GENOMICS IN ENDOCRINOLOGY

CONTEMPORARY ENDOCRINOLOGY

P. Michael Conn, SERIES EDITOR

Polycystic Ovary Syndrome: Current Controversies, from the Ovary to the Pancreas, edited by ANDREA DUNAIF, JEFFREY R. CHANG, STEPHEN FRANKS, AND RICHARD S. LEGRO, 2008

The Metabolic Syndrome: Epidemiology, Clinical Treatment, and Underlying Mechanisms, edited by BARBARA C. HANSEN AND GEORGE A. BRAY, 2008

Genomics in Endocrinology: DNA Microarray Analysis in Endocrine Health and Disease, edited by STUART HANDWERGER AND BRUCE J. ARONOW, 2008

Controversies in Treating Diabetes: Clinical and Research Aspects, edited by DEREK LEROITH AND AARON I. VINIK, 2007

Autoimmune Diseases in Endocrinology, edited by ANTHONY P. WEETMAN, 2007

When Puberty is Precocious: Scientific and Clinical Aspects, edited by ORA H. PESCOVITZ AND EMILY C. WALVOORD, 2007

Hypertension and Hormone Mechanisms, edited by ROBERT M. CAREY, 2007

Insulin Resistance and Polycystic Ovarian Syndrome: Pathogenesis, Evaluation and Treatment, edited by JOHN E. NESTLER, EVANTHIA DIAMANTI-KANDARAKIS, RENATO PASQUALI, AND D. PANDIS, 2007

The Leydig Cell in Health and Disease, edited by ANITA H. PAYNE AND MATTHEW PHILLIP HARDY, 2007

Treatment of the Obese Patient, edited by ROBERT F. KUSHNER AND DANIEL H. BESSESEN, 2007

Androgen Excess Disorders in Women: Polycystic Ovary Syndrome and Other Disorders, Second Edition, edited by RICARDO AZZIS, JOHN E. NESTLER, AND DIDIER DEWAILLY, 2006

Evidence-Based Endocrinology, edited by VICTOR M. MONTORI, 2006

Office Andrology, edited by PHILLIP E. PANTTON AND DAVID E. BATTAGLIA, 2005

Stem Cells in Endocrinology, edited by LINDA B. LESTER, 2005

Male Hypogonadism: Basic, Clinical, and Therapeutic Principles, edited by STEPHEN J. WINTERS, 2004

Androgens in Health and Disease, edited by CARRIE BAGATELL AND WILLIAM J. BREMNER, 2003

Endocrine Replacement Therapy in Clinical Practice, edited by A. WAYNE MEIKLE, 2003

Early Diagnosis of Endocrine Diseases, edited by ROBERT S. BAR, 2003

Type I Diabetes: Etiology and Treatment, edited by MARK A. SPERLING, 2003

Handbook of Diagnostic Endocrinology, edited by JANET E. HALL AND LYNNETTE K. NIEMAN, 2003

Pediatric Endocrinology: A Practical Clinical Guide, edited by SALLY RADOVICK AND MARGARET H. MACGILLIVRAY, 2003

Diseases of the Thyroid, Second Edition, edited by LEWIS E. BRAVERMAN, 2003

Developmental Endocrinology: From Research to Clinical Practice, edited by ERICA A. EUGSTER AND ORA HIRSCH PESCOVITZ, 2002

Osteoporosis: Pathophysiology and Clinical Management, edited by ERIC S. ORWOLL AND MICHAEL BLIZIOTES, 2002

Challenging Cases in Endocrinology, edited by MARK E. MOLITCH, 2002

Selective Estrogen Receptor Modulators: Research and Clinical Applications, edited by ANDREA MANNI AND MICHAEL F. VERDERAME, 2002

Transgenics in Endocrinology, edited by MARTIN MATZUK, CHESTER W. BROWN, AND T. RAJENDRA KUMAR, 2001

Assisted Fertilization and Nuclear Transfer in Mammals, edited by DON P. WOLF AND MARY ZELINSKI-WOOTEN, 2001

Adrenal Disorders, edited by ANDREW N. MARGIORIS AND GEORGE P. CHROUSOS, 2001

Endocrine Oncology, edited by STEPHEN P. ETHIER, 2000

Endocrinology of the Lung: Development and Surfactant Synthesis, edited by CAROLE R. MENDELSON, 2000

Sports Endocrinology, edited by MICHELLE P. WARREN AND NAAMA W. CONSTANTINI, 2000

Gene Engineering in Endocrinology, edited by MARGARET A. SHUPNIK, 2000

Endocrinology of Aging, edited by JOHN E. MORLEY AND LUCRETIA VAN DEN BERG, 2000

Human Growth Hormone: Research and Clinical Practice, edited by ROY G. SMITH AND MICHAEL O. THORNER, 2000

Hormones and the Heart in Health and Disease, edited by LEONARD SHARE, 1999

Menopause: Endocrinology and Management, edited by DAVID B. SEIFER AND ELIZABETH A. KENNARD, 1999

GENOMICS IN ENDOCRINOLOGY

DNA Microarray Analysis in Endocrine Health and Disease

Edited by

STUART HANDWERGER, MD

Cincinnati Children's Hospital Medical Center (CCHMC),
University of Cincinnati College of Medicine, Cincinnati, OH, USA

BRUCE J. ARONOW, PhD

University of Cincinnati College of Medicine, Cincinnati, OH, USA

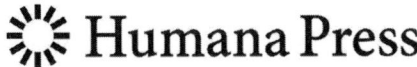 Humana Press

Editors

Stuart Handwerger
Cincinnati Children's Hospital Medical
    Center (CCHMC)
University of Cincinnati College of Medicine
3333 Burnett Avenue
Cincinnati, OH 45229, USA
stuart.handwerger@chmcc.org

Bruce Aronow
University of Cincinnati College of Medicine
Cincinnati, OH, USA

Series Editor

P. Michael Conn, MD
Division of Neuroscience
Oregon Health and Science University
Oregon National Primate Research Center
505 NW 185th Ave.
Beaverton, OR 97006, USA

ISBN: 978-1-58829-651-1   e-ISBN: 978-1-59745-309-7

Library of Congress Control Number: 2007931072

Cover illustration: A composite of a complex gene expression profile heatmap, a network view of gene interactions
that constitute the mechanisms for gene regulation, and a cartoon view of endocrine and paracrine relationships within
and between different endocrine tissues and somatic tissues responding to endocrine hormones. The heatmap depicts
clusters of genes (in rows) that are coordinately activated (red) or repressed (blue) in different tissues and as a function
of different influences (samples and treatments are in columns across the map) such as by different combinations of
hormonal agents. (Courtesy of Mike Zender and Bruce Aronow.)

Printed on acid-free paper

9 8 7 6 5 4 3 2 1

springer.com

PREFACE

During the last five years, the development of new technologies capable of monitoring genome function has gone from hopeful estimates to very solid depictions of genome output from individual samples. The application of this technology to the fields of endocrinology has resulted in new insights as well as the recognition that refinements in biological sampling, technology, experimentation, and data analysis and interpretation will yield a wealth of new understandings. In this volume, forty-five plus authors have contributed descriptions of a variety of investigations that reflect the beginnings of an integrated understanding of how different biological systems respond to endocrine signals. The development and application of focused studies that can reveal the mechanisms of endocrine action is a major and exciting challenge. Endocrine actions are divided over a broad range of mechanisms that encompass development, homeostasis, reproduction, and environmental adaptiveness of cells, tissues, systems, and organisms. Given adequate experimental design and analysis, gene expression profiling can provide considerable insight into endocrine system functions, responses, and interactions.

In our introductory chapter, we describe several issues in hypothesis formulation, experimental design, data analysis, and follow-up studies that may corroborate, validate, and extend hypotheses gained through microarray analyses. The contributed chapters span a variety of applications that we have divided into the areas of (1) genomic insights into molecular mechanisms responsible for hormone action, (2) genomic characterizations of endocrine producing tissues, and (3) genomic manifestations of diseases of hormonal systems.

We thank all of the authors for their efforts in summarizing some of their current understandings and to the members of our departments and institution who have shown patience and provided support for our efforts. We would also like to thank Humana Press for their enthusiasm on this project, and Allene Ford and Nancy Fehr for their invaluable assistance.

Stuart Handwerger
Bruce J. Aronow

CONTENTS

Preface .... *v*

Contributors .... *ix*

PART I   INTRODUCTION

1   Microarray-based Gene Expression Analysis of Endocrine
    Systems: *Principles of Experimental Design and Interpretation*
    Anil G. Jegga, DVM, Bruce J. Aronow, PhD,
    and Stuart Handwerger, MD *3*

PART II   HORMONE ACTION AND MOLECULAR MECHANISMS

2   Gene Expression Profiles and Transcription Factors Involved
    in Parathyroid Hormone Signaling in Osteoblastic Cells
    Ling Qin, PhD, Ping Qiu, PhD, and Nicola C.
    Partridge, PhD *25*

3   Analysis of Growth Hormone Effects on Hepatic Gene
    Expression in Hypophysectomized Rats
    Amilcar Flores-Morales, PhD and
    Leandro Fernández-Pérez, MD, PhD *41*

4   Gene Expression Profiling in Leiomyoma in Response to GnRH
    Therapy and TGF-β
    Nasser Chegini, PhD and Xiaoping Luo, MD *67*

5   Gene Profiling Analysis of Androgen Receptor Mediated
    Function
    Clay E. S. Comstock, PhD, Craig J. Burd, PhD,
    Walter J. Jessen, PhD, and Karen E. Knudsen, PhD *83*

6   Interrogating Estrogen Receptor α Signaling in Breast Cancer
    by Chromatin Immunoprecipitation Microarrays
    Alfred S. L. Cheng, PhD, Huey-Jen L. Lin, PhD,
    and Tim H. -M. Huang, PhD *115*

7   Gene Expression Analysis of the Adrenal Cortex in Health
    and Disease
    Anelia Horvath, PhD and Constantine A. Stratakis, MD,
    DMedSci *131*

PART III ENDOCRINE PRODUCING TISSUES

8   DNA Microarray Analysis of Human Uterine Decidualization
    Ori Eyal, MD and Stuart Handwerger, MD *147*

9   Large-scale DNA Microarray Data Analysis Reveals
    Glucocorticoid Receptor-mediated Breast Cancer Cell
    Survival Pathways
    Min Zou, PhD, Wei Wu, MD, PhD,
    and Suzanne D. Conzen, MD *165*

10 Application of Microarrays for Gene Transcript Analysis
    in Type 2 Diabetes
    R. Sreekumar, PhD, C. P. Kolbert, MS, Y. Asmann, PhD,
    and K. S. Nair, MD, PhD *185*

11 DNA Microarray Analysis of Effects of TSH, Iodide, Cytokines,
    and Therapeutic Agents on Gene Expression in Cultured
    Human Thyroid Follicles
    Kanji Sato, MD, PhD, Kazuko Yamazaki, BS,
    and Emiko Yamada, MD *207*

PART IV   DISEASES OF HORMONAL SYSTEMS

12 Genomics and Polycystic Ovary Syndrome (PCOS): *The Use*
    of Microarray Analysis to Identify New Candidate Genes
    Jennifer R. Wood, PhD and Jerome F. Strauss III, MD, PhD... *219*

13 Microarray Analysis of Alterations Induced by Obesity in White
    Adipose Tissue Gene Expression Profiling
    Julien Tirard, Ricardo Moraes, Danielle Naville, PhD,
    and Martine Bégeot, PhD *239*

14 Novel Molecular Signaling and Classification of Human
    Clinically Nonfunctioning Pituitary Adenomas Identified
    by Microarray and Reverse Transcription-Quantitative
    Polymerase Chain Reaction
    Chheng-Orn Evans, MS, Carlos S. Moreno, PhD,
    and Nelson M. Oyesiku, MD, PhD, FACS *263*

15 Gene Expression Studies of Prostate Hyperplasia in Prolactin
    Transgenic Mice
    Karin Dillner, PhD, Jon Kindblom, MD, PhD, Amilcar
    Flores-Morales, PhD, and Håkan Wennbo, MD, PhD *271*

Index *283*

CONTRIBUTORS

BRUCE J. ARONOW, PhD, *Division Biomedical Informatics, Cincinnati Children's Hospital Medical Center, Cincinnati, Ohio, USA Department of Pediatrics, University of Cincinnati College of Medicine, Cincinnati, OH*

YAN W. ASMANN, PhD, *Bioinformatics Application Scientist, Mayo Clinic College of Medicine, Rochester, MN*

MARTINE BÉGEOT, PhD, *Permanent Researcher from INSERM, Lyon, France*

CRAIG J. BURD, PhD, *Departments of Cell Biology and Pathology, University of Cincinnati, Cincinnati, OH*

NASSER CHEGINI, PhD, *Professor, Division of Reproductive Endocrinology and Infertility, Department of Obstetrics and Gynecology, University of Florida College of Medicine, Gainesville, FL*

ALFRED S. L. CHENG, PhD, *Post-doctoral Researcher, Division of Human Cancer Genetics, Department of Molecular Virology, Immunology, and Medical Genetics, Comprehensive Cancer Center, Ohio State University, Columbus, OH*

CLAY E. S. COMSTOCK, PhD, *Vontz Center for Molecular Studies, University of Cincinnati, Cincinnati, OH*

SUZANNE D. CONZEN, MD, *Associate Professor, Department of Medicine and The Ben May Institute of Cancer Research, The University of Chicago, Chicago, IL*

KARIN DILLNER, PhD, *Disease Biology, AstraZeneca R&D, Södertälje, Sweden*

CHHENG-ORN EVANS, MS, *Research Supervisor, Department of Neurosurgery, Emory University, Atlanta, GA*

ORI EYAL, MD, *The Edmond and Lily Safra Children's Hospital, The Chaim Sheba Medical Center, Sackler School of Medicine, Tel-Aviv University, Tel-Hashomer, Israel*

LEANDRO FERNÁNDEZ-PÉREZ, MD, PhD, *Associate Professor or Pharmacology, Department of Clinical Sciences, Molecular Pharmacology Group, University of Las Palmas de Gran Canaria–Canary Institute for Cancer Research, RTICCC, Canary Islands, Spain*

AMILCAR FLORES-MORALES, PhD, *Associate Professor of Experimental Endocrinology, Department of Molecular Medicine and Surgery, Karolinska Instututet, Karolinska Hospital, Stockholm, Sweden*

STUART HANDWERGER, MD, *Division of Endocrinology, Cincinnati Children's Hospital Medical Center, Cincinnati, Ohio, USA Department of Pediatrics, University of Cincinnati College of Medicine, Cincinnati, OH*

ANELIA HORVATH, PhD, *Senior Fellow, Section on Endocrinology and Genetics, (SEGEN)/DEB, NICHD, NIH, Bethesda, MD*

TIM H. -M. HUANG, PhD, *Professor, Division of Human Cancer Genetics, Department of Molecular Virology, Immunology, and Medical Genetics, Comprehensive Cancer Center, Ohio State University, Columbus, OH*

ANIL G. JEGGA, DVM, *Division of Biomedical Informatics, Cincinnati Children's Hospital Medical Center, Cincinnati, Ohio, USA Department of Pediatrics, University of Cincinnati College of Medicine, Cincinnati, OH*

WALTER J. JESSEN, PhD, *Division of Biomedical Informatics, Cincinnati Children's Hospital Medical Center, Cincinnati, OH*

JON KINDBLOM, MD, PhD, *Department of Oncology, Göteborg University, Göteborg, Sweden*

KAREN E. KNUDSEN, PhD, *Center for Environmental Genetics and Department of Cell Biology, University of Cincinnati, Cincinnati, OH*

CHRISTOPHER P. KOLBERT, MS, *Senior Research Technologist, Mayo Clinic College of Medicine, Rochester, MN*

HUEY-JEN L. LIN, PhD, *Assistant Professor, Division of Medical Technology, School of Allied Medical Professions, Ohio State University, Columbus, OH*

XIAOPING LUO, MD, *Research Assistant Professor, Division of Reproductive Endocrinology and Infertility, Department of Obstetrics and Gynecology, University of Florida College of Medicine, Gainesville, FL*

RICARDO MORAES, *Post-Graduate Student, Breast Center, Baylor College of Medicine, Houston, TX*

CARLOS S. MORENO, PhD, *Assistant Professor, Department of Pathology and Laboratory of Medicine and Cancer Institute, Emory University, Atlanta, GA*

K. SREEKUMARAN NAIR, MD, PhD, *David Murdock Dole Professor and Professor of Medicine, Division of Endocrinology, Mayo Clinic and Foundation, Rochester, MN*

DANIELLE NAVILLE, PhD, *Permanent Researcher from INSERM, Lyon, France*

NELSON M. OYESIKU, MD, PhD, FACS, *Professor and Vice-Chairman, Neurological Surgery; Director, Neurosurgery Residency Program; Director, Laboratory of Molecular Neurosurgery and Biotechnology, Department of Neurological Surgery, Emory University School of Medicine; Past-President, Congress of Neurological Surgeons; Past-President, Georgia Neurosurgical Society, Atlanta, GA*

NICOLA C. PARTRIDGE, PhD, *Professor and Chair, Department of Physiology and Biophysics, UMDNJ-Robert Wood Johnson Medical School, Piscataway, NJ*

LING QIN, PhD, *Assistant Professor, Department of Physiology and Biophysics, UMDNJ-Robert Wood Johnson Medical School, Piscataway, NJ*

PING QIU, PhD, *Principal Scientist, Bioinformatics Group and Discovery Technology Department, Schering-Plough Research Institute, Kenwilworth, NJ*

KANJI SATO, MD, PhD, *Professor, Field of Thyroid and Parathyroid Disease, Division of Internal Medicine, Graduate School of Medicine, Tokyo Women's Medical University, Shinjuku-ku, Tokyo, Japan*

RAGHAVAKAIMAL SREEKUMAR, PhD, *Associate Professor of Medicine, Mayo Clinic College of Medicine, Rochester, MN*

CONSTANTINE A. STRATAKIS, MD, DMedSci, *Chief, Heritable Disorders Branch, NICHD, NIH; Director, Pediatric Endocrinology Training Program, NICHD, NIH; Head, Section on Endocrinology and Genetics, (SEGEN)/DEB, NICHD, NIH, Bethesda, MD*

JEROME F. STRAUSS III, MD, PhD, *Executive Vice President for Medical Affairs, VCU Health System Dean, Professor of Obstetrics and Gynecology, Virginia Commonwealth University School of Medicine, Richmond, VA*

JULIEN TIRARD, *Graduate Student, INSERM, Lyon, France*

HAKAN WENNBO, MD, PhD, *Integrative Pharmacology, AstraZeneca R&D, Molndal, Sweden*

JENNIFER R. WOOD, PhD, *Assistant Professor, Department of Animal Science, University of Nebraska, Lincoln, NE*

WEI WU, MD, PhD, *Research Associate, Division of Molecular Oncology, Evanston Northwestern Healthcare Research Institute, Evanston, IL*

EMIKO YAMADA, MD, *Director, Thyroid Disease Institute, Kanaji Hospital, Kita-ku, Tokyo, Japan*

KAZUKO YAMAZAKI, BS, *Thyroid Disease Institute, Kanaji Hospital, Kita-ku, Tokyo, Japan*

MIN ZOU, PhD, *Post-doctoral Scholar, Department of Medicine, The University of Chicago, Chicago, IL*

I INTRODUCTION

1

Microarray-based Gene Expression Analysis of Endocrine Systems: Principles of Experimental Design and Interpretation

Anil G. Jegga, DVM, Bruce J. Aronow, PhD, and Stuart Handwerger, MD

CONTENTS

INTRODUCTION
GENE EXPRESSION PROFILING STUDIES IN ENDOCRINOLOGY
DESIGN OF EXPERIMENTS USING MICROARRAY ASSAYS
PRINCIPLES OF MICROARRAY DATA ANALYSIS
MICROARRAY DATA COMPLEXITY AND SELECTION OF APPROPRIATE
    STATISTICAL METHODS
MICROARRAY EXPERIMENTS AND EXTRACTION
    OF BIOLOGICAL INFORMATION
MICROARRAY DATA—STORAGE, STANDARDS AND EXCHANGE
MICROARRAY DATA VALIDATION
MICROARRAYS—PITFALLS AND LIMITATIONS
LITERATURE OVERVIEW—FURTHER READING
CONCLUSIONS
REFERENCES

Abstract

The fundamental rationale for the use of microarray-based gene expression profiling to characterize biological samples is based in part on the principle that cells, tissues, and perturbations applied to them can be characterized on the basis of their relative expression of genes and transcripts. Different biological states, cell types, and influences can be distinguished based on transcriptional profiles and the change in the relative levels of different genes and gene groups. This genomic expression profile-based discovery of biological states and effector-actions represents an essential element of a systems-based whole-genome approach to characterizing cells and tissues, and differs from the characterization of individual gene expression changes in isolation from one another, and has the potential to increase knowledge in all fields of biomedicine. The past two decades have seen a paradigm shift in which medical genetics has moved from being a tool of the basic investigator to play a role in the mainstream

From: *Contemporary Endocrinology: Genomics in Endocrinology:*
DNA Microarray Analysis in Endocrine Health and Disease
Edited by S. Handwerger and B. Aronow © Humana Press, Totowa, NJ

of medical practice. Identification of genetic causal agents of common endocrine disorders, deciphering underlying molecular pathophysiology of known conditions, development of new predictive tests for genetic abnormalities, and applications in the field of therapeutics are some of the implications of this shift. Endocrine systems, in particular, offer tremendous opportunities for the use of genomic analyses to understand physiological and pathological responses and effectors without being biased to a particular gene or set of genes. Therefore, the responses of diverse and potentially diversely affected systems can be broadly evaluated, constrained only by the limitation that there may be either a primary or secondary impact on transcript abundance. This emerging concept—endocrinomics—thus has the potential to significantly impact the field of endocrine research and clinical practice. However, advancements in the field are also limited by problems in collecting comprehensive datasets, the inherent complexity of multiple interacting systems, genetic variations between individuals, and some cumbersomeness associated with expression profiling technology and data analysis itself. This chapter discusses some of the issues to be considered in the design and analysis of microarray experiments for the characterization of endocrine-regulated systems.

INTRODUCTION

Investigators have observed correlations and cause-and-effect relationships between the expression and functions of known genes and biological phenotypes or human diseases for more than a half-century. Previously, this was done for one or a few genes at a time, relying on the use of nucleic acid (in situ hybridization, Northern blotting, etc.) or antibody (immunocytochemistry, Western blotting, etc.) probes. Descriptions of gene expression had been largely dependent on the availability of these probes and the intuition of researchers. However, continuing advances in the availability of genomic data associated with the Human Genome Project, the parallel development of microarray technology, and cross-species insights into conserved and evolved genome structures and functions have provided the means to perform global analyses of the expression of thousands of genes in a single assay. This has changed the landscape of biomedical research, but has created a gap in understanding of the general approaches that can be used by clinical investigators to map novel genes and biological states that occur in relatively unique patient populations and experimental conditions. The application of highly consistent and calibrated approaches to genomic expression profiling of clinical and experimental samples has considerable promise to generate new biological opportunities for both basic and clinical scientists. The conventional paradigm of clinical practice and endocrine research has typically involved the evaluation of single pathways and clinical indications, and the formulation of interventions that largely separate processes or phenomena, molecules and interactions. Although generally successful, this approach has limitations that are becoming increasingly evident as the frontiers of our understanding enter areas of higher complexity, e.g., the interactions of individual patient genetics, nutritional state, and co-morbidities in shaping a physiological or pathophysiological set of conditions and resulting disease processes. A full understanding of biological processes, including endocrine ones, can therefore only come from a systems approach in which interacting components are measured and understood with respect to their participation in normal and disease processes. Although current biomedical science is still far from this ideal, development and judicious use of unbiased assessment tools such as whole-genome expression profiles, combined with well-designed experimental approaches, have the potential to provide novel insights into biological systems.

A microarray or "gene chip" measures the relative expression levels of a gene by determining the amount of messenger RNA that is present (mRNA abundance) in one sample versus another. A microarray is composed of thousands of nucleotide sequences (probes), usually DNA, that can hybridize to complementary DNA or RNA that is usually fluorescently labeled. Nucleotide sequences are placed in specific arrangement, usually on a silicon microchip or a glass slide. Depending on the type of arrays, the DNA sequences can be short (oligonucleotide arrays) or long (cDNA arrays). The underlying principle of microarrays is that complementary sequences will bind to each other under proper conditions, whereas non-complementary sequences will fail to bind. Probes are usually designed to be gene- or transcript-specific. On a microarray, thousands of different nucleotide sequences are arranged in a grid or array, with each array position containing a DNA sequence from a particular gene. When labeled transcripts from a biological sample of interest are hybridized to the array, the level of binding to the array is proportional to the abundance of the labeled molecules that contain the corresponding sequence. High-intensity binding indicates that there is high abundance of the corresponding transcript.

To place these new developments in context for endocrinologists, this review will present an overview of issues to be considered in the experimental design of a microarray study and in data analysis. In addition, some of the opportunities and perils offered by public database data storage and data exchange will be discussed.

GENE EXPRESSION PROFILING STUDIES IN ENDOCRINOLOGY

When a target tissue is sampled, it is generally the case that different cell types in the tissue have different responses. Some cells may be affected directly, while others may respond secondarily based on an adjacent primary response. Thus, even a distal endocrine-regulated response may be composed of a combination of different endocrine mechanisms. Microarray technology has the power to dissect these different responses, provided that there is separation of different cell types from the target tissue. Thus, a fundamental consideration in experimental design involves the decision to profile the entire tissue or specific cell types. Variables include relatively simple factors such as time of sampling relative to the addition or inhibition of hormones or drugs, dose-response characteristics, combinatorial interactions, and different modes of administration. Biological systems are also highly responsive to many factors, such as genetic, nutritional, environmental perturbations that can lead to alterations in endocrine status and hormonal responsiveness. Conventional endocrine experiments usually test in isolation the effects of a single hormone, metabolic, or regulatory state or stimulus on the expression levels or functional state of one or more gene products. A basic drawback of this approach is that it disregards the complexity of biological systems. Some of the reductionism inherent in conventional single gene or protein analysis is overcome by the advent of microarray technology, but the need to control and evaluate other physiologic variables in a powerful experimental study design cannot be overcome by the use highly sensitive whole genome expression profile analysis. Indeed, the sensitivity to independent transcriptional effects of different individual stimuli poses a critical requirement for careful control and extensive replication and sampling in an experimental study design.

The sensitivity offered by whole-genome transcript profiling is particularly high with respect to its ability to identify new genes that are endocrine factor-responsive as well as genes that respond coordinately in characteristic behaviors. For example, DNA microarray-based gene expression studies have identified 45 genes not previously identified as thyroid hormone-responsive genes (1). Similarly, the specificity of insulin versus *IGF1* signaling (2) could be discriminated using cDNA microarrays, with 30 of 2221 genes tested increased in presence of *IGF1* but not by insulin. Of these, 27 genes were not previously reported as being *IGF1*-responsive genes. A good example of the application of microarrays to problems in clinical endocrinology is the characterization of Hurthle cell tumors and a determination of the clinical significance of identified phenotypes (3). Cancer subtype-specific signature profiling has the potential to help in treatment option selection, ranging from conservative to very aggressive, and for the selection of tumors that are likely to respond to endocrine-based therapies involving systems such as breast cancer, thyroid cancer, and prostate cancer. Thus, empirically defined relations between outcome and the specific "expression signature" of the particular tumor may permit more refined decision making, even before the functional significance of the different expression patterns is understood (3). In addition, microarray-based gene expression studies have the potential to unravel very fine and unexpected mechanisms modulated by hormones (see Table 1 for examples of endocrinology studies using DNA microarrays). As a first step towards an integrative understanding of biologic responses to hormonal agents, and the impact of this explosion of data, and to use this information towards an integrative understanding of biological processes, a wide variety of computational methods have been developed to detect co-expressed and co-regulated gene groups that define cellular states, activities, and responses. These state-specific and factor-regulated based signatures represent building blocks for the assembly of new views of biological systems.

Table 1
Examples of Some Studies in Endocrinology Using DNA Microarrays

Study	Tissue	Reference
Aldosterone vasopressin	Kidney cells	Robert-Nicoud et al. (50)
Androgen	Prostate cancer cells	Xu et al. (51)
Atherogenic stimulus	Human arterial endothelial cells	de Waard et al. (52)
Corticosteroid	Rat hippocampus	Datson et al. (53)
ER-alpha	Breast carcinoma cells	Gruvberger et al. (54)
Estrogen and Progesterone	Mouse lacrimal gland	Suzuki et al. (55)
Trophoblast differentiation	Placenta	Aronow et al. (56)
Insulin and IGF1	Mouse fibroblast NIH-3T3	Dupont et al. (2)
Progesterone and dexamethasone	Human endometrial cancer Ishikawa H cells	Davies et al. (57)
Vasopressin	Hypothalamoneurohypophyseal system in rats	Hindmarch et al. (58)

In the chapters that follow, the authors demonstrate profiling studies of a range of endocrine systems that reveal the magnitude of the challenge and opportunity to gain systems-level understanding of hormone actions and mechanisms in health and disease. In further sections of this chapter, we shall discuss the principles of experimental design that enable characterization of different endocrine-regulated systems.

DESIGN OF EXPERIMENTS USING MICROARRAY ASSAYS

In order to characterize biological systems based on their transcriptional states and responses, it is important to establish that there are reproducible differences between samples in one condition or state and those in another state. For example, differences between states that are determined or influenced by differential hormonal treatment effects. To establish hormonal factor influences on gene expression, one must show that hormone-induced expression alterations are sufficiently consistent in a base state and differ from those in the perturbed state (Fig. 1). Several experimental approaches can accomplish this, including the demonstration of dose-responsiveness, sensitivity to inhibition, and specificity of effects in a particular context or target cell. It is also important to use multiple trials for the occurrence of a treatment–effect test of the observations that demonstrate a significant effect on treated samples *(4,5)*. A variety of

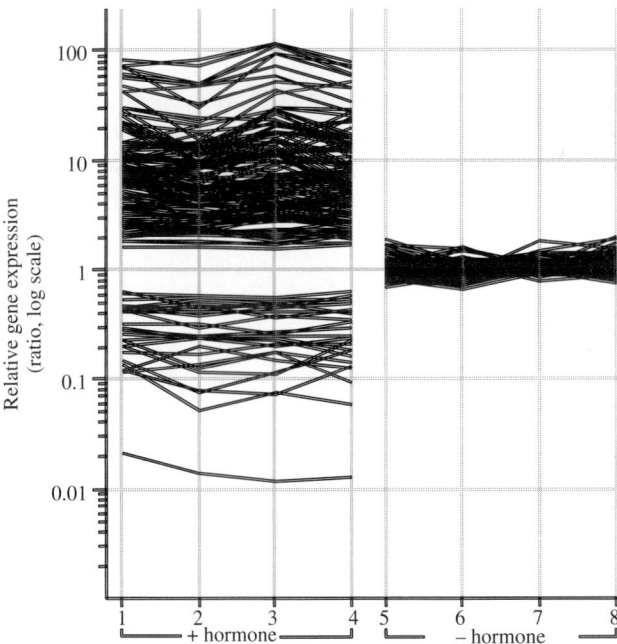

Fig. 1. Examples of genes whose expression is reproducibly altered in cells by hormone treatment. Gene expression was measured in four cultures of cells subjected to a single dose of hormone for a fixed time and was compared to that in four untreated cultures. Gene expression values in each isolate were referenced relative to the median of gene expression in the untreated samples and a Student's *t*-test corrected with Benjamini-Hochberg false discovery rate testing (FDR < 0.001) was used to identify the genes that differed in their expression between ± hormone-treated groups.

interesting phenomena can be observed based on the intrinsic variability of biological systems. The basic principle is that genes that function together tend to be regulated together. Thus, when many thousands of genes are measured over a large number of related samples (usually much greater than 5), coordinate patterns of regulation can be detected among small groups of genes. A number of factors influence one's ability to recognize clusters of coordinate regulation. These include the types of biological samples, their treatments or perturbations, and the technical accuracy of gene expression measurements. As increasing numbers of samples and related states are measured, one is able to detect improved correlations among groups of genes and among groups of samples. This allows both genes and samples to be clustered according to the genes and samples that share relative activation patterns. Biological systems vary in many complex manners. Occurrence of some variability among biological samples is inevitable, but some strong patterns cannot be fully anticipated prior to initial experimentation. Thus, pilot experiments can greatly improve insight into biological variation under different circumstances and can provide useful guidance into the range of samples necessary to achieve statistical significance. An excellent discussion of power calculations and a novel system for exploring baseline variation in microarray experiments can be found in Page et al. *(6)*. Careful control of experimental conditions can minimize variation due to secondary unknown factors, but stochastic, hidden, or intrinsic variables can cause significant within-group biological variation. Similarly, treated samples may differ in the extent of treatment effect, often for reasons that cannot be identified or stringently controlled.

The range of experimental variables that are able to impact critically on responses of endocrine systems is large. However, these modifying factors must therefore be considered carefully in experimental designs. Examples of modifying factors include hormone forms and combinations, dose, route of administration, cofactors that modify distribution, metabolism, degradation, elimination, host pre-treatment factors including environmental, other hormonal influences, target system sampling timings, consistency of target tissue dissection, and intrinsic compositional variability of target tissues. Tissue dissection and intrinsic compositional variability can impart very large effects and may be difficult to control at collection time. It is possible to detect and adjust for the effects of this following data collection. However, the presence of these variables in a sample series can seriously affect one's ability to identify genes that are regulated by the variable of interest (such as the effect of the hormonal influence or disease state). Examples include adjacent tissue that is hard to recognize at the time of dissection or the presence of variable compositions of cell types in blood in different individuals. Additional non-biologic factors associated with individual samples for microarray analysis include tissue preservation methods and RNA quality. RNA quality is critical to assure, particularly for post-mortem samples; and a variety of methods are available to do this *(7–12)*. Microarray platform and technology factors that impact on variability include cDNA amplification, probe labeling, hybridization conditions, and washing *(13)*. There are a variety of methods that can be employed to control, measure, and normalize these variables *(14–16)*.

As a general rule, experimental conditions should be standardized as much as possible and measurements should be extensively replicated in separate treatment trials. As discussed earlier, while there is no single answer to the question of how many replicates are necessary per treatment condition, various investigations have approached this as a "power analysis" problem *(6)*. In general, the number of samples that are necessary is proportional to the magnitude of the treatment effect relative to the magnitude of the sum of other factor effects within a biological group. There are two types of replicates. A technical replicate is a separate analysis of an identical RNA sample. Highly reproducible technical methods (i.e., labeling, hybridization, and measurement) require little or no technical replication. However, the accuracy of gene expression measurements from technologies that are somewhat noisy (i.e., home-built cDNA microarrays, some RNA amplification methods) can be substantially improved by the use of technical replication methods. The effect of technical variations is to increase the numbers of genes falsely changed in repeated measurements. A sample that is to be technically replicated should be separately sampled from a given RNA aliquot and subjected to separate labeling, hybridization, washing, and scanning. Conversely, biological replicates are independently treated samples that have been subjected to parallel dissection, purification, and quantifications as other samples within its group (Fig. 2).

If all these issues are taken into account in the experimental design, microarray experiments provide reliable data on gene expression at a systems level *(17)*. Until recently,

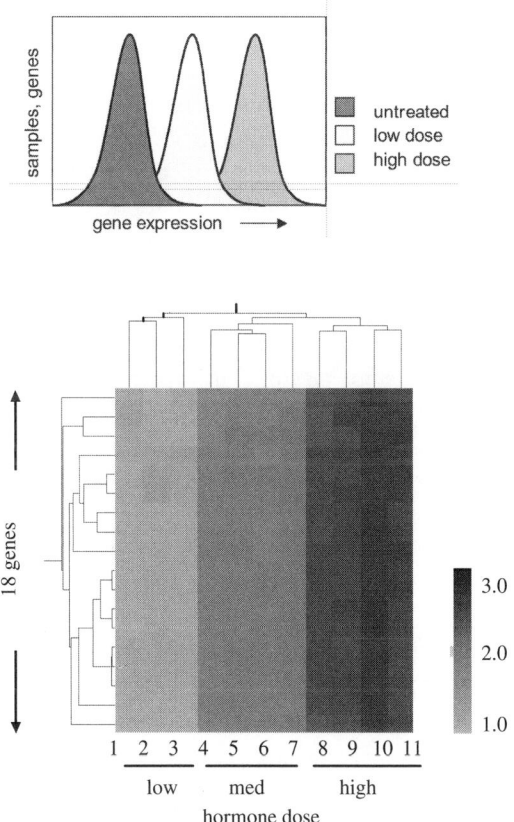

Fig. 2. Distributions of relative gene expression level changes imparted by two doses of a hormonal factor on independent samples. Even within highly replicated studies, there will be an intrinsic distribution of the magnitude of the effect in separate samples.

the amount of RNA required for reliable microarray experiments using conventional labeling of RNA has been in the range of 5–100 μg of total RNA *(18)*. However, recent advances in RNA labeling have allowed laser-capture microdissected samples to approach conventional labeling protocols with respect to signal reliability and data reproducibility (http://www.epibio.com/targetamp/TargetAmpPoster1005_lg.pdf). Before one proceeds to interpret microarray data related to a physiological or pathological state, the important question that needs to be addressed is "What between-samples and between-groups gene expression differences are signatures of the introduced perturbation?" For any system under study, it is important to understand both the magnitude and diversity of gene expression in the normal unperturbed state—i.e., the range of values presented by independently sampled entities—versus expression patterning that is introduced by treatments and other factors. For example, variability between cultured cells that are grown as a pool and then plated separately is generally of a smaller magnitude than that occurring between independent culture isolates obtained from genetically diverse individuals—e.g., in a tissue biopsy. Following treatment with a hormone, cultured cell systems demonstrate subgroups of genes that exhibit variable responses, with respect to magnitude, time and dose effects. However, a series of human subjects would be expected to show several additional levels of variability in their responses, including genetic, nutritional, and a range of other environmental or situational variables that may impact on the biological response to hormone treatment. In addition, some markers may respond to the hormone only if another variable is operative. The combination of these effects may lead to other secondary responses that in effect make for elaborate levels of complexity. Allison et al. have summarized the problems presented by biological variability in a recent review *(19)*. They point out that: (a) biological replication is essential; (b) there is strength in numbers—power and sample size; (c) pooling biological samples can sometimes be useful; and (d) avoiding confounding effects introduced by extraneous factors is crucial *(19)*. The occurrence of extraneous factors, however, may also be unavoidable. Consequently, an effective adaptation to biological variability is to undertake an experimental design strategy in which a large number of biologically similar samples are sampled. This approach allows many individuals within a group to corroborate the occurrence of expression patterns that are not necessarily shared by all members of the overall group. Another useful adaptation is to undertake the ascertainment of additional variables that may significantly influence independent sample responses.

PRINCIPLES OF MICROARRAY DATA ANALYSIS

Data handling, normalization, pattern detection, and biological network and ontology significance analyses are important issues that can greatly affect the interpretation of microarray-based gene expression studies. Optimal strategies for the analysis of microarray data are areas of intense investigation. Two basic approaches are usually adopted for microarray data analysis: (a) exploratory approaches in which data are analyzed to identify genes and biological samples within an experimental series that exhibit similar or correlated patterns of expression; and (b) classifier approaches in which gene expression variations among samples are related to known classes of samples, e.g., genes whose relative expression patterns distinguish normal, hormone-affected, hormone-deficient, or diseased tissues. The former approach is also referred to as unsupervised analyses because there is no constraint on which the samples should be treated as similar. The classification approach is a supervised analysis because the different samples are treated as separate classes.

The usual computational approach is to discover the optimal number of classes using an unsupervised learning procedure, i.e., cells or tissues with similar patterns of gene expression based on a correlation or distance score. There are various methods for partitioning samples into groups based on their gene expression profiles. A common unsupervised method is hierarchical pairwise clustering based on average-linkage between clusters to identify the most closely related classes in a tree representation of the relationships. Hierarchical clustering has successfully analyzed global gene expression that occurs in MEN1 (multiple endocrine neoplasia)-associated neuroendocrine tumors (20). Many genes were identified that are differentially expressed in neuroendocrine tumors arising in patients with the MEN1 syndrome as compared to normal human islet cells. Gene clusters are then subjected to manual or automated procedures to evaluate whether there is evidence for co-functionality of genes within the clusters and whether or not other clustering patterns provide additional data insights for the classification or shared perturbations among the sample series. Other unsupervised methods that have been used for microarray data analysis of endocrine disorders include principal component analysis to discover cardiovascular risk factors associated with the insulin resistance syndrome (21).

Supervised learning methods are a step more advanced than supervised analysis and require building and training a model to classify and then recognize unknown samples. Models are trained using a set of known samples that conform to predefined classes, and then a gene classifier is extracted from the training set that can be used to predict the occurrence of classes within the test set, thus validating the classifier genes. Alternatively, the same set can be used for both training and validation, by iteratively removing one or more members of the set from the training step and refining a list of genes that can validly predict all permutations (22). Genes can be classified as well. For example, Soinov et al. have employed supervised learning methods to reconstruct and extend known gene networks from expression data (23). A general problem with supervised learning methods is that of "overfitting the data" to the training samples. This leads to the derivation of classifiers that are only useful for classifying samples within a dataset, but are not able to classify samples that are more variable or diverse (22,24). This problem tends to result from inadequate numbers of independent experimental replications, but different circumstances and variance patterns significantly differ in the strength and reproducibility. This leads to more or less robustness of different expression profile classifiers.

MICROARRAY DATA COMPLEXITY AND SELECTION OF APPROPRIATE STATISTICAL METHODS

As listed before, there are many sources of noise and experimental variation in microarray-based gene expression data. In fact, every stage of the experiment—sample preparation, RNA labeling, hybridization, signal and background measurement, and variation between biological samples—is vulnerable to noise and signal variation. The net effect makes the task of extracting a list of significantly regulated or differentially expressed genes a formidable one—replete with both false-negative and false-positive predictions. Unfortunately, no simple solution can overcome these challenges, and finding the most suitable experimental design and data analysis for different questions remains a matter of contention among statisticians and biologists.

Microarray experiments require careful planning and execution as well as creative statistical and bioinformatics analyses. Experimental design and data analysis should be based on a clear understanding and statement of the experimental goals and overarching hypotheses. The experimental design should identify sources of variation and measure them. For example, the additional measurement of growth response of a tissue to a hormonal treatment can provide a strong basis for correlating gene expression changes. Nearly any measurement of biological response of a tissue to a treatment can provide an excellent basis for examining gene expression patterns. A good yardstick is to use, at minimum, three or four biological replicates. As cost is always a consideration, the number of treatments or time points can be reduced in favor of biological replication. However, discriminating early, middle and late responses can provide powerful insights into a biological process. In experiments that depend on tissues, samples from patients in groups that differ in drug sensitivity, clinical outcome, or clinical diagnosis can provide powerful bases for separating gene expression pattern differences between each group. Analysis of variance models allow for the detections of gene expression pattern differences that are greater between groups than within-group variations. Samples from single individuals should generally not be pooled but rather should be used to perform individual hybridization reactions. However, pooling samples can allow for an initially inexpensive approach to very expensive questions. However, there are also many problems with pooled sample analyses that diminish the usefulness of observations derived from them. In general, sample-to-sample reproducibility is a highly powerful significance measure, and sometimes genes that vary by larger magnitudes are not the ones that are most linked to the underlying difference in sample treatments. Thus, studies that contain small numbers of samples tend to generate lists of genes with expression changes that are not fully reproducible over multiple trials. Genes whose expression changes are not reproducible are termed false-positives. Genes not identified, whose expression changes are lower but more reproducible, are also characteristic of small sample experiments (termed false-negatives). Both improved statistical tests and more extensive experimental replications are necessary to overcome sensitivity and specificity issues.

A major contribution to microarray data analysis has been the application of formal statistical methods, including linear and non-linear models, ANOVA, and the application of sophisticated background and data normalization methods *(14–16)*. For instance, the use of linear models for microarray analysis in the BioConductor package (http://www.bioconductor.org) *(25)*, which is based on the open source R statistical programming language, supports many of these statistical methods and can allow for the assessment of differential expression in the presence of multiple interacting and non-interacting experimental variables.

R is a popular statistical programming language package used by researchers, and has been vigorously embraced by the microarray data analysis community. R itself is command-line driven and includes graphics utilities, and enables scripting and derivation of a wide variety of statistical methods and algorithms. One can, therefore, write programs to suit specific tasks or projects, making them ideal for microarray analysis. R includes both simple statistics, e.g., calculation of variance, standard deviations, chi-square tests, etc., and more complex permutation and machine learning-based algorithms.

Bioconductor is a collection of many tools that operate on genomic data for various purposes *(25)*. This is not just restricted to statistical tools, but includes annotation and database tools also. Each of these tools has a solid grounding on R and data can be dynamically interchanged between tools. The "Literature Overview—Further Reading" section later in this chapter lists some of the microarray data analysis packages and extended descriptions.

GeneSpring (Agilent, Inc. Palo Alto, CA, USA), Spotfire (Göteberg, Sweden), Partek (St Louis, MO, USA), and GeneSifter (VizXlabs, Seattle, WA, USA) are excellent commercial tools for microarray data analysis that incorporate lots of Swiss-army knife-like functions necessary to accomplish the first few stages of data analysis, including data input, normalization, referencing, statistical group and expression pattern filtering, and clustering and pattern significance evaluations. Interestingly, GeneSpring has developed methodologies to interact fully with R and BioConductor, and is quite versatile and popular.

MICROARRAY EXPERIMENTS AND EXTRACTION OF BIOLOGICAL INFORMATION

Once the sample-handling and bioinformatics aspects are appropriately managed, the investigator is still presented with the task of choosing among many potentially interesting differentially expressed genes for further study and the evaluation of the overall biological significance. This task can be simplified, e.g., by focusing on genes with known or putative regulatory functions, or on genes for which quality reagents (i.e., antibodies) exist. One of the most difficult tasks faced by biologists is what to do with a long list of genes from an expression microarray experiment. How relevant are these results biologically? Some of the analyses that can help answer questions surrounding the biological relevance include the examination of these lists for their relative enrichment of shared properties of the genes and proteins in the list. There are a number of emerging approaches to the detection of shared properties. For microarrays that cover the entire genome, the general approach is to annotate systematically each gene/transcript for the functions and properties of the corresponding protein and then to look for commonalities among the genes in the coordinately regulated gene list. Thus, similar annotations with respect to gene ontology, phenotype or disease-association, biochemical function, known biochemical and regulatory relationships, and known protein–protein and gene–gene interactions can all be identified and if the number of genes that share properties is greater than the number that might be chosen by chance from the genome, a probability can be derived that compares observed occurrences to expected occurrences of the shared property. Quantitative significance measures used to describe the relative enrichment of genes involved in a shared category such as a common pathway include Fisher's Exact test and the Hypergeometric Distribution test. For example, DAVID (Database for Annotation, Visualization, and Integrated Discovery—http://www.david.niaid.nih.gov) addresses this need via four web-based analysis modules: (a) Annotation tool rapidly appends descriptive data from several public databases to lists of genes; (b) GO charts assign genes to Gene Ontology functional categories based on user selected classifications and term specificity level; (c) KEGG charts assign genes to KEGG metabolic processes and enables users to view genes in the context of biochemical pathway maps; and (d) Domain charts group genes according to PFAM-conserved protein domains. The functionality provided by DAVID accelerates the analysis of genome-scale datasets by facilitating the transition from data collection to biological meaning *(26)*. Similarly, several tools are available to identify potential cis-regulatory signature modules in a group of co-expressed genes *(27–29)*.

Several complex microarray data analysis methods have been used to uncover previously unknown gene relationships. Molecular interactions and relevance networks have been extracted from clinical data using mutual information from genes *(30)*, genome data, and probabilistic graph models trained by integrating prior information about the relationships of genes in pathways and processes *(31)*. Inferring complex regulatory networks from microarray-based gene expression data, has been an active area of bioinformatics research *(27–29, 31, 32)*, but has had a tendency to yield mixed results. Frequently, the resulting network models are in agreement with experimental data but provide very little insight into novel gene regulatory relationships. A principal reason for this may be the underlying complexity of the regulatory interactions among genes and the presence of hidden variables that modify coordinate behaviors such as genetic polymorphisms in a population affecting the function of key regulatory genes. Nevertheless, these models are useful, and their refinement and eventual application should lead to a better understanding of the systemic response of cells or tissues to therapeutic agents.

The approach to extracting biological information from hormone-driven gene expression experiments should reveal both the individual genes and proteins, as well as the biochemical pathways and processes that are activated or suppressed in cells and tissues by hormone treatment. By comparing transcriptional and other biological responses of different cells and tissues to a range of hormones, maps and circuits can be generated, which reflect the dependence of diverse cells and tissues on single and multiple hormone agents. Whereas some actions may be apparent over brief periods of time, some actions may ensue over much longer periods. The interesting prospect for systems-based endocrinomics is that improved understanding of shared and specific gene expression patterns will lead to new biological insights, and that continued advances into the details of biological responses of target cells and tissues will improve our ability to extract additional information from the gene expression profile data.

MICROARRAY DATA—STORAGE, STANDARDS AND EXCHANGE

Microarray-based gene expression studies produce vast amounts of raw data necessitating standardization methods for data representation as well as analysis. An international collaboration between major research groups and companies, including the European Bioinformatics Institute (EBI), Affymetrix, Rosetta, and Agilent, has made major progress in the integration and standardization of microarray data and microarray experiment descriptions. Large-scale gene expression experiment databases include ArrayExpress in Britain and Gene Expression Omnibus (GEO) at the National Center for Biotechnology Information (NCBI). Submissions to expression array databases can be made via EBI's software tool, MIAMExpress. MIAME (Minimum Information About a Microarray Experiment) provides a logical framework for describing an array experiment, outlining the content, intent, and implications of a specific array experiment. MIAME was developed by the MGED Group at the EBI. It can be split into three types of information content: (a) Sample information—all relevant information regarding the origin of the hybridized RNA, physical and chemical parameters, etc.; (b) Gene information—all relevant information

Table 2
Downloadable Large Datasets of Microarray Measurements

CardioGenomics Program in Genomic Applications (http://cardiogenomics.med.harvard.edu/public-data.html): Microarrays involving mouse models of cardiac development and signal transduction, including measurements made in time-series.

Gene Expression Omnibus (GEO—http://www.ncbi.nlm.nih.gov/geo/): Currently holds over 30000 submissions representing approximately half a billion individual molecular abundance measurements, for over 100 organisms *(34)*.

Human Gene Expression Index (HugeIndex—http://www.hugeindex.org): Contains the results of gene expression experiments on human tissues.

Integrated Tumor Transcriptome Array and Clinical data Analysis (ITTACA—http://bioinfo-out.curie.fr/ittaca/): Centralizes public datasets containing both gene expression and clinical data and currently focuses on breast carcinoma, bladder carcinoma, and uveal melanoma *(59)*.

L2L (http://depts.washington.edu/l2l/): A database consisting of lists of differentially expressed genes compiled from published mammalian microarray studies, along with an easy-to-use application for mining the database with the user's own microarray data *(60)*.

Oncomine (http://www.oncomine.org/): Contains more than 65 gene expression datasets comprising nearly 48 million gene expression measurements form over 4700 microarray experiments. Differential expression analyses comparing most major types of cancer with respective normal tissues as well as a variety of cancer subtypes and clinical-based and pathology-based analyses are available for exploration *(61)*.

Public Expression Profiling Resource (PEPR—http://pepr.cnmcresearch.org): Permits gene-based queries of large Affymetrix array datasets without any specialized software.

Stanford Microarray Database (SMD—http://genome-www.stanford.edu/microarray): Currently supports the research of more than 1000 users in over 260 laboratories at Stanford and around the world. These users have entered data generated from more than 50 000 microarrays used to study the biology of 34 organisms, published more than 190 papers referring to data in SMD and have made the complete raw data from more than 7000 microarrays freely available via the SMD website. The public data can be selected, viewed, downloaded and analyzed by the public using most of the tools that are available to registered SMD users *(62)*.

Whitehead Institute Center for Genome Research (http://www.broad.mit.edu/cgi-bin/cancer/datasets.cgi): Microarrays from 12 publications involving many types of cancer, including some clinical measurements associated with each sample.

regarding tethered sequences on the array; and (c) Gene expression levels—actual numerical values obtained. For additional information, refer to the MIAME website (http://www.mged.org/Workgroups/MIAME/miame.html) or Brazma et al. *(33)*.

The GEO at the NCBI is the largest fully public repository for gene expression data *(34)*. The database has a flexible and open design that allows the submission, storage, and retrieval of many data types. These data include microarray-based experiments measuring the abundance of mRNA, genomic DNA and protein molecules, as well as non-array-based technologies, such as serial analysis of gene expression (SAGE) and mass spectrometry proteomic technology. As an open repository, the data in GEO have typically been analyzed and studied, and in most cases, the results have been published in journals. Nonetheless, pooling disparate data into one location and organizing them to be analyzable and cross-comparable using common interfaces add a valuable analytic layer not attainable when considering individual experiments. Mining GEO data can provide clues as to the function of uncharacterized genes and genetic networks by examining spatial and temporal expression patterns, and co-regulation with well-characterized markers *(34)*. Cross-comparison of independently generated but experimentally similar datasets can corroborate interesting gene expression trends that may be overlooked in one experiment alone *(35)*.The GEO database is publicly accessible through the World Wide Web at http://www.ncbi.nlm.nih.gov/geo. Table 2 lists some of the publicly available microarray data repositories.

MICROARRAY DATA VALIDATION

Two questions that are fundamental in validating microarray data are whether the data procured from a microarray experiment is valid in terms of the biological problem that it seeks to solve and whether the normalized values are biologically accurate and minimized for technical artifacts *(36)*. Good experimental design should include steps for ensuring quality control of the technical system, the integrity of RNA samples, cell viability, array quality, array element annotation, and the technical reproducibility between batches and samples. The same logic can be extended to "post-experimental" factors, such as normalization and array-scanning techniques. Data processing techniques remain under constant improvement. Thus, providing subsequent investigators access to original raw data files will allow others to take advantage of the original data for follow-up study. It can also be very useful to perform sampling of representative genes in the sample series subjected to microarray analysis using RTQ-PCR as well as large-scale between-platform validations such as by comparing differences of two very different samples on one microarray platform to the differences that are detected on another microarray platform. These approaches can be used to identify erroneous data or technical problems. A good sampling-based approach can verify whether the observed expression levels on the array are valid.

MICROARRAYS—PITFALLS AND LIMITATIONS

Notwithstanding the initial hype and excitement about microarrays, their pitfalls and limitations have led to several formal reevaluations, comparisons, and statements of limitations *(37–42)*. Kothapalli et al., following an examination of microarray data from two different systems, reported inconsistencies in sequence fidelity of the spotted microarrays, variability of differential expression, low specificity of cDNA probes, discrepancy in fold-change calculations, and lack of probe specificity for different isoforms of a gene *(43)*. Ntzani and Ioannidis examined 84 large-scale microarray expression datasets that address major clinical outcomes including death, metastasis, recurrence, and response to therapy. They found that these studies show variable prognostic performance *(44)*. Michiels et al. reanalyzed data from seven large published studies that have attempted to predict prognosis of patients with cancer on the basis of DNA microarray analysis. The results reveal that the lists of genes identified as predictors of prognosis were highly unstable and molecular signatures were strongly dependent on the selection of patients in the training sets *(45)*. Some of the problems in predictor pattern reproducibility may be attributable to technical variables that affect the quality of the datasets. Other problems relate to sample sizes that may be small relative to the complexity of human population genetics. Other reasons may include overprediction and a tendency that this causes a reliance on marginal signatures. It is possible that the identification of stronger signatures may provide somewhat less predictive power but more robust group classification reproducibility *(46)*. Indeed, using a subtype classification approach, Sorlie et al. *(47)* have shown the ability to recognize and classify breast cancer subtypes across separate populations and microarray platforms.

LITERATURE OVERVIEW—FURTHER READING

A large number of books dealing with experimental and statistical microarray data analysis have been published. It is beyond the scope of this chapter to provide a complete list, but some of the books that we found useful for statistical background *(1–5)* and for general background *(6–8)* are listed below:

1. Baldi, P., and G.W. Hatfield DNA Microarrays and Gene Expression, from Experiments to Data Analysis and Modeling, Cambridge University Press, 2002.
2. Parmigiani, G., E.S. Garett, R.A. Irizarry and S.L. Zeger The Analysis of Gene Expression Data, Springer, 2003.
3. Speed, T. (ed. and contributor) Statistical Analysis of Gene Expression Microarray Data, Chapman and Hall, 2003.
4. Ting Lee, M.L. Analysis of Microarray Gene Expression Data, Springer, 2004.
5. Wit, E. and J. McClure Statistics for Microarrays: Design, Analysis and Inference, Wiley, 2004.
6. Draghici, S. Data Analysis Tools for DNA Microarrays, Chapmann-Hall, 2003.
7. Knudsen, S. A Biologist's Guide to Analysis of DNA Microarray Data, Wiley, 2002.
8. Blalock, E.M. (ed. and contributor) A Beginner's Guide To Microarrays. Kluwer Academic Publishers, 2003.

Several excellent reviews dealing with microarray experimental design and analytical methods have also been published *(15, 19, 48, 49)*. The amount of software, both commercial and freeware, available for microarray analysis has exploded in recent years. When considering what software to use for microarray data analysis, some of the issues to be considered are: (a) ease of data import; (b) specific versus comprehensive packages; (c) strength for visualization; (d) freeware—although they are often not debugged; and (e) support of other statistical and mathematical functions, such as in SAS or MATLAB, and also the ability to use scripting languages like R statistical programming language. While it is beyond the scope of this chapter to list all available software, two websites that describe a wealth of microarray software resources are given:

1. http://genome-www5.stanford.edu/resources/restech.shtml
2. https://www.cs.tcd.ie/Nadia.Bolshakova/softwaretotal.html

CONCLUSIONS

DNA microarrays have given biomedical researchers access to a wealth of data. While there are significant problems, huge opportunities are also available. Microarray data analysis is far from easy, and the amount of effort required for critical analysis is nearly always underestimated. Importantly, it is difficult to standardize analysis steps because different datasets contain unique variance patterns that may need different approaches to optimize data and knowledge accrual. Much can be gained by thinking about the experimental design, statistical methods to be adopted, and the underlying biological processes likely to be reflected in the expression profiles and biological states. It is apt to quote the advice from the geneticist and statistician Ronald Fisher in 1938: "To call in the statistician after the experiment is done may be no more than asking him to perform a post-mortem examination: he may be able to say what the experiment died of." It is also important to prioritize the hypotheses and design the experiment so as to adequately power the possibility of observing repeated occurrences of a particular state and set of phenomena. It is wise to approach analysis with the same philosophy as the experiment itself: check after every step. Are the results obtained reasonable with respect to existing knowledge or intuition? With little prior support, there may be little to look forward to as well. Lastly, microarray results validation should be based on at least some additional measures that support principle comparisons and significant gene regulation occurrences. Other techniques (e.g., RTQ-PCR or Northern blotting) and database information can be particularly useful, and in situ hybridization evidence can confirm regulation occurring in specific cellular compartments. The integration of systems biology experimental approaches with genomics, transcriptomics, proteomics, molecular, and genetic knowledge will undoubtedly result in a much greater understanding of molecular, cellular, and systems processes that are subject to hormonal control.

REFERENCES

1. Feng, X., Y. Jiang, P. Meltzer, and P.M. Yen, Thyroid hormone regulation of hepatic genes in vivo detected by complementary DNA microarray. *Mol Endocrinol*, 2000. 14(7): p. 947–55.
2. Dupont, J., J. Khan, B.H. Qu, P. Metzler, L. Helman, and D. LeRoith, Insulin and IGF-1 induce different patterns of gene expression in mouse fibroblast NIH-3T3 cells: identification by cDNA microarray analysis. *Endocrinology*, 2001. 142(11): p. 4969–75.
3. Hoos, A., A. Stojadinovic, B. Singh, M.E. Dudas, D.H. Leung, A.R. Shaha, J.P. Shah, M.F. Brennan, C. Cordon-Cardo, and R. Ghossein, Clinical significance of molecular expression profiles of Hurthle cell tumors of the thyroid gland analyzed via tissue microarrays. *Am J Pathol*, 2002. 160(1): p. 175–83.
4. Lockhart, D.J. and E.A. Winzeler, Genomics, gene expression and DNA arrays. *Nature*, 2000. 405(6788): p. 827–36.
5. van't Veer, L.J., H. Dai, M.J. van de Vijver, Y.D. He, A.A. Hart, M. Mao, H.L. Peterse, K. van der Kooy, M.J. Marton, A.T. Witteveen, G.J. Schreiber, R.M. Kerkhoven, C. Roberts, P.S. Linsley,

R. Bernards, and S.H. Friend, Gene expression profiling predicts clinical outcome of breast cancer. *Nature*, 2002. 415(6871): p. 530–6.

6. Page, G.P., J.W. Edwards, G.L. Gadbury, P. Yelisetti, J. Wang, P. Trivedi, and D.B. Allison, The PowerAtlas: a power and sample size atlas for microarray experimental design and research. *BMC Bioinformatics*, 2006. 7: p. 84.

7. Shi, H. and R. Bressan, RNA extraction. *Methods Mol Biol*, 2006. 323: p. 345–8.

8. Staal, F.J., G. Cario, G. Cazzaniga, T. Haferlach, M. Heuser, W.K. Hofmann, K. Mills, M. Schrappe, M. Stanulla, L.U. Wingen, J.J. van Dongen, and B. Schlegelberger, Consensus guidelines for microarray gene expression analyses in leukemia from three European leukemia networks. *Leukemia*, 2006. 20(8): p. 1385–92.

9. Stangegaard, M., I.H. Dufva, and M. Dufva, Reverse transcription using random pentadecamer primers increases yield and quality of resulting cDNA. *Biotechniques*, 2006. 40(5): p. 649–57.

10. Verhaak, R.G., F.J. Staal, P.J. Valk, B. Lowenberg, M.J. Reinders, and D. de Ridder, The effect of oligonucleotide microarray data pre-processing on the analysis of patient-cohort studies. *BMC Bioinformatics*, 2006. 7: p. 105.

11. Walter, M.A., D. Seboek, P. Demougin, L. Bubendorf, M. Oberholzer, J. Muller-Brand, and B. Muller, Extraction of high-integrity RNA suitable for microarray gene expression analysis from long-term stored human thyroid tissues. *Pathology*, 2006. 38(3): p. 249–53.

12. Wang, H., J.D. Owens, J.H. Shih, M.C. Li, R.F. Bonner, and J.F. Mushinski, Histological staining methods preparatory to laser capture microdissection significantly affect the integrity of the cellular RNA. *BMC Genomics*, 2006. 7: p. 97.

13. Qian, X., B.W. Scheithauer, K. Kovacs, and R.V. Lloyd, DNA microarrays: recent developments and applications to the study of pituitary tissues. *Endocrine*, 2005. 28(1): p. 49–56.

14. Churchill, G.A., Fundamentals of experimental design for cDNA microarrays. *Nat Genet*, 2002. 32 Suppl: p. 490–5.

15. Quackenbush, J., Microarray data normalization and transformation. *Nat Genet*, 2002. 32 Suppl: p. 496–501.

16. Geller, S.C., J.P. Gregg, P. Hagerman, and D.M. Rocke, Transformation and normalization of oligonucleotide microarray data. *Bioinformatics*, 2003. 19(14): p. 1817–23.

17. Yuen, T., E. Wurmbach, R.L. Pfeffer, B.J. Ebersole, and S.C. Sealfon, Accuracy and calibration of commercial oligonucleotide and custom cDNA microarrays. *Nucleic Acids Res*, 2002. 30(10): p. e48.

18. Kacharmina, J.E., P.B. Crino, and J. Eberwine, Preparation of cDNA from single cells and subcellular regions. *Methods Enzymol*, 1999. 303: p. 3–18.

19. Allison, D.B., X. Cui, G.P. Page, and M. Sabripour, Microarray data analysis: from disarray to consolidation and consensus. *Nat Rev Genet*, 2006. 7(1): p. 55–65.

20. Dilley, W.G., S. Kalyanaraman, S. Verma, J.P. Cobb, J.M. Laramie, and T.C. Lairmore, Global gene expression in neuroendocrine tumors from patients with the MEN1 syndrome. *Mol Cancer*, 2005. 4(1): p. 9.

21. Zanolin, M.E., F. Tosi, G. Zoppini, R. Castello, G. Spiazzi, R. Dorizzi, M. Muggeo, and P. Moghetti, Clustering of cardiovascular risk factors associated with the insulin resistance syndrome: assessment by principal component analysis in young hyperandrogenic women. *Diabetes Care*, 2006. 29(2): p. 372–8.

22. Simon, R., M.D. Radmacher, K. Dobbin, and L.M. McShane, Pitfalls in the use of DNA microarray data for diagnostic and prognostic classification. *J Natl Cancer Inst*, 2003. 95(1): p. 14–8.

23. Soinov, L.A., M.A. Krestyaninova, and A. Brazma, Towards reconstruction of gene networks from expression data by supervised learning. *Genome Biol*, 2003. 4(1): p. R6.

24. Ambroise, C. and G.J. McLachlan, Selection bias in gene extraction on the basis of microarray gene-expression data. *Proc Natl Acad Sci U S A*, 2002. 99(10): p. 6562–6.

25. Gentleman, R.C., V.J. Carey, D.M. Bates, B. Bolstad, M. Dettling, S. Dudoit, B. Ellis, L. Gautier, Y. Ge, J. Gentry, K. Hornik, T. Hothorn, W. Huber, S. Iacus, R. Irizarry, F. Leisch, C. Li, M. Maechler, A.J. Rossini, G. Sawitzki, C. Smith, G. Smyth, L. Tierney, J.Y. Yang, and J. Zhang, Bioconductor: open software development for computational biology and bioinformatics. *Genome Biol*, 2004. 5(10): p. R80.

26. Dennis, G., Jr., B.T. Sherman, D.A. Hosack, J. Yang, W. Gao, H.C. Lane, and R.A. Lempicki, DAVID: Database for Annotation, Visualization, and Integrated Discovery. *Genome Biol*, 2003. 4(5): p. P3.

27. Berezikov, E., V. Guryev, and E. Cuppen, CONREAL web server: identification and visualization of conserved transcription factor binding sites. *Nucleic Acids Res*, 2005. 33(Web server issue): p. W447–50.

28. Jegga, A.G., A. Gupta, S. Gowrisankar, M.A. Deshmukh, S. Connolly, K. Finley, and B.J. Aronow, CisMols analyzer: identification of compositionally similar cis-element clusters in ortholog conserved regions of coordinately expressed genes. *Nucleic Acids Res*, 2005. 33(Web server issue): p. W408–11.

29. Sharan, R., I. Ovcharenko, A. Ben-Hur, and R.M. Karp, CREME: a framework for identifying cis-regulatory modules in human-mouse conserved segments. *Bioinformatics*, 2003. 19 Suppl 1: p. i283–91.

30. Butte, A.J. and I.S. Kohane, Mutual information relevance networks: functional genomic clustering using pairwise entropy measurements. *Pac Symp Biocomput*, 2000: p. 418–29.

31. Segal, E., H. Wang, and D. Koller, Discovering molecular pathways from protein interaction and gene expression data. *Bioinformatics*, 2003. 19 Suppl 1: p. i264–71.

32. Hartemink, A.J., D.K. Gifford, T.S. Jaakkola, and R.A. Young, Combining location and expression data for principled discovery of genetic regulatory network models. *Pac Symp Biocomput*, 2002: p. 437–49.

33. Brazma, A., P. Hingamp, J. Quackenbush, G. Sherlock, P. Spellman, C. Stoeckert, J. Aach, W. Ansorge, C.A. Ball, H.C. Causton, T. Gaasterland, P. Glenisson, F.C. Holstege, I.F. Kim, V. Markowitz, J.C. Matese, H. Parkinson, A. Robinson, U. Sarkans, S. Schulze-Kremer, J. Stewart, R. Taylor, J. Vilo, and M. Vingron, Minimum information about a microarray experiment (MIAME)-toward standards for microarray data. *Nat Genet*, 2001. 29(4): p. 365–71.

34. Barrett, T., T.O. Suzek, D.B. Troup, S.E. Wilhite, W.C. Ngau, P. Ledoux, D. Rudnev, A.E. Lash, W. Fujibuchi, and R. Edgar, NCBI GEO: mining millions of expression profiles – database and tools. *Nucleic Acids Res*, 2005. 33(Database issue): p. D562–6.

35. Lee, H.K., A.K. Hsu, J. Sajdak, J. Qin, and P. Pavlidis, Coexpression analysis of human genes across many microarray data sets. *Genome Res*, 2004. 14(6): p. 1085–94.

36. Chuaqui, R.F., R.F. Bonner, C.J. Best, J.W. Gillespie, M.J. Flaig, S.M. Hewitt, J.L. Phillips, D.B. Krizman, M.A. Tangrea, M. Ahram, W.M. Linehan, V. Knezevic, and M.R. Emmert Buck, Post-analysis follow-up and validation of microarray experiments. *Nat Genet*, 2002. 32 Suppl: p. 509–14.

37. Barnes, M., J. Freudenberg, S. Thompson, B. Aronow, and P. Pavlidis, Experimental comparison and cross-validation of the Affymetrix and Illumina gene expression analysis platforms. *Nucleic Acids Res*, 2005. 33(18): p. 5914–23.

38. Culhane, A.C., G. Perriere, and D.G. Higgins, Cross-platform comparison and visualisation of gene expression data using co-inertia analysis. *BMC Bioinformatics*, 2003. 4: p. 59.

39. de Reynies, A., D. Geromin, J.M. Cayuela, F. Petel, P. Dessen, F. Sigaux, and D.S. Rickman, Comparison of the latest commercial short and long oligonucleotide microarray technologies. *BMC Genomics*, 2006. 7: p. 51.

40. Irizarry, R.A., D. Warren, F. Spencer, I.F. Kim, S. Biswal, B.C. Frank, E. Gabrielson, J.G. Garcia, J. Geoghegan, G. Germino, C. Griffin, S.C. Hilmer, E. Hoffman, A.E. Jedlicka, E. Kawasaki, F. Martinez-Murillo, L. Morsberger, H. Lee, D. Petersen, J. Quackenbush, A. Scott, M. Wilson, Y. Yang, S.Q. Ye, and W. Yu, Multiple-laboratory comparison of microarray platforms. *Nat Methods*, 2005. 2(5): p. 345–50.

41. van Ruissen, F., J.M. Ruijter, G.J. Schaaf, L. Asgharnegad, D.A. Zwijnenburg, M. Kool, and F. Baas, Evaluation of the similarity of gene expression data estimated with SAGE and Affymetrix GeneChips. *BMC Genomics*, 2005. 6: p. 91.

42. Woo, Y., J. Affourtit, S. Daigle, A. Viale, K. Johnson, J. Naggert, and G. Churchill, A comparison of cDNA, oligonucleotide, and Affymetrix GeneChip gene expression microarray platforms. *J Biomol Tech*, 2004. 15(4): p. 276–84.

43. Kothapalli, R., S.J. Yoder, S. Mane, and T.P. Loughran, Jr., Microarray results: how accurate are they? *BMC Bioinformatics*, 2002. 3: p. 22.

44. Ntzani, E.E. and J.P. Ioannidis, Predictive ability of DNA microarrays for cancer outcomes and correlates: an empirical assessment. *Lancet*, 2003. 362(9394): p. 1439–44.

45. Michiels, S., S. Koscielny, and C. Hill, Prediction of cancer outcome with microarrays: a multiple random validation strategy. *Lancet*, 2005. 365(9458): p. 488–92.
46. Chang, H.Y., D.S. Nuyten, J.B. Sneddon, T. Hastie, R. Tibshirani, T. Sorlie, H. Dai, Y.D. He, L.J. van't Veer, H. Bartelink, M. van de Rijn, P.O. Brown, and M.J. van de Vijver, Robustness, scalability, and integration of a wound-response gene expression signature in predicting breast cancer survival. *Proc Natl Acad Sci U S A*, 2005. 102(10): p. 3738–43.
47. Sorlie, T., R. Tibshirani, J. Parker, T. Hastie, J.S. Marron, A. Nobel, S. Deng, H. Johnsen, R. Pesich, S. Geisler, J. Demeter, C.M. Perou, P.E. Lonning, P.O. Brown, A.L. Borresen-Dale, and D. Botstein, Repeated observation of breast tumor subtypes in independent gene expression data sets. *Proc Natl Acad Sci U S A*, 2003. 100(14): p. 8418–23.
48. D'Haeseleer, P., How does gene expression clustering work? *Nat Biotechnol*, 2005. 23(12): p. 1499–501.
49. Imbeaud, S. and C. Auffray, 'The 39 steps' in gene expression profiling: critical issues and proposed best practices for microarray experiments. *Drug Discov Today*, 2005. 10(17): p. 1175–82.
50. Robert-Nicoud, M., M. Flahaut, J.M. Elalouf, M. Nicod, M. Salinas, M. Bens, A. Doucet, P. Wincker, F. Artiguenave, J.D. Horisberger, A. Vandewalle, B.C. Rossier, and D. Firsov, Transcriptome of a mouse kidney cortical collecting duct cell line: effects of aldosterone and vasopressin. *Proc Natl Acad Sci U S A*, 2001. 98(5): p. 2712–6.
51. Xu, L.L., Y.P. Su, R. Labiche, T. Segawa, N. Shanmugam, D.G. McLeod, J.W. Moul, and S. Srivastava, Quantitative expression profile of androgen-regulated genes in prostate cancer cells and identification of prostate-specific genes. *Int J Cancer*, 2001. 92(3): p. 322–8.
52. de Waard, V., B.M. van den Berg, J. Veken, R. Schultz-Heienbrok, H. Pannekoek, and A.J. van Zonneveld, Serial analysis of gene expression to assess the endothelial cell response to an atherogenic stimulus. *Gene*, 1999. 226(1): p. 1–8.
53. Datson, N.A., J. van der Perk, E.R. de Kloet, and E. Vreugdenhil, Identification of corticosteroid-responsive genes in rat hippocampus using serial analysis of gene expression. *Eur J Neurosci*, 2001. 14(4): p. 675–89.
54. Gruvberger, S., M. Ringner, Y. Chen, S. Panavally, L.H. Saal, A. Borg, M. Ferno, C. Peterson, and P.S. Meltzer, Estrogen receptor status in breast cancer is associated with remarkably distinct gene expression patterns. *Cancer Res*, 2001. 61(16): p. 5979–84.
55. Suzuki, T., F. Schirra, S.M. Richards, N.S. Treister, M.J. Lombardi, P. Rowley, R.V. Jensen, and D.A. Sullivan, Estrogen's and progesterone's impact on gene expression in the mouse lacrimal gland. *Invest Ophthalmol Vis Sci*, 2006. 47(1): p. 158–68.
56. Aronow, B.J., B.D. Richardson, and S. Handwerger, Microarray analysis of trophoblast differentiation: gene expression reprogramming in key gene function categories. *Physiol Genomics*, 2001. 6(2): p. 105–16.
57. Davies, S., D. Dai, G. Pickett, and K.K. Leslie, Gene regulation profiles by progesterone and dexamethasone in human endometrial cancer Ishikawa H cells. *Gynecol Oncol*, 2005. 101(1): p. 62–70.
58. Hindmarch, C., S. Yao, G. Beighton, J. Paton, and D. Murphy, A comprehensive description of the transcriptome of the hypothalamoneurohypophyseal system in euhydrated and dehydrated rats. *Proc Natl Acad Sci U S A*, 2006. 103(5): p. 1609–14.
59. Elfilali, A., S. Lair, C. Verbeke, P. La Rosa, F. Radvanyi, and E. Barillot, ITTACA: a new database for integrated tumor transcriptome array and clinical data analysis. *Nucleic Acids Res*, 2006. 34(Database issue): p. D613–6.
60. Newman, J.C. and A.M. Weiner, L2L: a simple tool for discovering the hidden significance in microarray expression data. *Genome Biol*, 2005. 6(9): p. R81.
61. Rhodes, D.R., J. Yu, K. Shanker, N. Deshpande, R. Varambally, D. Ghosh, T. Barrette, A. Pandey, and A.M. Chinnaiyan, ONCOMINE: a cancer microarray database and integrated data-mining platform. *Neoplasia*, 2004. 6(1): p. 1–6.
62. Ball, C.A., I.A. Awad, J. Demeter, J. Gollub, J.M. Hebert, T. Hernandez-Boussard, H. Jin, J.C. Matese, M. Nitzberg, F. Wymore, Z.K. Zachariah, P.O. Brown, and G. Sherlock, The Stanford microarray database accommodates additional microarray platforms and data formats. *Nucleic Acids Res*, 2005. 33(Database issue): p. D580–2.

BMC Bioinformatics. 2006; 7: 84.
Published online 2006 February 22. doi: 10.1186/1471-2105-7-84.

The *PowerAtlas*: a power and sample size atlas for microarray experimental design and research

Reviewed by Grier P Page,[1] Jode W Edwards,[1,2] Gary L Gadbury,[3] Prashanth Yelisetti,[1] Jelai Wang,[1] Prinal Trivedi,[1] and David B Allison[1]

II   HORMONE ACTION AND MOLECULAR MECHANISMS

2

Gene Expression Profiles and Transcription Factors Involved in Parathyroid Hormone Signaling in Osteoblastic Cells

Ling Qin, PhD, Ping Qiu, PhD, and Nicola C. Partridge, PhD

CONTENTS

OSTEOPOROSIS AND PARATHYROID HORMONE TREATMENT
DNA MICROARRAY ANALYSES IDENTIFY NOVEL GENES REGULATED
  BY PTH IN OSTEOBLASTIC CELLS
THE IMPLICATIONS OF DNA MICROARRAY DATA ON DELINEATING
  THE MECHANISM OF PTH TREATMENT OF OSTEOPOROSIS
COMPUTATIONAL PROMOTER ANALYSIS OF PTH-REGULATED GENES
  REVEALS TRANSCRIPTION FACTORS INVOLVED IN PTH SIGNALING
CONCLUSIONS
ACKNOWLEDGMENTS
REFERENCES

Abstract

Parathyroid hormone (PTH), a peptide hormone regulating calcium homeostasis in humans, is also one of the most effective treatments for osteoporosis, a metabolic bone disease prevailing in the elderly. Therefore, studying PTH actions and its downstream signaling pathways in osteoblasts has been a focus of the bone research field. The recent advances in microarray technology have identified many novel PTH-regulated genes covering a wide range of biological functions and protein families. In this review, we summarize the implications of DNA microarray data on delineating the mechanism of PTH treatment of osteoporosis. We also describe a computational promoter analysis method to extract useful information about PTH-regulated transcription factors from the microarray dataset. The combination of microarray experiments and bioinformatics analyses will shed light on our final goal to reconstruct the global regulatory network established by PTH.

Key Words: parathyroid hormone, osteoporosis, bone metabolism, osteoblast, computational promoter analysis, transcription factor binding sites.

From: *Contemporary Endocrinology: Genomics in Endocrinology:*
DNA Microarray Analysis in Endocrine Health and Disease
Edited by S. Handwerger and B. Aronow © Humana Press, Totowa, NJ

OSTEOPOROSIS AND PARATHYROID HORMONE TREATMENT

Osteoporosis is a metabolic bone disease characterized by low bone mass and microarchitectural deterioration of bone tissue, leading to enhanced bone fragility and a consequent increase in fracture risk. Currently, it is a major public health threat for an estimated 44 million Americans, most of them postmenopausal women or the elderly. Osteoporotic fractures contribute substantially to morbidity and mortality in an aging world population, thus consuming considerable health resources. The cause of osteoporosis is an imbalance of bone remodeling. Bone, a highly mineralized tissue that provides mechanical support and metabolic functions, constantly undergoes remodeling. Bone remodeling occurs at discrete sites within the skeleton and proceeds in an orderly fashion, with bone resorption by osteoclasts always being followed by bone formation by osteoblasts, a phenomenon referred to as coupling. This physiological process is coordinated and tightly regulated by local and endocrine factors to ensure that the bone formation rate matches the bone resorption rate. However, in osteoporosis patients, osteoclast activity exceeds osteoblast activity, leading to irreversible bone loss.

The drugs currently available for osteoporosis treatment fall into two categories. The first, antiresorptives, inhibit osteoclastic bone resorption. This category includes bisphosphonates (alendronate, ibandronate, and risedronate) and oestrogenic compounds (estrogen, tamoxifen, and raloxifene). Although this category of agents dominates osteoporosis therapeutics, there are limitations to their efficacy. Even the most potent antiresorptive drugs only reduce the risk of osteoporotic fractures by about 50% (1) and at best increase bone density by about 10% over ten years' treatment (2). This is mainly due to the fact that bone remodeling is coupled and therefore the decrease in bone resorption usually correlates with a decrease in bone formation. The other category of therapy, anabolic therapy, promises to overcome this ineffectiveness by primarily targeting the osteoblast and directly promoting bone formation. To date, parathyroid hormone (PTH) 1–34 (teriparatide) is the only FDA-approved anabolic agent for osteoporosis treatment. Several clinical trials demonstrated that PTH is much more effective in stimulating the increase in bone mineral density (BMD) of the skeleton, especially in trabecular bone, than alendronate, the most popular bisphosphonate prescribed for osteoporosis patients (3, 4).

Parathyroid hormone is an 84 amino acid peptide secreted by the parathyroid glands and is one of the principal regulators of calcium homeostasis for humans and most likely all terrestrial vertebrates. The amino-terminal region of PTH (the first 34 amino acids) is associated with most of its known biologic actions and shows high homology among the different vertebrate species. PTH's main targets in the body are bone and kidney, leading to an increase in serum calcium concentrations. Interestingly, administration of either the full length or N-terminal peptide (1–34) of PTH is a double-edged sword for bone metabolism, since continuous infusion of PTH causes bone loss (catabolic action), while intermittent administration induces bone formation (anabolic action). The disparity of these two actions has attracted intensive studies over a long period, and a major effort has been made to investigate the mechanism of its anabolic action because of its use for the osteoporotic patient.

The bone-forming cells, osteoblasts, originate from bone marrow stromal stem cells. These precursors undergo proliferation and differentiation into preosteoblasts and then into mature osteoblasts. The latter cells are the primary PTH-responsive cells because they express higher levels of the PTH type I receptor (PTH1R) than preosteoblasts (5,6,7). Although there is some emerging evidence that osteoclasts express PTH1R and may respond directly to the PTH signal (8, 9), the prevailing view is that PTH stimulates osteoclast activity indirectly through the osteoblast by upregulating RANKL expression by the osteoblast (10). Therefore, understanding the PTH signaling pathway and downstream gene regulation in the osteoblast is the most crucial step toward revealing the mechanism of PTH treatment of osteoporosis.

DNA MICROARRAY ANALYSES IDENTIFY NOVEL GENES REGULATED BY PTH IN OSTEOBLASTIC CELLS

At the tissue and cellular levels, daily PTH injection exerts its anabolic effects by promoting osteoblast differentiation (11,12,13,14) and inhibiting osteoblast apoptosis (15). At a molecular level, PTH binds to the PTH1R, a G-protein-coupled receptor, on the osteoblast and strongly activates the $G\alpha s$/cAMP/protein kinase A (PKA) pathway and weakly activates the $G\alpha q$/phospholipase C (PLC)/protein kinase C (PKC) pathway. Recently, it was found that the increased level of cAMP by PTH treatment also leads to activation of Epac (also known as cAMP–guanine nucleotide exchange factor, cAMP-GEF)/Rap1/B-raf pathway in the osteoblast and results in phosphorylation of extracellular signal-regulated kinases (ERKs) (16). Two decades of studies have revealed a number of genes that are regulated by PTH, such as c-fos, interleukin-6, collagenase-3, tissue inhibitors of metalloproteinases, RANKL, type 1 collagen, regulator of G-protein signaling 2, etc. [reviewed in (17)]. However, how G-protein activation leads to changes in the above genes' expression, whether there are other genes regulated by PTH, when expression of these genes are changed, and how these genes regulate each other, are all largely unknown. A recent advance in studying global gene expression patterns using DNA microarray makes it possible to address the above questions and therefore to reveal PTH's actions. To date, there are two published reports adopting this strategy, and each of those results yielded important insights into the mechanism of PTH action and opens up avenues for future research.

In the first report (18), UMR 106-01, a strongly PTH-responsive rat osteosarcoma osteoblastic cell line, was chosen for this study. A total of 125 genes and 30 expressed sequence tags (ESTs) out of 8,799 known rat transcripts and ESTs on the chip (RG-U34A, Affymetrix) were found to have at least twofold expression changes at 4, 12, 24 h after rPTH (1–34, 10^{-8} M) treatment. Among those 125 genes, 14 were previously known to be PTH-regulated, but this was the first report of regulation of the remainder of the genes. Quantitative reverse transcriptase-PCR indicated that 90% of those genes are truly regulated more than twofold by PTH (1–34) in UMR 106-01 cells, demonstrating the validity of the microarray data. Interestingly, the genes cover a wide range of different biological functions and most of them belong to the following protein families: hormones, growth factors, and receptors; signal transduction pathway proteins; transcription factors; proteases; metabolic enzymes; structural and matrix proteins; transporters, etc. However, because no in vitro cell culture system can replicate the dual actions of PTH, it is difficult to know which of the genes seen to be regulated in vitro contribute to PTH's anabolic action and which to PTH's catabolic action. It is possible that these genes contribute to both of PTH's actions because a recent hypothesis proposed that the different kinetics of PTH anabolic or catabolic administration leads to different temporal and level of expression of the same set of genes, rather than different sets of genes being expressed, and it is this difference which results in opposite outputs *(19)*.

To elucidate PTH's in vivo function, a second report investigated gene expression changes in the osteoblast-rich metaphyseal area of the rat distal femur after either intermittent PTH injection or continuous PTH infusion for a week *(20)*. It was found that both PTH treatments co-regulated 22 genes including known bone formation genes (i.e., collagens, osteocalcin, decorin, and osteonectin), while intermittent PTH injection regulated an additional 19 genes and continuous PTH infusion regulated an additional 173 genes. The genes also covered a wide range of functions, including adhesion, matrix-associated proteins, signal transduction, catalysis and metabolism, nucleic acid binding and transcription factors, cell cycle, transporters, etc. It is surprising to find that compared to continuous treatment (195 genes) intermittent PTH treatment only regulated 41 genes, 33 of them with less than twofold change. This is probably due to the fact that, in the intermittent treatment, RNA samples were harvested 24 h after the last PTH injection. PTH injections mostly transiently regulate gene expression in bone with a peak change at about 1 h and return to baseline after 4 h *(13,21,22)*. This may also partially explain why only a total of 14 of the same genes are regulated in the UMR 106-01 dataset and the in vivo bone dataset. It is possible that the UMR 106-01 dataset more closely resembles the gene expression changes after 1 h of PTH injection in vivo. Due to the heterogeneity of in vivo bone samples, it should be noted that some of the genes in the in vivo bone dataset are actually not expressed by the osteoblast and their response to PTH is not a direct effect. A subsequent study of PDGFα, a cytokine regulated by continuous PTH treatment, found that it was expressed by the mast cells in the bone marrow *(23)*.

THE IMPLICATIONS OF DNA MICROARRAY DATA
ON DELINEATING THE MECHANISM OF PTH TREATMENT
OF OSTEOPOROSIS

Taken together, there are about 300 PTH-regulated genes revealed by the above two microarray assays. Those genes cover nine (binding, catalytic activity, chaperone regulator, enzyme regulator, motor, signal transducer, structural molecule, transcription regulator, and transporter) out of fourteen categories grouped by molecular function according to gene ontology (GO). Several striking features about PTH signaling emerged after reviewing these gene lists. First, PTH significantly changes the expression levels of many hormones, cytokines, and growth factors produced by osteoblasts and, therefore, may influence the behaviors of osteoblasts or their neighboring cells. For example, PTH regulates the expression of secreted frizzled-related protein 4 (Wnt signaling), amphiregulin and TGF-α (epidermal growth factor receptor signaling), Jagged1 (Notch signaling), bone morphogenetic proteins (BMPs, BMP receptor signaling), etc. Second, several genes (PTH1R, adenylyl cyclase, cAMP phosphodi-esterases, and protein phosphatase inhibitor-1) are regulated in such ways that will inhibit the PTH1R/Gαs/cAMP/PKA pathway, the major route for PTH action. This feedback mechanism may explain the mechanism of PTH's dual functions *(19)*. Third, many neuron-specific genes that were not expected to be present in the osteoblast are actually activated by PTH treatment, revealing brain signaling pathways that may be important in bone metabolism. Fourth, a substantial number of PTH-regulated genes are extracellular matrix-associated genes, such as collagens, matrix metalloproteinases, various proteinases and peptidases, suggesting that PTH significantly regulates the microenvironment of osteoblasts. In this section, several genes revealed in the microarray data and studied in terms of their functions in PTH treatment and in bone metabolism are discussed.

Amphiregulin

Amphiregulin belongs to a family of epidermal growth factor (EGF)-like ligands, which also includes EGF, transforming growth factor alpha (TGF-α), heparin-binding EGF (HB-EGF), betacellulin, and epiregulin *(24)*. All of these ligands bind to the EGF receptor (EGFR/ErbB1), while the latter three also bind to ErbB4. The EGFR (ErbB1) is a receptor tyrosine kinase and lies at the beginning of a complex signal transduction cascade that modulates cell proliferation, survival, adhesion, migration, and differentiation *(25)*. Upon ligand binding, the EGFR undergoes dimerization and phosphorylation at tyrosine residues in its intracellular domain, thus activating several important cellular signal transduction pathways. The major signaling routes are the Ras-Raf-MAP-kinase *(26)* and PI-3-kinase-Akt pathways *(27)*.

Amphiregulin was first isolated from conditioned medium of MCF-7 human breast carcinoma cells exposed to phorbol 12-myristate-13-acetate (PMA) *(28, 29)*. It is bifunctional since it inhibits the growth of many human tumor cells but stimulates the proliferation of other cells, such as normal fibroblasts and keratinocytes *(30, 31)*. Previously, EGF has been shown to have several effects on bone cells or on bone: it stimulates osteoblast proliferation *(32)*, decreases alkaline phosphatase *(33)* and collagen production *(34)*, changes bone nodule formation *(35)*, and yet has catabolic effects on bone *(36)*, i.e., similar to amphiregulin, bifunctional effects. Nevertheless, the production, detailed mechanism, and the significance of the EGF signaling pathway in bone are not well understood.

DNA microarray analysis and subsequent RT-PCR data indicated that amphiregulin is a general immediate response gene for PTH action, both in osteoblastic cell lines and in bone *(22)*. In UMR 106-01 cells, the peak regulation occurs at 1 h after 10^{-8} M rPTH (1–34) treatment with a 23-fold increase in mRNA. Similar results were obtained with rat primary osteoblastic cells (23-fold at 1 h) and mouse preosteoblastic MC3T3 cells (4-fold at 1 h). More interestingly, the level of amphiregulin mRNA in rat osteoblast-enriched femoral metaphyseal primary spongiosa was dramatically elevated to about 12-fold after 1 h of hPTH (1–38) injection, and quickly decreased to about twofold after 4 h. Additional promoter analysis demonstrated that the amphiregulin promoter strongly responds to rPTH (1–34) treatment in UMR cells *(37)*. Functional studies demonstrated that amphiregulin is a potent growth factor for preosteoblastic cells in an EGFR-dependent manner. However, continuous treatment of rat primary osteoblastic cultures with amphiregulin completely inhibited osteogenic differentiation, with no bone nodule formation, reduced alkaline phosphatase activity, and very low expression of osteocalcin, a late osteoblast marker. Moreover, amphiregulin-null mice displayed significantly lesser tibial trabecular bone than wild-type mice, by microCT analysis. Accordingly, a model has been proposed that PTH acts on the mature or differentiating osteoblast inducing expression of amphiregulin, and this growth factor then increases the preosteoblast population by stimulating their proliferation and inhibiting their differentiation. Since this is a transient regulation, the overall outcome will be an increase in the osteoblast population, therefore contributing to PTH's anabolic effect. Since bone marrow mesenchymal stem cells express EGFR (38) and can proliferate and form colonies in a serum-deprived medium as long as EGF is present (39), it is possible that AR has a role in regulating the mesenchymal stem cell pool and, therefore, influencing several other cell lineages derived from mesenchymal stem cells, such as adipocytes and chondrocytes.

Jagged-1

The trans-membrane Jagged and Delta protein families are ligands for Notch receptors (Notch 1–4) in vertebrates (40). Notch signaling activated by those ligands plays an important role in cell fate decisions in many different tissues in multicellular organisms; for example, the nervous system, the vascular system, the hematopoietic system, somites, muscle, skin, and pancreas. The UMR DNA microarray data revealed that Jagged-1 is a PTH-regulated gene. Further, RT-PCR studies confirmed that 2 h of 10^{-8} M rPTH (1–34) increased Jagged-1 mRNA expression 8-fold and 5 days of intermittent injections of hPTH (1–34) increased Jagged-1 mRNA levels 4-fold in the rat osteoblast-rich femoral metaphyses (41), while continuous infusion did not affect Jagged-1 expression (20).

A separate elegant investigation also found that Jagged-1 protein was overexpressed in the osteoblasts of transgenic mice having a constitutively active PTH1R under the control of an osteoblast-specific $\alpha1(I)$ collagen promoter (42). This overexpression of Jagged-1 led to elevated Notch signaling in the hematopoietic stem cell population and resulted in expansion of the stem cell compartment. Although this may not directly be involved in PTH's anabolic effect and its treatment of osteoporosis, it has significant pharmacological impact because intermittent PTH injection could be a potent tool for reducing the damage inflicted on bone marrow by cytotoxic cancer chemotherapeutic drugs and ionizing radiation (43).

Previous studies have shown that osteoblastic cells express Notch receptors (44,45). To investigate a potential role of Jagged-1 in PTH's anabolic effect, rat primary osteoblastic cells were treated with dish-tethered Jagged-1 protein (41). It was found that, similar to amphiregulin, Jagged-1 stimulated preosteoblast proliferation but inhibited its further differentiation. Therefore, the role of Jagged-1 in PTH treatment of osteoporosis might be similar to that of amphiregulin, increasing the preosteoblast or even mesenchymal stem cell pool, leading to an increase in osteoblasts and increased bone formation.

Mitogen-Activated Protein Kinase (MAPK) Phosphatase-1 (Mkp-1), Cyclin D1 and p21^{cip1}

A prevailing hypothesis to explain PTH's anabolic effect is that PTH promotes osteoblast differentiation. As a terminally differentiated cell, the osteoblast must exit the cell cycle. Therefore, arresting cell cycle progression of the osteoblast could be part of the mechanism elicited by PTH to facilitate differentiation. While the effect of PTH injection on osteoblast proliferation is not clear in vivo, in vitro studies were also contradictory: while PTH could stimulate the proliferation of osteoblastic cells under some circumstances, such as low concentrations of PTH (10^{-11} M) *(46)*, high cell density *(47)*, short exposure to PTH *(48)*, or in a particular cell line (TE-85) *(49)*, at a high-concentration (10^{-8} M) PTH consistently inhibits the growth of several well-established osteoblastic cell lines [UMR 106-01 *(50)*, MC3T3-E1 *(51)*, SaOS-2 *(52)*, and rat calvarial primary cultures *(22)*].

The UMR microarray data revealed that two genes involved in cell cycle progression (MKP-1 and cyclin D1) were regulated by PTH. MKP-1 is a dual specificity phosphatase capable of removing both phosphotyrosine and phosphothreonine from the Thr-X-Tyr motif of MAP kinases (reviewed in *(53,54)*). Evidence strongly suggests that one family of the MAP kinases (ERK1/2, p42/p44^{MAPK}), whose dual phosphorylation correlates with cell proliferation in most cell types, are substrates for MKP-1 *(55)*. Cell cycle analyses indicated that overexpression of MKP-1 arrested UMR 106-01 cells in G1 phase *(21)*. In mammalian cells, the cell cycle is governed by the activities of the cyclin-dependent kinases (CDKs) and their regulatory cyclin partners. Progression from G1 to S phase requires cyclin D/cdk4(6) and cyclin E/cdk2, and is regulated by CDK inhibitors (CDKIs): INK4 proteins interfere with cyclin D/cdk4(6), and CIP/KIP members (p21^{Cip1}, p27^{Kip1}, and p57^{Kip2}) bind to both complexes. Further experiments confirmed that PTH strongly upregulated MKP-1 and downregulated cyclin D1 in osteoblastic cell lines at both mRNA and protein levels *(21)*. Screening all cyclin-related proteins additionally identified p21^{cip1} as an immediate response gene for PTH *(21)*. Furthermore, intermittent injections of hPTH (1–38) increased MKP-1 and p21^{cip1} and decreased cyclin D1 mRNA levels in the distal femoral metaphyses. These data strongly suggest that arresting the cell cycle progression of the mature osteoblast is a physiologically relevant event for PTH's anabolic actions. Since cell cycle arrest is required for terminal cell differentiation, this could be one mechanism for PTH to exert its anabolic function by promoting osteoblast differentiation in its target cells.

COMPUTATIONAL PROMOTER ANALYSIS OF PTH-REGULATED GENES REVEALS TRANSCRIPTION FACTORS INVOLVED IN PTH SIGNALING

The DNA microarray dataset conveys much more information than just identifying genes regulated at the mRNA level. The use of DNA microarrays to study global gene expression profiles is emerging as a pivotal technology in functional genomics. Comparison of gene expression profiles under different biological conditions reveals corresponding modifications in cellular transcriptional programs. Microarray measurements do not, however, directly reveal the regulatory networks that underlie the observed transcriptional modulation. While it successfully identifies transcription factors regulated at the transcriptional level, it cannot reveal those regulated at post-transcriptional levels. For example, PTH signaling in the osteoblast quickly results in phosphorylation of cAMP-response element (CRE)-binding protein (CREB) at Ser-133, and this phosphorylated CREB is required for the enhanced transcription of the AP-1 family member, c-fos *(56)*, receptor activator of NFκB-ligand (RANKL), an inducer for osteoclastogenesis *(57)*, and amphiregulin *(37)*. However, DNA microarray is unable to identify CREB as an important mediator for PTH function because its mRNA and protein levels remain the same after PTH treatment. Due to the current rapid accumulation of complete genome sequences, it is feasible to use both microarray results and computational promoter analysis to deduce the transcription factors involved in signaling pathways. In this section, we will use the UMR microarray dataset as an example to explain how to identify transcription factors playing an important role in PTH signaling *(18)*.

Methods

The underlying assumption of using high-throughput expression profiling technology in computational promoter analysis is that a set of co-regulated genes usually share a similar set of regulatory motifs, which, in most cases, are the transcription factor binding sites (TFBSs). Therefore, the principle of computational promoter analysis is to identify the over-represented TFBSs in the promoter regions of genes in the microarray dataset. When we performed these analyses, the rat genomic sequences had yet to be completed, but the human draft was finished. Hence, we mapped all the PTH-regulated genes in the UMR (a rat cell) dataset to their human orthologs and used promoters of human orthologs for the subsequent promoter analysis. Among the 125 genes, 73 genes had a human ortholog.

The first step toward analysis was to find the promoter, especially the transcription start site (TSS). While the translation start site is easily identified by examining the coding region, it is much more difficult to find the TSS in eukaryotic genes because it is not always close to the translation start site, sometimes occurring several kilobases away from the translational start site. To ensure that the 5′ end of the mRNA is close to the TSS, only mRNAs that encode the N-terminus of the protein were used for transcript mapping. Furthermore, only sequences in the Genbank Refseq database were used to reduce gene redundancy. A local alignment software package, AAT, was used for alignment of the 5′ end of the cDNA with the human genome draft sequence. To reduce the number of undesirable matches due to interspersed repeats, the DNA sequence was screened for interspersed repeats using RepeatMasker. Only genes having 5,000 bp of continuous regions upstream of the first exon start sites were kept for further transcription factor site analysis. We term this PTH-regulated gene list as the sample list. It contains 63 genes. Meanwhile, a reference list was constructed by collecting human genes in the Genbank Refseq section that have more than 5000 bp continuous regions upstream of the first exon start sites. It contains 4221 genes.

Because most transcription factors bind to short (5–25 bp), degenerate sequence motifs, a position weight matrix (PWM) or International Union of Pure and Applied Chemistry (IUPAC) string is often used to summarize the binding specificity of these factors. TRANSFAC (http://transfac.gbf.de/TRANSFAC) is a database of transcription factors, their binding sites, and TFBS sequence profiles generated from the published literature and represented in PWM and IUPAC string *(58,59)*. The consensus sequences represented by the IUPAC string or PWM in TRANSFAC can be used to search the promoters for known TFBSs. Several programs have been developed to perform searches based on PWM and IUPAC: SIGNAL SCAN *(60)*, MATRIX SEARCH *(61)*, Matinspector *(62)*, ConsInspector *(63)*, TFSearch *(64)*, etc.

In order to reveal over-represented TFBSs, we calculated their occurrence frequency in the sample list and the reference list, respectively. For each gene in the list, 500, 1000, and 2000 bp upstream of the first exon start site were retrieved as possible promoter sequence. Three different lengths of promoter were used for later analysis in order to avoid possible contamination of the wrong annotation of the first exon. For each distance, all promoter sequences within a list were pooled together and then checked for TFBSs using the TRANSFAC database. Matrix similarity score (a value to measure the similarity between the matrix and the actual DNA sequence; complete match has a value of 1) of 0.8 was used as cut-off score. The matrix's normal occurrence frequency is equal to total number of matrix occurrences in the reference database/total promoter number. Using statistical analyses, we were able to identify those matrices with significantly higher occurrence frequency in the sample list than in the reference list. Their corresponding transcription factors are likely to play a role in PTH signaling.

Results and Validation

Using the above computational method, 12 groups of transcription factors and their matrices were identified from the UMR microarray dataset (Table 1). The reason that some groups contain more than one transcription factor is that one transcription factor's matrix may tolerate another matrix. For example, Bach1's matrix (SATGAGTCATGNY) is a stringent version of AP-1's matrix (TGAGTCAKC). Therefore, a Bach1 binding site must be also an AP-1 binding site. But an AP-1 binding site may not necessarily be a Bach1 binding site. In this case, the significance of a Bach1 binding site occurring in a promoter region may be a consequence of Bach1 playing a role in PTH signaling, or may just reflect an AP-1 binding site.

The fact that previously known PTH-regulated transcription factors (AP-1, CREB, and c-myc) and transcription factors identified in the UMR microarray dataset (C/EBPs) are present in Table 1 demonstrates that the above computational approach does indeed correctly identify transcription factors that may mediate PTH signaling. It is comforting to see that CREB, a transcription factor regulated by PTH at the post-translational level and therefore not identifiable in microarray experiments, is pinpointed by this computational method. Recently, another transcription factor in this list, activated protein 2 (AP2), was found to be regulated by PTH at the post-translational level and this regulation is important for PTH activation of insulin-like growth factor binding protein-5 (IGFBP-5) *(65)*, further demonstrating the validity of this list. Other transcription factors in this list, such as AP-4, SP1, FoxD3 (also known as Hfh2, a member of the forkhead family of transcription factors), and MEF2 (myocyte enhancer factor-2), etc., points out an interesting direction for future study of the PTH regulation network.

Certainly, there will be false-positive results from this method for several reasons. First, mammalian promoters are very diverse and hard to predict. The transcript mapping approach (finding TSS by aligning full-length mRNA with their counterpart genomic sequence) described above usually achieves over 80% accuracy in promoter localization *(66–68)*. The incompleteness of full-length transcripts and 5′ end splice variants are the main reasons for the false-positive results. This will be improved by future advances in gene annotation. Second, although a PWM-based search is sensitive, it is not very specific because not every putative TFBS binds with its corresponding transcription factor in vivo and is, therefore, functionally significant. Comparative genomics (also referred to as phylogenetic footprinting) is a widely used method by many researchers to reduce the false-positive rate based on the assumption that functional sequences such as coding regions and regulatory modules are more

Table 1
The Transcription Factors that May Play a Role in PTH Signaling, as Revealed
by computational Promoter Analysis

Group	Transcription factor	No. of genes containing matrix	Identifier	Consensus binding sequence
1	AP-1	30	V$AP1_01	TGAGTCAKC
1	Bach1	3	V$BACH1_01	SATGAGTCATGNY
2	AP-2	38	V$AP2_Q6	MKCCCSCNGGCG
3	AP-4	19	V$AP4_Q5	CAGCTG
3	AP-4	39	V$AP4_Q6	CWCAGCTGG
4	ATF4	12	V$ATF4_Q2	SVTGACGYMABG
4	CREB	8	V$CREB_Q2	STGACGTAA
5	C/EBP beta	39	V$CEBPB_02	KNTTGCNYAAY
5	C/EBP	59	V$CEBP_Q2	TTGCNNAA
6	FoxD3	37	V$FOXD3_01	AWTGTTTRTTT
7	GC box	39	V$GC_01	RGGGGCGGGGCNK
7	MazR	16	V$MAZR_01	SGGGGGGGGGMC
7	SP1	54	V$SP1_01	GRGGCRGGGW
7	SP1	39	V$SP1_Q6	GGGGGCGGGGY
8	MEF2	8	V$MEF2_02	KCTAWAAATAGM
8	MEF2	4	V$MEF2_03	WKCTAWAAATAGM
8	RSRFC4	4	V$RSRFC4_Q2	RNKCTATTTWTAGMWN
9	MYCMAX	35	V$MYCMAX_B	ANCACGTGNNW
10	ZF5	46	V$ZF5_B	RNRNRCGCGCW
11	EVI1	12	V$EVI1_04	GATANGANWAGATA
12	MTATA	7	V$MTATA_B	BNTWTAAANCNBNVSS
12	TATA	27	V$TATA_01	STATAAAWR

The consensus binding sequences were written in IUPAC 15-letter code.
ATF4, activating transcription factor 4; FoxD3, fork head box D3; MazR, MAZ related factor;
MEF2, myocyte enhancer factor-2; RSRFC4, related to serum response factor C4; EVI1, ectopic
viral integration site 1 encoded factor; MTATA, muscle TATA box, C/EBP: CAAT/enhancer binding
protein.

conserved evolutionarily than nonfunctional regions. Third, this computational method
is also likely to omit some true TFBSs because in multicellular organisms, regulatory
elements are not necessarily located immediately upstream of the TSS when compared
with single-celled organisms, and can be found upstream or downstream of the gene, in
introns, or even spread over tens or even hundreds of kilobase pairs. Lastly, this method
is unable to extract unknown transcription factors or transcription factors with unknown
binding sites. One approach to solve this problem is searching for over-represented
motifs in the collection of co-regulated promoter regions (69,70). Popular programs that
can perform this task are Consensus, MEME, Gibbs Sampler, DIALIGN, ANN-Spec,
AlignACE, PROJECTION, MDScan, YMF, etc. (71). The biggest challenge remaining
is to match those motifs with the corresponding transcription factors.

Despite the limitations mentioned above, the aforementioned computational promoter
analysis is general and can be applied to the analysis of transcriptional networks
controlling any biological process in combination with gene expression microarrays.

Previously, such methodologies have been successfully demonstrated mainly in prokaryotes and lower eukaryotic organisms (72,73,74). Our computational promoter analysis is one of several examples that have been successful with mammalian cells. In another study, a similar computational approach was used to reveal transcription factors in E2F target genes and in each phase of cell cycle progression using several sets of microarray data from human cells (75).

CONCLUSIONS

An ultimate goal of studying hormone regulation is delineating the transcriptional regulation of the entire genome and reconstructing the global regulatory network. The rapid accumulation of genome sequences and microarray technology make the former task a reality. For the latter task, microarray analysis is not enough because it only captures a small portion of the overall picture. It must be combined with cis-regulatory element analysis to decipher the regulatory network. The improvements in various bioinformatics tools make this task feasible. PTH signaling in the osteoblast provides an excellent example for this kind of study. The data obtained can identify co-regulated genes, as well as novel transcription factors not known to be involved in a signaling network. Nevertheless, these microarray and bioinformatics analyses must be confirmed by further experimentation, either at the single-gene level or by newly developed techniques, such as ChIP-on-chip.

ACKNOWLEDGMENTS

N.C.P. was supported by National Institutes of Health (DK48109 and DK47420). L.Q. was supported by National Institutes of Health (DK071988-01) and by a National Osteoporosis Foundation Research Grant.

REFERENCES

1. Cranney A, Guyatt G, Griffith L, Wells G, Tugwell P, Rosen C 2002. Meta-analyses of therapies for postmenopausal osteoporosis. IX: Summary of meta-analyses of therapies for postmenopausal osteoporosis. Endocr Rev 23:570–8.
2. Bone HG, Hosking D, Devogelaer JP, Tucci JR, Emkey RD, Tonino RP, Rodriguez-Portales JA, Downs RW, Gupta J, Santora AC, Liberman UA 2004. Ten years' experience with alendronate for osteoporosis in postmenopausal women. N Engl J Med 350:1189–99.
3. Black DM, Greenspan SL, Ensrud KE, Palermo L, McGowan JA, Lang TF, Garnero P, Bouxsein ML, Bilezikian JP, Rosen CJ 2003. The effects of parathyroid hormone and alendronate alone or in combination in postmenopausal osteoporosis. N Engl J Med 349:1207–15.
4. Finkelstein JS, Hayes A, Hunzelman JL, Wyland JJ, Lee H, Neer RM, Black DM, Greenspan SL, Ensrud KE, Palermo L, McGowan JA, Lang TF, Garnero P, Bouxsein ML, Bilezikian JP, Rosen CJ 2003. The effects of parathyroid hormone, alendronate, or both in men with osteoporosis. N Engl J Med 349:1216–26.
5. Aubin JE, Heersche JNM 2001. Cellular actions of parathyroid hormone on osteoblast and osteoclast differentiation. In Bilezikian J, Marcus R, Levine M (eds): "The parathyroids." Academic Press, pp 199–212, San Diego, CA.
6. Kalajzic I, Staal A, Yang WP, Wu Y, Johnson SE, Feyen JH, Krueger W, Maye P, Yu F, Zhao Y, Kuo L, Gupta RR, Achenie LE, Wang HW, Shin DG, Rowe DW 2005. Expression profile of osteoblast lineage at defined stages of differentiation. J Biol Chem 280:24618–26.

7. Qi H, Aguiar DJ, Williams SM, La Pean A, Pan W, Verfaillie CM 2003. Identification of genes responsible for osteoblast differentiation from human mesodermal progenitor cells. Proc Natl Acad Sci U S A 100:3305–10.

8. Dempster DW, Hughes-Begos CE, Plavetic-Chee K, Brandao-Burch A, Cosman F, Nieves J, Neubort S, Lu SS, Iida-Klein A, Arnett T, Lindsay R 2005. Normal human osteoclasts formed from peripheral blood monocytes express PTH type 1 receptors and are stimulated by PTH in the absence of osteoblasts. J Cell Biochem 95:139–48.

9. Gay CV, Zheng B, Gilman VR 2003. Co-detection of PTH/PTHrP receptor and tartrate resistant acid phosphatase in osteoclasts. J Cell Biochem 89:902–8.

10. Teitelbaum SL 2000. Bone resorption by osteoclasts. Science 289:1504–8.

11. Dobnig H, Turner RT 1995. Evidence that intermittent treatment with parathyroid hormone increases bone formation in adult rats by activation of bone lining cells. Endocrinology 136:3632–8.

12. Hodsman AB, Fraher LJ, Ostbye T, Adachi JD, Steer BM 1993. An evaluation of several biochemical markers for bone formation and resorption in a protocol utilizing cyclical parathyroid hormone and calcitonin therapy for osteoporosis. J. Clin. Invest. 91:1138–48.

13. Onyia JE, Bidwell J, Herring J, Hulman J, Hock JM 1995. In vivo, human parathyroid hormone fragment (hPTH 1–34) transiently stimulates immediate early response gene expression, but not proliferation, in trabecular bone cells of young rats. Bone 17:479–84.

14. Schmidt IU, Dobnig H, Turner RT 1995. Intermittent parathyroid hormone treatment increases osteoblast number, steady state messenger ribonucleic acid levels for osteocalcin, and bone formation in tibial metaphysis of hypophysectomized female rats. Endocrinology 136:5127–34.

15. Jilka RL, Weinstein RS, Bellido T, Roberson P, Parfitt AM, Manolagas SC 1999. Increased bone formation by prevention of osteoblast apoptosis with parathyroid hormone. J. Clin. Invest. 104: 439–46.

16. Fujita T, Meguro T, Fukuyama R, Nakamuta H, Koida M 2002. New signaling pathway for parathyroid hormone and cyclic AMP action on extracellular-regulated kinase and cell proliferation in bone cells. Checkpoint of modulation by cyclic AMP. J Biol Chem 277:22191–200.

17. Swarthout JT, D'Alonzo RC, Selvamurugan N, Partridge NC 2002. Parathyroid hormone-dependent signaling pathways regulating genes in bone cells. Gene 282:1–17.

18. Qin L, Qiu P, Wang L, Li X, Swarthout JT, Soteropoulos P, Tolias P, Partridge NC 2003. Gene expression profiles and transcription factors involved in parathyroid hormone signaling in osteoblasts revealed by microarray and bioinformatics. J Biol Chem 278:19723–31.

19. Qin L, Raggatt LJ, Partridge NC 2004. Parathyroid hormone: a double-edged sword for bone metabolism. Trends Endocrinol Metab 15:60–5.

20. Onyia JE, Helvering LM, Gelbert L, Wei T, Huang S, Chen P, Dow ER, Maran A, Zhang M, Lotinun S, Lin X, Halladay DL, Miles RR, Kulkarni NH, Ambrose EM, Ma YL, Frolik CA, Sato M, Bryant HU, Turner RT 2005. Molecular profile of catabolic versus anabolic treatment regimens of parathyroid hormone (PTH) in rat bone: an analysis by DNA microarray. J Cell Biochem 95: 403–18.

21. Qin L, Li X, Ko JK, Partridge NC, Tamasi J, Raggatt L, Feyen JH, Lee DC, Dicicco-Bloom E 2005. Parathyroid hormone uses multiple mechanisms to arrest the cell cycle progression of osteoblastic cells from G1 to S phase. J Biol Chem 280:3104–11.

22. Qin L, Tamasi J, Raggatt L, Li X, Feyen JH, Lee DC, Dicicco-Bloom E, Partridge NC 2005. Amphiregulin is a novel growth factor involved in normal bone development and in the cellular response to parathyroid hormone stimulation. J Biol Chem 280:3974–81.

23. Lotinun S, Sibonga JD, Turner RT 2003. Triazolopyrimidine (trapidil), a platelet-derived growth factor antagonist, inhibits parathyroid bone disease in an animal model for chronic hyperparathyroidism. Endocrinology 144:2000–7.

24. Mendelsohn J, Baselga J 2000. The EGF receptor family as targets for cancer therapy. Oncogene 19:6550–65.

25. Yarden Y 2001. The EGFR family and its ligands in human cancer. Signaling mechanisms and therapeutic opportunities. Eur J Cancer 37:S3–8.

26. Alroy I, Yarden Y 1997. The ErbB signaling network in embryogenesis and oncogenesis: signal diversification through combinatorial ligand-receptor interactions. FEBS Lett 410:83–6.

27. Burgering BM, Coffer PJ 1995. Protein kinase B (c-Akt) in phosphatidylinositol-3-OH kinase signal transduction. Nature 376:599–602.

28. Shoyab M, McDonald VL, Bradley JG, Todaro GJ 1988. Amphiregulin: a bifunctional growth-modulating glycoprotein produced by the phorbol 12-myristate 13-acetate-treated human breast adeno-carcinoma cell line MCF-7. Proc Natl Acad Sci U S A 85:6528–32.

29. Shoyab M, Plowman GD, McDonald VL, Bradley JG, Todaro GJ 1989. Structure and function of human amphiregulin: a member of the epidermal growth factor family. Science 243:1074–6.

30. Cook PW, Mattox PA, Keeble WW, Pittelkow MR, Plowman GD, Shoyab M, Adelman JP, Shipley GD 1991. A heparin sulfate-regulated human keratinocyte autocrine factor is similar or identical to amphiregulin. Mol Cell Biol 11:2547–57.

31. Plowman GD, Green JM, McDonald VL, Neubauer MG, Disteche CM, Todaro GJ, Shoyab M 1990. The amphiregulin gene encodes a novel epidermal growth factor-related protein with tumor-inhibitory activity. Mol Cell Biol 10:1969–81.

32. Ng KW, Partridge NC, Niall M, Martin TJ 1983. Stimulation of DNA synthesis by epidermal growth factor in osteoblast-like cells. Calcif Tissue Int 35:624–8.

33. Kumegawa M, Hiramatsu M, Hatakeyama K, Yajima T, Kodama H, Osaki T, Kurisu K 1983. Effects of epidermal growth factor on osteoblastic cells in vitro. Calcif Tissue Int 35:542–8.

34. Hata R, Hori H, Nagai Y, Tanaka S, Kondo M, Hiramatsu M, Utsumi N, Kumegawa M 1984. Selective inhibition of type I collagen synthesis in osteoblastic cells by epidermal growth factor. Endocrinology 115:867–76.

35. Antosz ME, Bellows CG, Aubin JE 1987. Biphasic effects of epidermal growth factor on bone nodule formation by isolated rat calvaria cells in vitro. J Bone Miner Res 2:385–93.

36. Raisz LG, Simmons HA, Sandberg AL, Canalis E 1980. Direct stimulation of bone resorption by epidermal growth factor. Endocrinology 107:270–3.

37. Qin L, Partridge NC 2005. Stimulation of amphiregulin expression in osteoblastic cells by parathyroid hormone requires the protein kinase A and cAMP response element-binding protein signaling pathway. J Cell Biochem 96:632–40.

38. Satomura K, Derubeis AR, Fedarko NS, Ibaraki-O'Connor K, Kuznetsov SA, Rowe DW, Young MF, Gehron Robey P 1998. Receptor tyrosine kinase expression in human bone marrow stromal cells. J Cell Physiol 177:426–38.

39. Gronthos S, Simmons PJ 1995. The growth factor requirements of STRO-1-positive human bone marrow stromal precursors under serum-deprived conditions in vitro. Blood 85:929–40.

40. de La Coste A, Freitas AA 2006. Notch signaling: Distinct ligands induce specific signals during lymphocyte development and maturation. Immunol Lett 102:1–9.

41. Bevelock LM, Li X, Boumah C, Partridge NC 2005. Parathyroid hormone regulation of Notch signaling during osteoblastic maturation. J Bone Miner Res 20 Suppl. 1:s30.

42. Calvi LM, Adams GB, Weibrecht KW, Weber JM, Olson DP, Knight MC, Martin RP, Schipani E, Divieti P, Bringhurst FR, Milner LA, Kronenberg HM, Scadden DT 2003. Osteoblastic cells regulate the haematopoietic stem cell niche. Nature 425:841–6.

43. Whitfield JF 2005. Parathyroid hormone (PTH) and hematopoiesis: new support for some old observations. J Cell Biochem 96:278–84.

44. Dallas DJ, Genever PG, Patton AJ, Millichip MI, McKie N, Skerry TM 1999. Localization of ADAM10 and Notch receptors in bone. Bone 25:9–15.

45. Schnabel M, Fichtel I, Gotzen L, Schlegel J 2002. Differential expression of Notch genes in human osteoblastic cells. Int J Mol Med 9:229–32.

46. Swarthout JT, Doggett TA, Lemker JL, Partridge NC 2001. Stimulation of extracellular signal-regulated kinases and proliferation in rat osteoblastic cells by parathyroid hormone is protein kinase C-dependent. J Biol Chem 276:7586–92.

47. MacDonald BR, Gallagher JA, Russell RG 1986. Parathyroid hormone stimulates the proliferation of cells derived from human bone. Endocrinology 118:2445–9.

48. Scutt A, Duvos C, Lauber J, Mayer H 1994. Time-dependent effects of parathyroid hormone and prostaglandin E2 on DNA synthesis by periosteal cells from embryonic chick calvaria. Calcif Tissue Int 55:208–15.

49. Finkelman RD, Mohan S, Linkhart TA, Abraham SM, Boussy JP, Baylink DJ 1992. PTH stimulates the proliferation of TE-85 human osteosarcoma cells by a mechanism not involving either increased cAMP or increased secretion of IGF-I, IGF-II or TGF beta. Bone Miner 16:89–100.

50. Partridge NC, Opie AL, Opie RT, Martin TJ 1985. Inhibitory effects of parathyroid hormone on growth of osteogenic sarcoma cells. Calcif Tissue Int 37:519–25.

51. Tam VK, Clemens TL, Green J 1998. The effect of cell-matrix interaction on parathyroid hormone (PTH) receptor binding and PTH responsiveness in proximal renal tubular cells and osteoblast-like cells. Endocrinology 139:3072–80.

52. Nasu M, Sugimoto T, Kaji H, Chihara K 2000. Estrogen modulates osteoblast proliferation and function regulated by parathyroid hormone in osteoblastic SaOS-2 cells: role of insulin-like growth factor (IGF)-1 and IGF-binding protein-5. J Endocrinol 167:305–13.

53. Camps M, Nichols A, Arkinstall S 2000. Dual specificity phosphatases: a gene family for control of MAP kinase function. FASEB J 14:6–16.

54. Keyse SM 2000. Protein phosphatases and the regulation of mitogen-activated protein kinase signalling. Curr Opin Cell Biol 12:186–92.

55. Sun H, Charles CH, Lau LF, Tonks NK 1993. MKP-1 (3CH134), an immediate early gene product, is a dual specificity phosphatase that dephosphorylates MAP kinase in vivo. Cell 75:487–93.

56. Tyson DR, Swarthout JT, Partridge NC 1999. Increased osteoblastic c-fos expression by parathyroid hormone requires protein kinase A phosphorylation of the cyclic adenosine 3',5'-monophosphate response element-binding protein at serine 133. Endocrinology 140:1255–61.

57. Fu Q, Jilka RL, Manolagas SC, O'Brien CA 2002. Parathyroid hormone stimulates receptor activator of NFkappa B ligand and inhibits osteoprotegerin expression via protein kinase A activation of cAMP-response element-binding protein. J Biol Chem 277:48868–75.

58. Wingender E, Chen X, Fricke E, Geffers R, Hehl R, Liebich I, Krull M, Matys V, Michael H, Ohnhauser R, Pruss M, Schacherer F, Thiele S, Urbach S 2001. The TRANSFAC system on gene expression regulation. Nucleic Acids Res 29:281–3.

59. Wingender E, Chen X, Hehl R, Karas H, Liebich I, Matys V, Meinhardt T, Pruss M, Reuter I, Schacherer F 2000. TRANSFAC: an integrated system for gene expression regulation. Nucleic Acids Res 28:316–9.

60. Prestridge DS 1996. SIGNAL SCAN 4.0: additional databases and sequence formats. Comput Appl Biosci 12:157–60.

61. Chen QK, Hertz GZ, Stormo GD 1995. MATRIX SEARCH 1.0: a computer program that scans DNA sequences for transcriptional elements using a database of weight matrices. Comput Appl Biosci 11:563–6.

62. Quandt K, Frech K, Karas H, Wingender E, Werner T 1995. MatInd and MatInspector: new fast and versatile tools for detection of consensus matches in nucleotide sequence data. Nucleic Acids Res 23:4878–84.

63. Frech K, Herrmann G, Werner T 1993. Computer-assisted prediction, classification, and delimitation of protein binding sites in nucleic acids. Nucleic Acids Res 21:1655–64.

64. Frech K, Quandt K, Werner T 1997. Finding protein-binding sites in DNA sequences: the next generation. Trends Biochem Sci 22:103–4.

65. Erclik MS, Mitchell J 2005. Activation of the insulin-like growth factor binding protein-5 promoter by parathyroid hormone in osteosarcoma cells requires activation of an activated protein-2 element. J Mol Endocrinol 34:713–22.

66. Marino-Ramirez L, Spouge JL, Kanga GC, Landsman D 2004. Statistical analysis of over-represented words in human promoter sequences. Nucleic Acids Res 32:949–58. Print 2004.

67. Trinklein ND, Aldred SJ, Saldanha AJ, Myers RM 2003. Identification and functional analysis of human transcriptional promoters. Genome Res 13:308–12.

68. Wang L, Wu Q, Qiu P, Mirza A, McGuirk M, Kirschmeier P, Greene JR, Wang Y, Pickett CB, Liu S 2001. Analyses of p53 target genes in the human genome by bioinformatic and microarray approaches. J Biol Chem 276:43604–10.

69. Roth FP, Hughes JD, Estep PW, Church GM 1998. Finding DNA regulatory motifs within unaligned noncoding sequences clustered by whole-genome mRNA quantitation. Nat Biotechnol 16:939–45.

70. van Helden J, Andre B, Collado-Vides J 1998. Extracting regulatory sites from the upstream region of yeast genes by computational analysis of oligonucleotide frequencies. J Mol Biol 281:827–42.

71. Qiu P 2003. Recent advances in computational promoter analysis in understanding the transcriptional regulatory network. Biochem Biophys Res Commun 309:495–501.

72. Liu XS, Brutlag DL, Liu JS 2002. An algorithm for finding protein-DNA binding sites with applications to chromatin-immunoprecipitation microarray experiments. Nat Biotechnol 20:835–9.

73. Sinha S, Tompa M 2003. YMF: A program for discovery of novel transcription factor binding sites by statistical overrepresentation. Nucleic Acids Res 31:3586–8.

74. Workman CT, Stormo GD 2000. ANN-Spec: a method for discovering transcription factor binding sites with improved specificity. Pac Symp Biocomput 5:464–475.

75. Elkon R, Linhart C, Sharan R, Shamir R, Shiloh Y 2003. Genome-wide in silico identification of transcriptional regulators controlling the cell cycle in human cells. Genome Res 13:773–80.

3

Analysis of Growth Hormone Effects on Hepatic Gene Expression in Hypophysectomized Rats

Amilcar Flores-Morales, PhD and Leandro Fernández-Pérez MD, PhD

Contents

Introduction
Transcriptional regulation by GH
GH-regulation of hepatic gene expression
  in hypophysectomized male rats
Summary
Acknowledgments
References

Abstract

As indicated by its name, the Growth hormone is the main regulator of longitudinal growth in mammals. GH is mainly produced in the pituitary gland and acts distantly on target tissues through the activation of the transmembrane GH receptor. The liver expresses the highest content of GH receptor. Accordingly, the liver is of key importance for the physiological actions of GH. Global expression analysis of the hepatic GH actions using microarrays clearly indicate that most of the known physiological effects of GH can be explained, at least in part, through its effects on the transcription of specific genes. To this end, GH is known to activate a network of transcription factors in liver that include among others the nuclear receptor such as PPARα, CAR or SHP, buts also SREBP CRBP, and STAT5b. The latest is of particular importance in the regulation of body growth through its regulation of the expression of IGF-I and the ALS of the IGFBP3 binding protein which in turn mediate many of the GH actions in extra hepatic tissues. STAT5b and presumably the members of the nuclear receptor family under GH control also influence genes involve in xenobiotic metabolism. GH actions in liver lead to increase lipogenesis and decreased aminoacid catabolism, thereby promoting anabolic growth in bone and muscle tissue. These effects can be explained by increase expression of lipogeneic genes as consequence of the activation of the key lipogenic transcription factor SREBP1 as well as diminished expression of PPARα, a transcription factor regulating genes involved in lipid oxidation. In addition, reduced aminoacid catabolism correlates to diminished expression of aminotransferases in liver upon GH treatment. Microarray-based expression profiling of GH actions has not only provided molecular correlates to its known physiological action buts has identified a large number of regulated genes for which physiological function is unknown or poorly understood. This opens the opportunity to increase our understanding of liver physiology and the characterization of novel GH effects.

From: *Contemporary Endocrinology: Genomics in Endocrinology:*
DNA Microarray Analysis in Endocrine Health and Disease
Edited by S. Handwerger and B. Aronow © Humana Press, Totowa, NJ

Key Words: Growth hormone (GH); GH receptor (GHR); Janus kinase (JAK); acid labile subunit (ALS); signal transducer and activator of transcription (STAT); mitogen activated protein kinase (MAPK); insulin receptor substrate 1 (IRS-1); focal adhesion kinase (FAK); protein kinase C (PKC); suppressors of cytokine signaling (SOCS); histone acetyltransferases (HAT); carbonic anhydrase III (CAIII); n-Myc interactor (Nmi); hypophysectomized (Hx); sterol regulatory element-binding protein-1c (SREBP-1c); small heterodimer partner (SHP); farnesoid X receptor (FXR); pregnane X receptor (PXR); hepatocyte nuclear factor-4 alpha (HNF-4α); liver X receptor (LXR); PPAR alpha; free fatty acids (FFA); major urinary protein (MUP); apolipoprotein (Apo); constitutive androstane receptor (CAR)

INTRODUCTION

The best known physiological effect of growth hormone (GH) is the regulation of postnatal longitudinal bone growth *(1, 2)*. GH promotes growth through diverse and pleiotropic effects on cellular metabolism and differentiation. It regulates carbohydrate, lipid, nitrogen, and mineral metabolism *(3–5)*, and stimulates DNA synthesis, differentiation, and mitogenesis in a variety of cell types in different tissues *(6–9)*. GH is also important in the maintenance of the immune system *(10, 11)*, heart development *(12)* and has been shown to act on the brain in modulating emotion, stress response, and behavior *(13)*.

Responsiveness to GH in target cells is primarily dependent upon the expression of the GH receptor (GHR) *(14, 15)*. GHR is a single membrane-spanning cell surface protein member of the Class I cytokine receptor superfamily *(16)*. Like other members of the family, GHR lacks intrinsic kinase activity, and signal transduction is mediated by Janus Kinase 2 (JAK2), a cytoplasmic tyrosine kinase that is associated to the so-called box 1 in the membrane proximal region of the GHR cytoplasmic domain *(17, 18)*. JAK2 activation is triggered by GH-induced receptor dimerization, which induces conformational changes resulting in JAK2 transphosphorylation and catalytic activation. Subsequently, the receptor and several signaling proteins are phosphorylated on key tyrosine residues, resulting in the activation of multiple signaling pathways.

In humans, GHR-inactivating mutations cause the Laron syndrome, which is characterized by severe postnatal growth retardation *(19)*. This phenotype is recapitulated in GHR$^{-/-}$ mice, which show a 50% reduction in normal adult body weight *(20)*. JAK2 is also essential for GHR signaling. Mutation of box 1 in the receptor or deletion of JAK2 renders the GH receptor inactive *(21, 22)*.

GHR is found in multiple tissues, including muscle, bone, kidney, mammary gland, adipose, and embryonic stem cells. The highest concentration is found in the liver, specifically in the hepatocytes. Accordingly, GH regulates multiple aspects of liver metabolism, including urea cycle, bile acid synthesis, cholesterol and lipoprotein metabolism, glucose homeostasis, xenobiotic metabolism, etc. *(23)*.

The hepatic effects of GH are essential for its physiological actions. This is clearly illustrated by the mechanisms that control somatic growth. Circulating levels of IGF-1 are under GH control through its actions on hepatic gene expression *(24)*. IGF-1 mediates many of the systemic GH effects and its importance for longitudinal growth is illustrated by the fact that IGF-1 knockout mice reach 40% of the growth observed in wild-type mice *(25, 26)*. Interestingly, liver specific deletion of the IGF-1 gene reduces circulating IGF-1 levels by 65% but does not affect growth *(27)*. Further reduction to 15% of the normal levels can be achieved by the deletion of the Acid Labile Subunit (ALS), another hepatic GH-regulated gene, resulting in a significant reduction of body length *(28)*. We will not discuss in detail the mechanisms governing GHR expression [reviewed in *(29, 30)*]. Nevertheless, it is important to mention that factors acting at the transcriptional, translational, and posttranslational levels can influence GHR synthesis and, thereby, regulate GH sensitivity. These factors, some of which are listed in Table 1, include nutritional status, endocrine context, developmental stage (ontogeny), and various tissue/cell-specific control mechanisms.

A long history of GH studies has resulted in a large body of knowledge related to issues of experimental design. Many aspects that would be left aside when studying other systems become highly important in the GH field, since they have a direct impact on the interpretation of experimental data. For example, given that recombinant human and bovine GHs are readily available, it is common to exogenously administer recombinant GH to animal or tissue cultures as a way to study GH effects. In this case, the choice of hormone in relation to the animal models in use can alter the results of the experiment. Human GH is able to bind and activate both GH and prolactin receptors in rodents, which is important to consider when analyzing tissues that express both receptors. On the other hand, bovine recombinant GH is just able to activate the GHR and therefore constitutes a better choice to study GHR-mediated actions in rodent models *(31)*.

The way the hormone is administered also has important implications. Male and female rats show dramatically different GH secretion patterns. Male rats show intermittent peaks of GH secretion every 3–4 hours, with undetectable serum levels between

Table 1
Regulation of GHR Expression

Factor	Effect on GHR expression	System	Reference
Nutrition			
Undernutrition and fasting	↓ GHR (mRNA)	Liver, rat	*(123,124)*
	↓ GHR (mRNA)	Hepatocytes, rat	*(125)*
Glucose starvation	↓ GHR (mRNA)	Hepatocytes, pig	*(126)*
Endocrine System			
Chronic GH treatment	↑GHR (binding)	Liver, rat, pig and sheep	*(127,128)*
GH deficiency (hypophysectomy)	↓ GHR (number)	Liver, rabbit	*(129)*
Acute GH treatment	↑GHR, 1h (binding)	Liver, rat	*(130)*
	↓ GHR, 6h (binding)		
GH overexpression	↑GHR (binding)	Liver, transgenic mice	*(131)*
Pregnancy	↑GHR (mRNA and binding)	Liver, mouse	*(132)*
Estrogen	↑GHR (mRNA)	Liver, rat	*(133)*
Dexamethasone	↓ GHR (mRNA)	Liver, rat	*(133)*
Insulin	↑GHR (mRNA and protein)	HuH7 cells	*(134)*
	↓ GHR surface		
	↓ GHR (binding)	H4 cells	*(135)*
T3	↑GHR (mRNA)	HuH7 cells	*(136)*
IGF-1	↑GHR (mRNA)	Liver, rat	*(137)*

The arrow indicates the effect on GHR expression: ↑: increase; ↓: decrease.

peaks. On the contrary, female rats show a more continued pattern of secretion with smaller and more frequent GH secretory peaks. The different GH secretion patterns are the key determinant of gender differences in liver gene expression. Therefore, continued GH secretion as found in females (and imitated by GH treatment of male rats with minipumps) is known to feminize the expression of several genes involved in sterol and drug metabolism. Classical examples are the female specific CYP2C12 and the male enriched CAIII (carbonic anhydrase III), which are respectively induced and repressed by GH treatment with minipumps while oppositely regulated by the pulsatile GH pattern as found in males *(32)*.

It is also important to notice that several mechanisms of negative feedback exist in GH-responsive tissues that downregulate GHR signaling upon GH binding *(33)*. Consequently, GH-deficient animals, either because of genetic causes or experimentally generated by hypophysectomy, are more sensitive than GH-secreting animals to the actions of exogenously administered GH. On the other hand, GH-transgenic mice have much higher circulating levels of GH than its normal physiological levels, and the GH-regulated mechanism described using this model may have more relation with disease states such as acromegaly than with normal physiological actions of GH *(34)*). Finally, GH-activated signaling is a transient phenomenon; therefore, the choice of treatment time can determine which of the GH effects can be measured and studied.

Although the physiological characterization of GH actions started more than 90 years ago, it is not until recent times that the molecular mechanisms behind these effects are starting to be revealed. The human, mouse, and rat genome sequencing projects are almost completed which promises to dramatically increase our understanding of the genetic basis of GH actions. To fulfill this promise one must rely on the application of high throughput technologies to speed up the acquisition of experimental data at a genomic scale. The use of DNA microarrays for the parallel determination of the expression of tenths of thousands of genes in an easy-to-perform assay makes this technology the most widely used source of genome-scale experimental data in life sciences *(35)*. Microarray analysis of gene expression has already been used to study GH actions in different models. These include measurements of hepatic changes in gene expression in livers of male rats upon hypophysectomy and after treatment with GH in continued (minipump-based, female-like) fashion *(36)*. GH effects have also been measured in old male rats liver muscle and fat upon daily injections of GH *(37)*. Several mouse genetic models have been studied, as well, including GH deficient, Ghrhr[lit/lit] mice in comparison to wild-type (WT) littermates and after short term GH treatment *(38)*; also, the hepatic changes detected in liver of GH transgenic mice *(39)* and in mice expressing truncated versions of the GHR *(40)*. With the use of microarrays, our knowledge of GH-responsive genes has greatly increased. It is now possible to generate and test novel hypotheses regarding some of the GH effects that lack adequate molecular explanations *(36)*.

TRANSCRIPTIONAL REGULATION BY GH

The transcriptional actions of GH are exerted in a timely fashion by a network of transcription factors. Several transcription factors are known to be regulated by GH, including STATs 1, 3 and 5, the α, β, δ isoforms of the C/EBP family *(41–43)*, HNF-1 α *(44)*, HNF-6 *(45)*, IRF-1 *(46)*, mtTF1 *(47)*, c-jun, c-fos *(48,49)*, Foxm1b *(50)*,

Runx2 *(51)*, Hoxa1 *(52)*, ear-2, TAFII28 *(36)*, etc. As shown in Table 2, the number of DNA binding proteins known to be regulated by GH has been dramatically expanded by the utilization of microarray technology. How this transcriptional network orchestrates the physiological actions of GH is until now poorly understood. The discovery that the Signal Transducer and Activator of Transcription 5 (STAT5a/b) is activated by GH *(53)* represented a breakthrough in the understanding of the mechanisms behind GH transcriptional regulation.

STAT5 is a cytosolic latent transcription factor. Upon GH stimulation, STAT5 associates to GHR and rapidly becomes phosphorylated in tyrosine residues *(54)*. Once STAT5 proteins are phosphorylated, they dissociate from the receptor, dimerize (homo- or hetero-dimerization) through the SH2 domains in each STAT molecule, and migrate to the nucleus, where they bind specific DNA-response elements with the consensus sequence TTCNNNGAA *(53)* and activate gene transcription. A multitude of signaling molecules have been described to be activated by GH, including Mitogen Activated Protein Kinase (MAPK), Insulin Receptor Substrate 1 (IRS-1), Focal Adhesion Kinase (FAK), Protein Kinase C (PKC), Ras-like GTPases, and other STATs [recently reviewed in Zhu et al. *(55)*].

The contribution of many of these pathways to the physiological actions of GH in the liver and extrahepatic tissues, and more specifically to its transcriptional effects, remains unclear in most cases. In many instances the data has been obtained only from in vitro studies in cultured cells. Moreover, many of these pathways are also activated by several additional growth factors and cytokines that have limited functional similarity to GH. This is not the case with STAT5b, where a clear role in GH specific physiological actions is well demonstrated. Analysis of STAT5b deficient mice shows that it is directly involved in sexually dimorphic longitudinal growth and hepatic gene expression in mice. STAT5b deficient mice have a 27% reduction in body growth in males but not in females, elevated GH plasma levels, and reduced circulating IGF-1, as well as an inhibition of GH-induced lipolysis in adipose tissue *(56)*. Importantly, STAT5b mutations have also been found in humans, resulting in a severe growth deficiency comparable to that of patients with GHR-inactivating mutations (57). The analysis of GH-induced gene expression shows that STAT5b is required for the hepatic expression of several GH-regulated genes which are important for somatic growth such as IGF-1, IGFBP3, ALS, and SOCS2 *(58–60)*.

The molecular mechanisms that regulate STAT5b actions illustrate the complexity of the GH-regulated transcriptional network. Several mechanisms act to modulate STAT5b actions in response to GH. First, the activation of STAT5b is transient in nature due to the rapid action of negative regulatory mechanisms. STAT5b is subject to negative regulation by the actions of tyrosine phosphatases, such as Shp-2, the cytoplasmic protein-tyrosine phosphatase 1B (PTP1B), and possibly the T-cell protein-tyrosine phosphatase (TC-PTP) *(33)*). GHR activity leading to STAT5b activation is also negatively regulated by members of the SOCS protein family. As a consequence, sustained effects of GH on target tissues would necessarily require secondary effector molecules. Accordingly, STAT5 is known to influence the expression of other transcription factors (see Table 2 and Table 3). An example is the transcriptional regulation of HNF-6 by GH. STAT5b is known to bind the promoter and to regulate the transcription of HNF-6, which in turn activates some of the female-enriched GH-regulated genes *(45, 61)*.

Table 2
Genes with DNA-Binding Activity Regulated by GH in Liver and Skeletal Muscle, Identified by Microarray Analysis and Classified According to Current Gene Ontology Terms

Tissue	Gene Bank	Name	Exp
Liver	**Upregulated**		
	AB025017	Zinc finger protein 36	1, 3
	AF015953	Aryl hydrocarbon receptor (AhR) nuclear translocator-like	1
	AY004663	Nuclear factor, interleukin 3, regulated	1
	NM_012591	Interferon regulatory factor 1	1
	NM_012603	v-myc avian myelocytomatosis viral oncogene homolog	1
	NM_012747	**Signal transducer and activator of transcription (STAT) 3**	1
	NM_012912	Activating transcription factor 3	1
	NM_022671	One cut domain, family member 1	1
	NM_022858	HNF-3/forkhead homolog-1	1
	U67083	KRAB-zinc finger protein KZF-2	1
	AF092840	Nibrin	2, 3
	BC029197	X-box binding protein 1	2, 3
	AK016949	SWI/SNF related, actin dependent regulator of chromatin, subfamily a, member 5	3
	NM_008416	Jun-B oncogene	3
	NM_009883	**CCAAT/enhancer binding protein (C/EBP), beta**	3
	NM_010591	Jun oncogene	3
	NM_019963	Signal transducer and activator of transcription (STAT) 2	3
	Downregulated		
	NM_022380	**Signal transducer and activator of transcription (STAT) 5B**	1
	NM_023090	Endothelial PAS domain protein 1	1
	U09229	Cut (Drosophila)-like 1	1
	U20796	Nuclear receptor subfamily 1, group D, member 2	1
	AF009328	Nuclear receptor subfamily 1, group I, member 3	2, 3
	AF374266	**Sterol regulatory element binding factor (SREBP) 1**	2
	BC036982	**Thyrotroph embryonic factor**	2
	X89577	**Peroxisome proliferator activated receptor (PPAR) alpha**	2, 3
	AK009739	Kruppel-like factor 15	3
	NM_008260	**Forkhead box A3 (Foxa3) (HNF-3γ)**	3
	NM_010902	Nuclear factor, erythroid derived 2, like 2	3
	NM_016974	D site albumin promoter binding protein	3
Skeletal Muscle	**Upregulated**		
	AW144714	Nuclear receptor subfamily 2, group F, member 6	4
	NM_008241	Forkhead box G1 (Foxg1)	4
	D45210	Zinc finger protein 260	2, 3
	M22326	Early growth response 1	2, 3
	NM_007564	Zinc finger protein 36, C3H type-like 1	2, 3
	NM_007679	**CCAAT/enhancer binding protein (C/EBP), delta**	2, 3
	NM_008416	Jun-B oncogene	2, 3

NM_010234	**FBJ osteosarcoma oncogene (c-fos)**	2, 3
NM_013597	Myocyte enhancer factor 2A (Mef2a)	2
AK018370	Bcl6 interacting corepressor	3
BC010786	cAMP responsive element binding protein 3-like 3 (Creb3l3)	3
BC020042	Core promoter element binding protein (Copeb)	3
NM_008390	Interferon regulatory factor 1	3
NM_009029	Retinoblastoma 1	3
NM_010028	DEAD/H (Asp-Glu-Ala-Asp/His) box polypeptide 3, X-linked	3
NM_010902	Nuclear factor, erythroid derived 2, like 2	3
NM_011498	Basic helix-loop-helix domain containing, class B2	3
NM_013468	Ankyrin repeat domain 1 (cardiac muscle)	3
NM_013692	TGFB inducible early growth response 1	3
NM_013842	X-box binding protein 1	3
NM_020033	Ankyrin repeat domain 2 (stretch responsive muscle)	3
NM_020558	Nuclear DNA binding protein	3
Y15163	Cbp/p300-interacting transactivator, Glu/Asp-rich carboxy-terminal domain, 2	3
Downregulated		
BC013461	Downregulator of transcription 1	4
NM_010866	Myogenic differentiation 1	2
BC011118	**CCAAT/enhancer binding protein (C/EBP), alpha**	2
NM_009557	Zinc finger protein 46	3
NM_016974	D site albumin promoter binding protein	3

Experiments analyzed (Exp): 1. Hypophysectomized rats injected + GH (short term); 2.Ghrhr$^{\text{lit/lit}}$ mouse + GH 2h; 3. SOCS2$^{-/-}$ Ghrhr$^{\text{lit/lit}}$ mouse + GH 2h; 4. Old male rats + GH (injection twice per day for 1 week). In bold: transcription factors that were previously known to be regulated by GH.

Table 3
Hepatic Genes Regulated by Growth Hormone Treatment in Hypophysectomized Male Rats

Gene ID	Gene name	Fold change
Lipid and bile acid metabolism		
D85189	Acyl-CoA synthetase long-chain family member 4	4.4
J02585	Stearoyl-Coenzyme A desaturase 1	3.9
U13253	Fatty acid binding protein 5, epidermal	2.8
AB071986	ELOVL family member 6, elongation of long chain fatty acids	2.7
NM_024143	Bile acid CoA ligase	2.5
NM_017332	Fatty acid synthase	2.4
AB005743	Fatty acid transporter	2.1
AF157026	Solute carrier family 34 (sodium phosphate), member 2	2.0
NM_012703	Thyroid hormone responsive protein, spot 14	1.9
NM_012738	Apolipoprotein A-I	1.9
AF202887	Apolipoprotein A-V	1.8
NM_016994	Complement component 3 (C3)	1.8
Y12517	Cytochrome b5, mitochondrial isoform	1.8

(Continued)

Table 3
(Continued)

Gene ID	Gene name	Fold change
NM_031712	PDZ domain containing 1	1.8
AB012933	Acyl-CoA synthetase 5	1.7
NM_031976	Protein kinase, AMP-activated, beta 1 non-catalytic subunit	1.6
AF286470	Sterol regulatory element binding factor 1 (SREBP-1)	1.6
NM_016987	ATP citrate lyase	1.6
AF095449	3-hydroxyacyl-CoA dehydrogenase	0.7
AB004329	Acetyl-Coenzyme A carboxylase beta	0.7
AF309558	SEC14-like 2	0.7
NM_013084	Acyl-Co A dehydrogenase, short/branched chain	0.7
AJ132352	SEC14-like 3	0.6
D00569	2,4-dienoyl-CoA reductase	0.6
NM_017222	Solute carrier family 10, member 2 (ileal sodium/bile acid cotransporter)	0.5
NM_031559	Carnitine palmitoyltransferase 1, liver	0.5
Steroid and endo-xenobiotic metabolism		
M31363	Sulfotransferase, hydroxysteroid preferring 2	10.5
NM_022258	Alpha-1-B glycoprotein	9.9
AF182168	Aldo-keto reductase family 1, member B7	9.1
D50580	Carboxylesterase 5	7.3
NM_031572	P450 2c40 (CYP2C12)	6.8
NM_017158	P450 2c7 (CYP2C7)	6.0
M31031	P450 2f	4.1
J05035	Steroid 5 alpha-reductase 1	3.9
X61233	Gluthatione-S-transferase class pi	3.0
U89280	Hydroxysteroid 17-beta dehydrogenase 9	2.3
NM_012753	P45017a1	1.5
NM_012531	Catechol-O-methyltransferase	1.5
M33313	CYP2A2 (steroid hydroxylase IIA2)	0.7
J02868	P450 2d13	0.6
U66322	Leukotriene B4 12-hydroxydehydrogenase	0.6
AB037424	Aldo-keto reductase family 7, member A2	0.6
NM_017080	Hydroxysteroid 11-beta dehydrogenase 1	0.6
U70825	5-oxoprolinase (ATP-hydrolysing)	0.6
NM_022407	Aldehyde dehydrogenase family 1, member A1	0.6
NM_013215	Aldo-keto reductase family 7, member A3	0.5
U09742	P450 3A2 (CYP3A2)	0.5
NM_022224	Phosphotriesterase related protein	0.5
NM_020540	Glutathione *S*-Transferase M4	0.4
NM_012693	P450 2A1	0.4
X79991	P450 3a18	0.4
U46118	P450 3a13	0.4
NM_013105	P450 3A3	0.4
NM_032082	Hydroxyacid oxidase 2 (long chain)	0.4
NM_012584	Steroid delta-isomerase, 3 beta	0.3

J02861	P450 2c13	0.3
L32601	20 alpha-hydroxysteroid dehydrogenase	0.3
U33500	Retinol dehydrogenase type II (RODH II)	0.2
NM_012883	Estrogen sulfotransferase	0.1
NM_019184	Cytochrome P450, subfamily IIC (mephenytoin 4-hydroxylase)	0.1
M13646	P450 3a11	0.1

Transcriptional Regulation

NM_022671	One cut domain, family member 1	3.8
AI407707	Strongly similar to tripartite motif protein 24 [Mus musculus]	3.0
U56241	V-maf musculoaponeurotic fibrosarcoma oncogene family, protein B (avian)	2.0
NM_138838	POU domain, class 3, transcription factor 1	0.7
NM_013864	N-myc downstream regulated 2 (Ndr2)	0.7
M25804	Nuclear receptor subfamily 1, group D, member 1 (Rev-ErbA-alpha)	0.6
NM_133583	N-myc downstream regulated gene 2	0.6
D25233	Retinoblastoma 1	0.6
L25674	Transcription factor ear-21	0.6
NM_013196	PPAR-alpha	0.6
AF311875	Period homolog 3 (Drosophila)	0.5
NM_012543	D site albumin promoter binding protein	0.5
AW918049	Pirin	0.5
D86580	Nuclear receptor subfamily 0, group B, member 2 (SHP)	0.4
NM_022941	Nuclear receptor subfamily 1, group I, member 3 (CAR)	0.4
BF567013	Transcription factor CP2 (predicted)	0.4
AW918049	Pirin	0.3

Iron metabolism, redox

U76714	Solute carrier family 39 (iron-regulated transporter), member 1	2.4
AF106944	PRx III (peroxiredoxin)	2.1
L38615	Glutathione synthetase	1.8
AW142278	Ubiquinol-cytochrome c reductase hinge protein	1.7
NM_017055	Transferrin (Tf)	1.7
NM_133304	Hephaestin	1.6
NM_012833	ATP-binding cassette, sub-family C (CFTR/MRP), member 2	1.6
NM_030826	Glutathione peroxidase 1	1.6
NM_022500	Ferritin light chain 1 (liver, iron-storage protein)	1.5
NM_013059	Alkaline phosphatase, tissue-nonspecific	0.6
NM_017154	Xanthine dehydrogenase	0.6
AY027527	NADPH oxidase 4(kidney superoxide - producing NADPH oxidase)	0.4
NM_019292	Carbonic anhydrase 3	0.3

Secreted, membrane proteins

AB039827	Major urinary protein 5 (alpha-2u globulin)	3.9
AF006203	Insulin-like growth factor binding protein complex acid-labile subunit	2.5
NM_012630	Prolactin receptor	2.4
X70369	Collagen 1 type III	2.1

(Continued)

Table 3
(Continued)

Gene ID	Gene name	Fold change
M15481	Insulin-like growth factor 1	2.1
BG671562	Replication factor C (activator 1) 4 (predicted)	2.1
M81785	Syndecan 1	2.1
J00734	Fibrinogen -chain	1.9
AF532184	Coagulation factor VII	1.9
NM_012959	Glial cell line derived neurotrophic factor family receptor alpha 1	1.9
NM_007467	Amyloid precursor-like protein (mouse)	1.9
NM_031699	Claudin 1	1.9
NM_024486	Activin A receptor, type 1	1.7
Y13413	Amyloid beta (A4) precursor protein-binding, family B, member 3	1.6
NM_017155	Adenosine A1 receptor	1.6
AF169313	Hepatic fibrinogen/angiopoietin-related protein (HARP)	1.6
X70369	Collagen, type III, alpha 1	1.6
NM_013156	Cathepsin L	1.6
NM_012898	Alpha-2-HS-glycoprotein	1.5
M27883	Serine protease inhibitor, Kazal type 1 (pancreatic secretory trypsin inhibitor)	0.6
AJ236873	Collagen 1 type XVIII (endostatin precursor)	0.6
NM_031780	Amyloid beta (A4) precursor protein-binding, family A, member 2	0.5
Signal transduction, motility and transport.		
AB075973	Down syndrome critical region homolog 1 (human)	2.0
NM_017317	RAM (low Mr GTP-binding protein)	1.9
NM_022866	Solute carrier family 13 (sodium-dependent dicarboxylate transporter), member 3	1.9
NM_134457	Seven in absentia 2	1.7
U37252	14-3-3 protein, -subtype	1.7
NM_019378	SNAP-25-interacting protein (delays onset cell spreading in the early stage of cell adhesion to fibronectin; exocytosis)	1.6
NM_013005	Phosphatidylinositol 3-kinase, regulatory subunit, polypeptide 1 (p85)	1.6
M14050	Heat shock 70kD protein 5	1.6
NM_022382	Myomegalin (may play a role in cAMP-dependent intracellular signaling)	1.6
AJ303456	Wiskott-Aldrich syndrome protein interacting protein	1.5
NM_019196	Multiple PDZ domain protein (may play a role in G-protein coupled receptors)	1.5
AF306457	RAN, member RAS oncogene family	1.5
AW534340	Deleted in malignant brain tumors 1 (putative scavenger receptor)	0.6
NM_022693	SH3-domain binding protein 4	0.6
NM_139082	BMP and activin membrane-bound inhibitor	0.6
AF033355	Phosphatidylinositol-4-phosphate 5-kinase, type II, beta	0.6

AF218575	Nibrin (activation cell cycle checkpoints and telomere maintance)	0.6
NM_022849	Deleted in malignant brain tumors 1 (putative scavenger receptor)	0.6
AW141364	Mitogen-activated protein kinase associated protein 1	0.6
AF199504	Casitas B-lineage lymphoma b (ubiquitin-protein ligase; regulate autoimmunity; implicated as a susceptibility gene for type I Diabetes (IDDM))	0.6
X89603	Metallothionein 3 (growth inhibitory factor)	0.6
AY029283	Caspase 11	0.5
NM_031970	Heat shock 27kDa protein 1	0.4
NM_031548	Sodium channel, nonvoltage-gated, type I, alpha polypeptide (regulates salt and fluid transport in the kidney)	0.3
NM_031332	Solute carrier family 22 (Oatp 8, organic anion transporter), member 8	0.3

Intermediary Metabolism

NM_030656	Alanine-glyoxylate aminotransferase (dual metabolic roles of gluconeogenesis(mitochondria) and glyoxylate detoxification (peroxisomes)	2.3
Y07744	UDP-N-acetylglucosamine-2-epimerase/N-acetylmannosamine kinase	1.9
AW140979	Transketolase	1.8
NM_012624	Pyruvate kinase, liver and RBC (catalyzes conversion of ATP and pyruvate to ADP and phosphoenolpyruvate in glycolisis)	1.7
AF100154	Cysteine conjugate-beta lyase; cytoplasmic (Class-I pyridoxal-phosphate-dependent aminotransferase)	1.7
NM_017084	Glycine N-methyltransferase	0.6
NM_019370	Ectonucleotide pyrophosphatase/phosphodiesterase 3	0.6
NM_022635	N-acetyltransferase 8 (camello like)	0.6
NM_030850	Betaine-homocysteine methyltransferase (homocystein metabolism)	0.6
D28560	Ectonucleotide pyrophosphatase/phosphodiesterase 2	0.5
NM_031039	Glutamic pyruvic transaminase 1, soluble (GPT, glutamic-piruvate transaminase (alanine transaminase), gluconeogenesis)	0.5

Miscellaneous

AF169636	Paired-Ig-like receptor A3 (predicted)	3.5
H34199	Similar to chromosome 16 open reading frame 33; minus -99 protein (predicted)	2.0
NM_144743	Carboxylesterase isoenzyme gene	1.9
BF395809	Strongly similar to poptosis-inducing factor (AIF)-like mitochondrion-associated inducer of death [Mus musculus]	1.9
NM_053714	Progressive ankylosis homolog	1.9
AI229654	Similar to DIP13 alpha (predicted)	1.6
M86870	Protein disulfide isomerase related protein (calcium-binding protein, intestinal-related)	1.6
NM_012789	Dipeptidylpeptidase 4	1.6

(Continued)

Table 3
(Continued)

Gene ID	Gene name	Fold change
D49434	Arylsulfatase B	1.6
Gene ID	Gene name	Fold change
NM_138530	MAWD binding protein	0.6
Y17328	Crystallin, mu	0.6
X13231	Sulfated glycoprotein 2	0.6
AF023657	Mannosidase, endo-alpha	0.5
NM_022297	Dimethylarginine dimethylaminohydrolase 1	0.5
NM_031073	Neurotrophin 3	0.4

The transcriptional regulation of eukaryotic genes involves the interaction of transcription factors with chromosomal remodeling complexes and components of the basal transcriptional machinery. STAT5b is known to interact with different cofactors such as histone acetyltransferases (HATs), especially CBP/p300 and NCoA1/Src1, to potentiate STAT5-mediated transactivation *(62)*. In the case of STAT5a, the interaction occurs with the C-terminal transactivation domain in the STAT molecule. n-Myc inter-actor (Nmi) interacts with the coiled-coil domain of STAT5 and has been proposed to facilitate the association of STAT5 with p300/CBP, resulting in the enhanced STAT5-dependent transcription of STAT5 *(63)*. The co-repressor SMRT also interacts with the coiled-coil domain of STAT5, but, in contrast with Nmi, it inhibits the expression of STAT5 target genes *(64)*.

STAT5 also interacts with other transcription factors such as the glucocorticoid receptor (GR), the specificity protein (Sp)1 *(65)*, Ying Yang-1 (YY1) *(66)*, and CCAAT/enhancer binding protein (C/EBPβ) *(67)*. The GR-STAT5 interaction is of special importance for the hepatic actions of STAT5b. This interaction is independent of the DNA binding activity of the GR and results in an increase of the transcrip-tional activity of STAT5b-regulated genes *(68)*. The physiological importance of the co-activator function of GR on STAT5b is evident in the liver-specific GR knockout mice, which show a dramatic reduction in body size and reduced hepatic expression of STAT5b-dependent genes, such as IGF-1, ALS, and SOCS2 *(69)*. On the other hand, STAT5b is also known to interact with PPAR alpha, but in this case the interaction results in the mutual inhibition of the transactivation activity of both transcription factors *(36, 70, 71)*.

GH-REGULATION OF HEPATIC GENE EXPRESSION IN HYPOPHYSECTOMIZED MALE RATS

Hypophysectomy in combination with hormonal replacement in rats has been classi-cally used to study hormonal regulation of different physiological processes. In an early example of an application of microarray technology in the endocrine field, we used an array containing approximately 3000 gene probes to study the effects of hypophy-sectomy on gene expression in three different tissues: liver, kidney, and heart. The effects of chronic (three weeks) hGH replacement therapy on hypophysectomized (Hx) animals were also analyzed in the same tissues *(36)*. An interesting observation was that hypophysectomy causes a marked decrease in RNA content in multiple tissues. In our hands, and in agreement with previous results, the content of RNA per cell is reduced by approximately 40% in liver by hypophysectomy, and the fraction of mRNA in the total pool follows this pattern *(72)*,. Accordingly, a large number of genes downregulated by hypophysectomy were identified in comparison to relatively few upregulated genes, and there was a large overlapping among genes regulated by hypophysectomy in the three tissues analyzed *(36)*. Among the upregulated genes in the liver, the biggest group corresponded to ribosomal proteins, thus suggesting that ribosomal content is less sensitive to downregulation by hypophysectomy. On the other hand, the study of GH-induced changes in gene expression showed that a relatively small fraction of the genes which are regulated in Hx rats are reciprocally regulated by GH treatment in the tissues analyzed. This clearly indicates the importance of pituitary hormones other than GH for the transcriptional activity of the tissues under study. In contrast with the changes induced by hypophysectomy, most of the GH effects are highly tissue specific (see Fig. 1).

Although the state of hypophysectomy creates a physiological situation unique to the model itself (of multiple hormonal deficiency), chronic hGH treatment of Hx male rats results in clear restoration of body growth, among other well documented physiological and metabolic changes *(73)*. The Hx animal has a higher proportion of fat and a

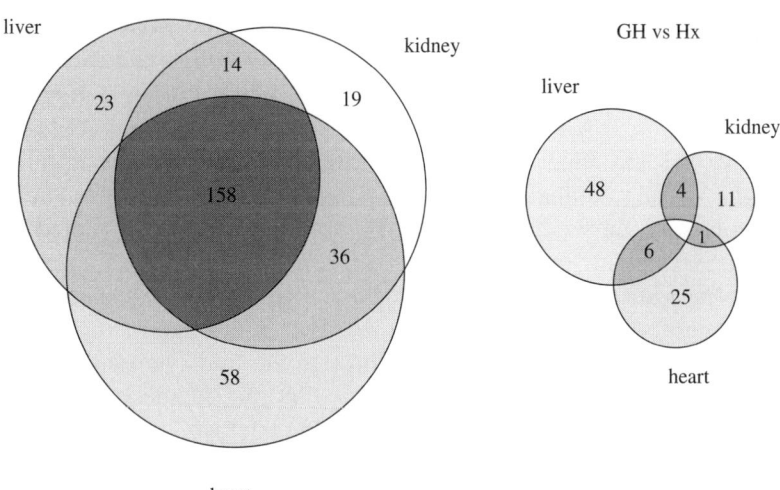

Fig. 1. Summary of the results obtained in the microarray analysis of hypophysectomy and GH treatment in the liver, kidney, and heart of the male rat. Each *circle* represents one organ, and the area is proportional to a number of transcripts. Numbers indicate the amount of different mRNAs differentially regulated in the comparison between normal and Hx tissues and in the comparison between Hx and GH-treated Hx liver, kidney, and heart.

lower proportion of protein in its carcass. Chronic hGH administration improves the nitrogen balance and reverses the changes in body composition found in Hx animals *(73,74)*. The protein anabolic effects of GH include increased uptake of amino acids in peripheral tissues, decreased protein breakdown and usage of amino acids for energy expenditure, and increased protein synthesis *(75)*.

These anabolic effects are energy costly and are partly accomplished by increasing the utilization of FFA in peripheral tissues *(3, 76)*. Thus, GH treatment promotes lipolysis and prevents lipogenesis in adipose tissue, which increases the availability of FFA for energy expenditure. The liver expresses high levels of GHR and subsequently plays a key role in the physiological actions induced by the hormone in the Hx animals. The identification of hepatic GH-regulated genes in this model can help to identify molecular mechanisms responsible for the physiological changes described earlier. We have now extended our initial findings regarding GH regulation of hepatic gene expression in Hx male rats *(36)* to the analysis of 7000 gene probes using a microarray technology. The results are shown in Table 3. In this model, GH induced changes in the expression levels of approximately 170 different genes. As expected, genes with a known role in the regulation of body growth, such as IGF-1 and ALS, were induced by the treatment. A functional classification of the regulated genes using Gene Ontology terms revealed that most identified genes are involved in regulating several aspects of cellular intermediary metabolism, specifically lipid, amino acid, and sterol/xenobiotic metabolism.

In contrast with its lipolytic effects in adipose tissue, GH exerts lipogenic actions in the liver. At the physiological level, these actions are characterized by increased hepatic lipoprotein and triglyceride production, increased secretion of very low density lipoprotein (VLDL), and increased expression of low-density lipoprotein (LDL) receptors and hepatic cholesterol transport *(77–79)*. The analysis of microarray gene expression profiles provides some clues about how these effects are achieved. In the first place, there seems to be a stimulatory effect of GH on lipogenesis that is evident in the induction of genes for lipogenic enzymes FAS, ATP-citrate lyase, SCD-1 and fatty acid elongase 2, Acetyl CoA synthase *(80)*, and for genes involved in lipid transport: FABP-5, Apolipoprotein A-I, Apolipoprotein A-V, and PDZ domain-containing protein 1. Importantly, GH stimulates the expression of SREBP1c, a transcription factor that constitutes the master regulator of lipogenesis through its effects in the expression of genes involved in fatty acid synthesis (Fig. 2). Indeed, GH-regulated FAS, ATP-cytrate lyase, SCD-1 and fatty acid elongase 2, Acetyl CoA synthase, and spot 14 are known to be transcriptionally regulated by SREBP1c *(80)*. Another mechanism whereby GH might promote lipogenesis in the liver is through the downregulation of lipid oxidation. This seems to be achieved through the transcriptional downregulation of PPAR alpha, a key regulator of genes involved in lipid catabolism *(36,81)*. Accordingly, the expression of gene carnitine palmitoyltransferase 1, liver 2,4-dienoyl CoA reductase 1, and Acyl-CoA dehydrogenase, which are involved in mitochondrial fatty acid beta oxidation, are repressed by GH.

The expression of small heterodimer partner (SHP) is oppositely regulated by hypophysectomy and GH treatment. GH downregulates the expression of SHP, a nuclear receptor family member that negatively affects the transcriptional activity of several liver-enriched transcription factors (e.g., GR, LXR, TR, HNF-4β LRH-1, PXR) that are involved in bile acid and cholesterol homeostasis, as well as other import

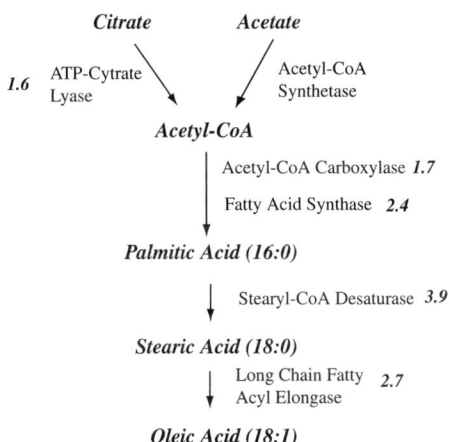

Fig. 2. Effect of GH treatment of hypophysectomized male rats in the hepatic expression of genes involved in fatty acid synthesis. Fold changes in gene expression after GH treatment are indicated in italics.

liver functions *(82)*. SHP is known to negatively regulate LXR stimulatory actions on SREBP1c gene expression *(83)*. Consequently, negative regulation of SHP by GH can contribute to its lipogenic actions. SHP is also known to inhibit PXR in addition to LXR and LRH-1. PXR is involved in cholesterol catabolism and excretion of cholesterol, as well as in the metabolism of xenobiotics. GH induction of the known PXR regulated gene Mrp2 *(84)*, responsible for the excretion of bilirubin glucoronides and glucoronidated bile salts, may also be secondary to SHP downregulation. Upregulation of Mrp2 and also bile Acid CoA ligase *(85)*, which catalyzes the first step in the conjugation of bile acids with taurin and glycine, possibly contributes to GH stimulation of bile flow in Hx rats *(86,87)*. A confirmation of the importance of SHP as a mediator of GH actions comes from microarray analysis of gene expression in mice overexpressing SHP *(88)*, which shows that many of the GH-regulated genes (e.g., CD36, mouse major urinary protein (MUP), serum amyloid A2, complement component 3, etc.) also change expression in this model.

Despite the stimulation of genes involved in fatty acid synthesis, chronic GH treatment is known to reduce triglyceride content in liver *(77, 87)*, an effect that is probably related to an increased capacity for liver VLDL secretion concomitant to increased utilization of fat in muscle tissue. GH increases triglyceride secretion in rat liver through the increased expression of the microsomal triglyceride transfer protein *(89)*, and by promoting RNA processing from ApoB100 into apoB48, thereby promoting VLDL lipoprotein assembly and secretion *(77,78)*. Upregulation of FAS and downregulation of ACCβ could also contribute to reducing liver triglyceride content, a hypothesis supported by the phenotype of liver specific FAS$^{-/-}$ and ACC beta$^{-/-}$. In the first case, the mice show fatty liver, while ACCβ $^{-/-}$ mice show reduced hepatic triglyceride levels in comparison to controls *(90, 91)*. On the other hand, upregulation of Apo A-V can contribute to reducing circulating levels of triglycerides in GH-treated rats. Although its mechanism of action remains unclear, Apo A-V overexpression reduces triglyceride levels in mice, while Apo A-V$^{-/-}$ mice have hypertriglyceridemia.

Moreover, several polymorphisms of Apo A-V have been described in humans associated to increased circulating levels of triglycerides *(92)*. Hx animals have increased LDL and reduced HDL levels, an effect that is normalized upon GH treatment *(87)*. These effects could be related with the increased production of Apo A-I, the major apolipoprotein in HDL particles and a cofactor for LCAT, which is responsible for the formation of most cholesteryl esters in plasma *(93)*.

GH treatment in animal models and in humans causes positive changes in the serum lipid profile *(87)*. Nevertheless, the pharmacological use of GH has been restricted for concerns regarding its anti-insulinic actions *(3)*. How and in which tissues GH antagonizes insulin activated pathways is a poorly understood matter. Some of the GH-regulated genes in the liver can negatively impinge on insulin signaling. The stimulation of phosphatidylinositol PI3-kinase, regulatory subunit, polypeptide 1 gene, which encodes for the p85 alpha regulatory subunit of PI3 kinase, could negatively influence insulin signaling. The catalytic subunit (p110) of PI3-kinase is essential in the activation by insulin of several downstream signaling pathways. The p85 alpha regulatory subunit of PI 3-kinase is required to mediate the insulin-dependent recruitment of PI3-kinase to the plasma membrane, yet mice with reduced p85 expression have increased insulin sensitivity *(94)*. Moreover, overexpression of p85 alpha results in the sequestration of IRS-1 into cytosolic complexes that are defective in phosphatidylinositol-3,4,5-trisphosphate production *(95)*. Another GH-induced gene alpha-2-HS-glycoprotein (fetuin), is also known to have an inhibitory activity on the insulin receptor tyrosine kinase in vitro and in vivo *(96)*. On the other hand, Cbl-b, which is downregulated by GH, is being proposed to exert positive actions on insulin signaling *(97, 98)*, although other reports deny these initial findings *(98)*.

GH treatment in animal model and humans causes dramatic changes in endo/xenobiotic-metabolizing enzyme activities *(99, 100)*. The analysis of GH effects in Hx rats has traditionally been used to study GH effects in xenobiotic metabolism and has resulted in the identification of genes such as CYPs, sulfotransferases, and glutathione-*S*-transferases as being GH-regulated. In keeping with the key role of GH secretory patterns to establish and maintain gender dimorphism in rat liver, GH treatment of Hx rats affect the expression of several genes known to be sexually dimorphic. Given that GH was given in a continued (female-like) fashion, we could identify many of the induced genes to be specific or predominant in female rat liver. These include the induction of CY2C12 *(101)*, steroid 5α-reductase *(102)*, CYP2C7 *(103)*, and several other genes *(104–106)* that are activated by the female-specific GH pattern or are female-predominant. Conversely, continuous administration of GH to Hx male rats dramatically decreased the transcription of CYP2C11, CYP2C13, carbonic anhydrases as well as other genes known to be male specific and inhibited by the female-specific GH pattern.

CYP2C11 is the most abundant member of the P450 xenobiotic-inducible super-family in non-stimulated liver that metabolizes testosterone (steroid 16α-hydroxylase) and also xenobiotics (e.g., mephenytoin). Unlike 2C12 and 2C11 genes, the expression of the male-specific 2C13 is not dependent on GH pulsation, as judged by its high level of expression in Hx rats of both sexes. The male-specific expression of 2C13 gene is primarily derived from the lack of a major suppressive influence of the continuous pattern of plasma GH seen in females. Other CYP genes that were markedly downreg-ulated by the continuous administration of GH to Hx male rats include members of the CYP3A subfamily of P450 enzymes (e.g., CYP3A3, CYP3A11, CYP3A13, CYP3A18) (see Table 3). CYP3A enzymes are very active in steroid and bile acid 6β-hydroxylation and the oxidation of several xenobiotics *(100, 107)*. Because of their wide substrate specificity and prominent expression in the liver (and gut), they are the most important group of enzymes involved in xenobiotic (drug) metabolism *(108)*.

The nuclear receptor PXR is the master regulator of CYP3A genes. On the other hand, the constitutive androstane receptor (CAR) can also transactivate CYP3A genes. Similarly to PXR, CAR is a nuclear receptor abundantly expressed in the liver and regulates the metabolism of xenobiotics. CAR is best known for its role as an inducer of the phenobarbital-responsive CYP2B gene family, which metabolizes phenobarbital *(109)*. Sixty-eight years ago, female rats were already observed to be more sensitive to the effects of barbiturates (i.e., a more prolonged sleeping time) than male rats *(110)*. A later study conducted by Shapiro et al. showed that male, but not female, rats respond to the effects of phenobarbital by increasing hexobarbital hydroxylase activity, and aminopyrine N-demethylation *(111)*. Interestingly, we have now shown that continuous GH administration to Hx male rats has no effects on PXR expression but downregulates CAR, thus suggesting that GH control of some of the male predominant genes can be exerted through its effect on CAR. Sex-dependent differences have also been observed in the expression of conjugative enzymes (phase II drug metabolizing enzymes), such as sulfotransferases *(112)*, glutathione S-transferases *(113)*, and glucuronyl-transferases *(114)*. Using microarray-based gene expression profiling to study the feminizing effects of GH in Hx rats, we and others have reported sexual dimorphism of phase II drug metabolizing enzymes *(105, 106)*. For example, hydroxysteroid sulfotransferase (Sult2A1), a female-predominant enzyme that catalyses the metabolic activation of potent carcinogenic polycyclic arylmethanols *(112, 115)* is upregulated by continuous GH administration to Hx male rats. In contrast, estrogen sulfotransferase and glutathione S-transferase M4 (Gstm4) are downregulated by the same treatment.

In human subjects, sex differences in metabolism and pharmacokinetics have been observed with several clinically important drugs *(116)*. With some drugs, shorter plasma average-lives (associated with greater clearance rates) are seen in male humans, while with other drugs the female is more active in drug metabolism. Unfortunately, we do not know yet if the mechanisms that operate to regulate P450-dependent drug metabolism in rodent models are directly applicable to human drug metabolism. However, the relevance of GH signaling pathways in humans, particularly STAT5b regulatory mechanisms, in relation to liver sex dimorphism is supported by some experimental evidence. GH secretion in humans is also sexually dimorphic and qualitatively similar to the pattern observed in rodents *(117, 118)*. Moreover, several clinical studies indicate that GH treatment can alter P450-dependent drug metabolism in humans *(99)*. In the near future, additional microarray experiments may be useful to further characterize the sex dependence of liver gene expression and to identify the role that GH plays in establishing and maintaining this sexual dimorphism in endo/xenobiotic metabolism both in rodents and humans.

GH treatment stimulates growth by promoting the conservation of amino acids. This effect is achieved by increasing nitrogen retention and reducing ureagenesis and amino acid oxidation. The amino groups are removed from nearly all α-amino acids by transamination to α-ketoglutarate in order to form L-glutamate. Oxidative deamination of L-glutamate in the mitochondria by glutamate dehydrogenase releases ammonia, which is transformed into urea by the enzymes involved in the urea cycle. An inhibitory effect of GH on the activity of the urea cycle enzymes has been reported *(119)*. GH can also affect this process by inhibiting glutamate production through downregulation of transaminases. Indeed, we detected that glutamate pyruvate transaminase gene expression was induced in Hx compared with normal rats and was repressed upon hGH treatment.

Another aspect of GH action that can influence somatic growth is its effect on erythropoiesis. Erythrocyte numbers are in proportion to body mass during the growth period in mammals, which guarantees that increased oxygen consumption during growth is matched by an increased oxygen transport capacity. GH treatment of Hx rats is known to increase erythropoiesis in proportion to increased blood volume without affecting erythrocyte concentration (121). These effects are achieved by a complex mechanism that includes the stimulation of erythropoietin (Epo) synthesis, as well as direct effects of IGF-1 on erythropoietic cell precursors. Erythropoiesis requires a balanced supply of iron for the synthesis of the haemoglobin in the erythroid cell. The hepatocyte plays a key role in this process as the main site of iron storage. In line with its erythropoietic actions, GH stimulates the expression of ferritin, transferrin, and haphaestin and iron-regulated transporter 1 (IREG1) in liver (Table 3). Most of the iron circulates as a diferric anion bound to transferrin. This complex is recognized by the transferrin receptor in immature erythroid cells, other body cells, and internalized and utilized. The excess of iron is stored in the liver in conjunction with ferritin. Iron transported to the circulation through the actions of IREG1 acts in coordination with haphaestin. The IREG1/haphaestin system is active both in the basal membrane of the enterocyte and the hepatocyte, thereby regulating the availability of dietary and liver stored iron *(121)*. The actions of GH seem then to enhance liver storage capacity and the mobilization of hepatic iron into circulation, as well as increased circulating iron through its action on transferrin expression

SUMMARY

More than 100 years of GH research have resulted in the accumulation of a body of knowledge regarding its physiological actions that is unparalleled by most of the known proteins. This knowledge is not restricted to a few animal models but based on studies on numerous species, from fish to mammals, representing a big portion of the recent evolutionary history. Importantly, the use of recombinant human GH as a pharmacological treatment for growth disorders has allowed the development of in-depth studies of its physiological effects in humans.

Despite the dramatic advances in recent years in the understanding of the basic molecular mechanisms of GH actions, many of the physiological actions of GH still lack molecular explanation. The use of high throughput technologies applied to genomic, proteomic and metabolic research can help us to clarify those mechanisms. The study of GH physiological actions at a molecular level is specially challenging, since it includes coordinated effects of the hormone in multiple tissues. Moreover, those effects may vary in relation to the time of treatment, mode of GH administration, and some effects may be specific for certain species.

The study described here represents an early step in the application of gene expression profiling to GH research and uses a relatively simple experimental design. Even with this limited scope, gene expression profiling offers multiple novel insights at the molecular level that can be readily related to physiological changes. As the technology evolves and the data analysis tools are developed to deal with the large amount of information generated by this kind of experiments, one would expect that more complex experiments are performed. Future experiments in GH research should have sophisticated experimental designs to match the complex and pleotropic actions of GH. Those studies will include time courses, simultaneous assessment of GH actions in multiple tissues, in various species, and in isolated cell types. The challenge, with this information at hand, is to build a model of GH actions that link molecular events to physiological actions at a systemic level. This model will not only be valuable to understand the GH system, but it will also help us to identify molecular targets for pharmacological intervention with the goal of mimicking many of the positive effects associated to GH treatment in humans without the secondary effects associated to direct hormonal treatment.

ACKNOWLEDGMENTS

This work was supported in part by Vetenskapsrådet and WCN (Sweden) and by grants from Ministerio de Sanidad y Consumo (FIS 1/1000 to L.F.), Ministerio de Ciencia y Tecnología (PETRI1995-0711 and SAF2003-02117 to L.F.) and ULPGC-Pfizer Spain (CN-78/02-05045 to L.F).

REFERENCES

1. Isaksson, O. G., Jansson, J. O. & Gause, I. A. (1982). Growth hormone stimulates longitudinal bone growth directly. *Science* **216**, 1237–9.
2. Isgaard, J., Nilsson, A., Lindahl, A., Jansson, J. O. & Isaksson, O. G. (1986). Effects of local administration of GH and IGF-1 on longitudinal bone growth in rats. *Am J Physiol* **250**, E367–72.
3. Davidson, M. B. (1987). Effect of growth hormone on carbohydrate and lipid metabolism. *Endocr Rev* **8**, 115–31.
4. Kostyo, J. L. (1968). Rapid effects of growth hormone on amino acid transport and protein synthesis. *Ann N Y Acad Sci* **148**, 389–407.
5. Kostyo, J. L. & Nutting, D. F. (1974). Growth hormone and protein metabolism. In *Handbook of physiology* (Sawyer, W. H. & Knobil, E., eds.), Vol. IV, part 2, pp. 187–210. American Physiology Society, Washington DC.
6. Madsen, K., Friberg, U., Roos, P., Eden, S. & Isaksson, O. (1983). Growth hormone stimulates the proliferation of cultured chondrocytes from rabbit ear and rat rib growth cartilage. *Nature* **304**, 545–7.
7. Billestrup, N. & Nielsen, J. H. (1991). The stimulatory effect of growth hormone, prolactin, and placental lactogen on beta-cell proliferation is not mediated by insulin-like growth factor-I. *Endocrinology* **129**, 883–8.
8. Barnard, R., Ng, K. W., Martin, T. J. & Waters, M. J. (1991). Growth hormone (GH) receptors in clonal osteoblast-like cells mediate a mitogenic response to GH. *Endocrinology* **128**, 1459–64.
9. Slootweg, M. C., Swolin, D., Netelenbos, J. C., Isaksson, O. G. & Ohlsson, C. (1997). Estrogen enhances growth hormone receptor expression and growth hormone action in rat osteosarcoma cells and human osteoblast-like cells. *J Endocrinol* **155**, 159–64.
10. Chen, H. T., Schuler, L. A. & Schultz, R. D. (1998). Growth hormone receptor and regulation of gene expression in fetal lymphoid cells. *Mol Cell Endocrinol* **137**, 21–9.
11. Jeay, S., Sonenshein, G. E., Postel-Vinay, M. C., Kelly, P. A. & Baixeras, E. (2002). Growth hormone can act as a cytokine controlling survival and proliferation of immune cells: New insights into signaling pathways. *Mol Cell Endocrinol* **188**, 1–7.
12. Lombardi, G., Colao, A., Ferone, D., Marzullo, P., Orio, F., Longobardi, S. & Merola, B. (1997). Effect of growth hormone on cardiac function. *Horm Res.* **48 Suppl 4**:38–42. Review.

13. Yoshizato, H., Fujikawa, T., Soya, H., Tanaka, M. & Nakashima, K. (1998). The growth hormone (GH) gene is expressed in the lateral hypothalamus: Enhancement by GH-releasing hormone and repression by restraint stress. *Endocrinology* **139**, 2545–51.

14. Tsushima, T. & Friesen, H. G. (1973). Radioreceptor assay for growth hormone. *J Clin Endocrinol Metab* **37**, 334–7.

15. Leung, D. W., Spencer, S. A., Cachianes, G., Hammonds, R. G., Collins, C., Henzel, W. J., Barnard, R., Waters, M. J. & Wood, W. I. (1987). Growth hormone receptor and serum binding protein: Purification, cloning and expression. *Nature* **330**, 537–43.

16. Cosman, D. (1993). The hematopoietin receptor superfamily. *Cytokine* **5**, 95–106.

17. Argetsinger, L. S., Campbell, G. S., Yang, X., Witthuhn, B. A., Silvennoinen, O., Ihle, J. N. & Carter Su, C. (1993). Identification of JAK2 as a growth hormone receptor-associated tyrosine kinase. *Cell* **74**, 237–44.

18. Carter-Su, C., Argetsinger, L. S., Campbell, G. S., Wang, X., Ihle, J. & Witthuhn, B. (1994). The identification of JAK2 tyrosine kinase as a signaling molecule for growth hormone. *Proc Soc Exp Biol Med* **206**, 210–5.

19. Rosenfeld, R. G., Rosenbloom, A. L. & Guevara-Aguirre, J. (1994). Growth hormone (GH) insensitivity due to primary GH receptor deficiency. *Endocr Rev* **15**, 369–90.

20. Zhou, Y., Xu, B. C., Maheshwari, H. G., He, L., Reed, M., Lozykowski, M., Okada, S., Cataldo, L., Coschigamo, K., Wagner, T. E., Baumann, G. & Kopchick, J. J. (1997). A mammalian model for Laron syndrome produced by targeted disruption of the mouse growth hormone receptor/binding protein gene (the Laron mouse). *Proc Natl Acad Sci U S A* **94**, 13215–20.

21. Sotiropoulos, A., Perrot Applanat, M., Dinerstein, H., Pallier, A., Postel Vinay, M. C., Finidori, J. & Kelly, P. A. (1994). Distinct cytoplasmic regions of the growth hormone receptor are required for activation of JAK2, mitogen-activated protein kinase, and transcription [see comments]. *Endocrinology* **135**, 1292–8.

22. Frank, S. J., Gilliland, G., Kraft, A. S. & Arnold, C. S. (1994). Interaction of the growth hormone receptor cytoplasmic domain with the JAK2 tyrosine kinase. *Endocrinology* **135**, 2228–39.

23. Mathews, L. S., Enberg, B., Norstedt, G., (1989) Regulation of rat growth hormone receptor gene expression. *J Biol Chem.* **264**(17):9905–10.

24. Sjogren, K., Liu, J. L., Blad, K., Skrtic, S., Vidal, O., Wallenius, V., LeRoith, D., Tornell, J., Isaksson, O. G., Jansson, J. O. & Ohlsson, C. (1999). Liver-derived insulin-like growth factor I (IGF-I) is the principal source of IGF-I in blood but is not required for postnatal body growth in mice. *Proc Natl Acad Sci U S A* **96**, 7088–92.

25. Powell-Braxton, L., Hollingshead, P., Giltinan, D., Pitts-Meek, S. & Stewart, T. (1993). Inactivation of the IGF-I gene in mice results in perinatal lethality. *Ann N Y Acad Sci* **692**, 300–1.

26. Powell-Braxton, L., Hollingshead, P., Warburton, C., Dowd, M., Pitts-Meek, S., Dalton, D., Gillett, N. & Stewart, T. A. (1993). IGF-I is required for normal embryonic growth in mice. *Genes Dev* **7**, 2609–17.

27. Yakar, S., Liu, J. L., Stannard, B., Butler, A., Accili, D., Sauer, B. & LeRoith, D. (1999). Normal growth and development in the absence of hepatic insulin-like growth factor I. *Proc Natl Acad Sci U S A* **96**, 7324–9.

28. Haluzik, M., Yakar, S., Gavrilova, O., Setser, J., Boisclair, Y. & LeRoith, D. (2003). Insulin resistance in the liver-specific IGF-1 gene-deleted mouse is abrogated by deletion of the acid-labile subunit of the IGF-binding protein-3 complex: Relative roles of growth hormone and IGF-1 in insulin resistance. *Diabetes* **52**, 2483–9.

29. Goodyer, C. G., Zogopoulos, G., Schwartzbauer, G., Zheng, H., Hendy, G. N. & Menon, R. K. (2001). Organization and evolution of the human growth hormone receptor gene 5′-flanking region. *Endocrinology* **142**, 1923–34.

30. Schwartzbauer, G. & Menon, R. K. (1998). Regulation of growth hormone receptor gene expression. *Mol Genet Metab* **63**, 243–53.

31. Caron, R. W., Jahn, G. A. & Deis, R. P. (1994). Lactogenic actions of different growth hormone preparations in pregnant and lactating rats. *J Endocrinol* **142**, 535–45.

32. Mode, A., Tollet, P., Strom, A., Legraverend, C., Liddle, C. & Gustafsson, J. A. (1992). Growth hormone regulation of hepatic cytochrome P450 expression in the rat. *Adv Enzyme Regul* **32**, 255–63.

33. Flores-Morales, A., Greenhalgh, C. J., Norstedt, G. & Rico-Bautista, E. (2005). Negative regulation of GH receptor signaling. *Mol Endocrinol.* **20**(2):241–253.

34. Palmiter, R. D., Norstedt, G., Gelinas, R. E., Hammer, R. E. & Brinster, R. L. (1983). Metallothionein-human GH fusion genes stimulate growth of mice. *Science* **222**, 809–14.

35. Holloway, A. J., van Laar, R. K., Tothill, R. W. & Bowtell, D. D. (2002). Options available—from start to finish—for obtaining data from DNA microarrays II. *Nat Genet* **32 Suppl**, 481–9.

36. Flores-Morales, A., Stahlberg, N., Tollet-Egnell, P., Lundeberg, J., Malek, R. L., Quackenbush, J., Lee, N. H. & Norstedt, G. (2001). Microarray analysis of the in vivo effects of hypophysectomy and growth hormone treatment on gene expression in the rat. *Endocrinology* **142**, 3163–76.

37. Tollet-Egnell, P., Parini, P., Stahlberg, N., Lonnstedt, I., Lee, N. H., Rudling, M., Flores-Morales, A. & Norstedt, G. (2004). Growth hormone-mediated alteration of fuel metabolism in the aged rat as determined from transcript profiles. *Physiol Genomics* **16**, 261–7.

38. Greenhalgh, C. J., Rico-Bautista, E., Lorentzon, M., Thaus, A. L., Morgan, P. O., Willson, T. A., Zervoudakis, P., Metcalf, D., Street, I., Nicola, N. A., Nash, A. D., Fabri, L. J., Norstedt, G., Ohlsson, C., Flores-Morales, A., Alexander, W. S. & Hilton, D. J. (2005). SOCS2 negatively regulates growth hormone action in vitro and in vivo. *J Clin Invest* **115**, 397–406.

39. Olsson, B., Bohlooly, Y. M., Brusehed, O., Isaksson, O. G., Ahren, B., Olofsson, S. O., Oscarsson, J. & Tornell, J. (2003). Bovine growth hormone-transgenic mice have major alterations in hepatic expression of metabolic genes. *Am J Physiol Endocrinol Metab* **285**, E504–11.

40. Rowland, J. E., Lichanska, A. M., Kerr, L. M., White, M., d'Aniello, E. M., Maher, S. L., Brown, R., Teasdale, R. D., Noakes, P. G. & Waters, M. J. (2005). In vivo analysis of growth hormone receptor signaling domains and their associated transcripts. *Mol Cell Biol* **25**, 66–77.

41. Rastegar, M., Rousseau, G. G. & Lemaigre, F. P. (2000). CCAAT/enhancer-binding protein-alpha is a component of the growth hormone-regulated network of liver transcription factors. *Endocrinology* **141**, 1686–92.

42. Piwien-Pilipuk, G., Van Mater, D., Ross, S. E., MacDougald, O. A. & Schwartz, J. (2001). Growth hormone regulates phosphorylation and function of CCAAT/enhancer-binding protein beta by modulating Akt and glycogen synthase kinase-3. *J Biol Chem* **276**, 19664–71.

43. Liao, J., Piwien-Pilipuk, G., Ross, S. E., Hodge, C. L., Sealy, L., MacDougald, O. A. & Schwartz, J. (1999). CCAAT/enhancer-binding protein beta (C/EBPbeta) and C/EBPdelta contribute to growth hormone-regulated transcription of c-fos. *J Biol Chem* **274**, 31597–604.

44. Meton, I., Boot, E. P., Sussenbach, J. S. & Steenbergh, P. H. (1999). Growth hormone induces insulin-like growth factor-I gene transcription by a synergistic action of STAT5 and HNF-1alpha. *FEBS Lett* **444**, 155–9.

45. Lahuna, O., Rastegar, M., Maiter, D., Thissen, J. P., Lemaigre, F. P. & Rousseau, G. G. (2000). Involvement of STAT5 (signal transducer and activator of transcription 5) and HNF-4 (hepatocyte nuclear factor 4) in the transcriptional control of the hnf6 gene by growth hormone. *Mol Endocrinol* **14**, 285–94.

46. Le Stunff, C. & Rotwein, P. (1998). Growth hormone stimulates interferon regulatory factor-1 gene expression in the liver. *Endocrinology* **139**, 859–66.

47. Yoshioka, S., Okimura, Y., Takahashi, Y., Iida, K., Kaji, H., Matsuo, M. & Chihara, K. (2004). Up-regulation of mitochondrial transcription factor 1 mRNA levels by GH in VSMC. *Life Sci* **74**, 2097–109.

48. Ashcom, G., Gurland, G. & Schwartz, J. (1992). Growth hormone synergizes with serum growth factors in inducing c-fos transcription in 3T3-F442A cells. *Endocrinology* **131**, 1915–21.

49. Gurland, G., Ashcom, G., Cochran, B. H. & Schwartz, J. (1990). Rapid events in growth hormone action. Induction of c-fos and c-jun transcription in 3T3-F442A preadipocytes. *Endocrinology* **127**, 3187–95.

50. Krupczak-Hollis, K., Wang, X., Dennewitz, M. B. & Costa, R. H. (2003). Growth hormone stimulates proliferation of old-aged regenerating liver through forkhead box m1b. *Hepatology* **38**, 1552–62.

51. Ziros, P. G., Georgakopoulos, T., Habeos, I., Basdra, E. K. & Papavassiliou, A. G. (2004). Growth hormone attenuates the transcriptional activity of Runx2 by facilitating its physical association with Stat3beta. *J Bone Miner Res* **19**, 1892–904.

52. Zhang, X., Zhu, T., Chen, Y., Mertani, H. C., Lee, K. O. & Lobie, P. E. (2003). Human growth hormone-regulated HOXA1 is a human mammary epithelial oncogene. *J Biol Chem* **278**, 7580–90.

53. Wood, T. J., Sliva, D., Lobie, P. E., Pircher, T. J., Gouilleux, F., Wakao, H., Gustafsson, J. A., Groner, B., Norstedt, G. & Haldosen, L. A. (1995). Mediation of growth hormone-dependent transcriptional activation by mammary gland factor/Stat 5. *J Biol Chem* **270**, 9448–53.

54. Xu, B. C., Wang, X., Darus, C. J. & Kopchick, J. J. (1996). Growth hormone promotes the association of transcription factor STAT5 with the growth hormone receptor. *J Biol Chem* **271**, 19768–73.

55. Zhu, T., Goh, E. L., Graichen, R., Ling, L. & Lobie, P. E. (2001). Signal transduction via the growth hormone receptor. *Cell Signal* **13**, 599–616.

56. Fain, J. N., Ihle, J. H. & Bahouth, S. W. (1999). Stimulation of lipolysis but not of leptin release by growth hormone is abolished in adipose tissue from Stat5a and b knockout mice. *Biochem Biophys Res Commun* **263**, 201–5.

57. Kofoed, E. M., Hwa, V., Little, B., Woods, K. A., Buckway, C. K., Tsubaki, J., Pratt, K. L., Bezrodnik, L., Jasper, H., Tepper, A., Heinrich, J. J. & Rosenfeld, R. G. (2003). Growth hormone insensitivity associated with a STAT5b mutation. *N Engl J Med* **349**, 1139–47.

58. Davey, H. W., McLachlan, M. J., Wilkins, R. J., Hilton, D. J. & Adams, T. E. (1999). STAT5b mediates the GH-induced expression of SOCS-2 and SOCS-3 mRNA in the liver. *Mol Cell Endocrinol* **158**, 111–6.

59. Davey, H. W., Xie, T., McLachlan, M. J., Wilkins, R. J., Waxman, D. J. & Grattan, D. R. (2001). STAT5b is required for GH-induced liver IGF-I gene expression. *Endocrinology* **142**, 3836–41.

60. Woelfle, J. & Rotwein, P. (2004). In vivo regulation of growth hormone-stimulated gene transcription by STAT5b. *Am J Physiol Endocrinol Metab* **286**, E393–401.

61. Lahuna, O., Fernández, L., Karlsson, H., Maiter, D., Lemaigre, F. P., Rousseau, G. G., Gustafsson, J. & Mode, A. (1997). Expression of hepatocyte nuclear factor 6 in rat liver is sex-dependent and regulated by growth hormone. *Proc Natl Acad Sci U S A* **94**, 12309–13.

62. Litterst, C. M., Kliem, S., Marilley, D. & Pfitzner, E. (2003). NCoA-1/SRC-1 is an essential coactivator of STAT5 that binds to the FDL motif in the alpha-helical region of the STAT5 transactivation domain. *J Biol Chem* **278**, 45340–51.

63. Zhu, M., John, S., Berg, M. & Leonard, W. J. (1999). Functional association of Nmi with Stat5 and Stat1 in IL-2- and IFNgamma-mediated signaling. *Cell* **96**, 121–30.

64. Nakajima, H., Brindle, P. K., Handa, M. & Ihle, J. N. (2001). Functional interaction of STAT5 and nuclear receptor co-repressor SMRT: Implications in negative regulation of STAT5-dependent transcription. *Embo J* **20**, 6836–44.

65. Martino, A., Holmes, J. H. T., Lord, J. D., Moon, J. J. & Nelson, B. H. (2001). Stat5 and Sp1 regulate transcription of the cyclin D2 gene in response to IL-2. *J Immunol* **166**, 1723–9.

66. Bergad, P. L., Towle, H. C. & Berry, S. A. (2000). Yin-yang 1 and glucocorticoid receptor participate in the Stat5-mediated growth hormone response of the serine protease inhibitor 2.1 gene. *J Biol Chem* **275**, 8114–20.

67. Wyszomierski, S. L. & Rosen, J. M. (2001). Cooperative effects of STAT5 (signal transducer and activator of transcription 5) and C/EBPbeta (CCAAT/enhancer-binding protein-beta) on beta-casein gene transcription are mediated by the glucocorticoid receptor. *Mol Endocrinol* **15**, 228–40.

68. Stocklin, E., Wissler, M., Gouilleux, F. & Groner, B. (1996). Functional interactions between Stat5 and the glucocorticoid receptor. *Nature* **383**, 726–8.

69. Tronche, F., Opherk, C., Moriggl, R., Kellendonk, C., Reimann, A., Schwake, L., Reichardt, H. M., Stangl, K., Gau, D., Hoeflich, A., Beug, H., Schmid, W. & Schutz, G. (2004). Glucocorticoid receptor function in hepatocytes is essential to promote postnatal body growth. *Genes Dev* **18**, 492–7.

70. Zhou, Y. C. & Waxman, D. J. (1999). STAT5b down-regulates peroxisome proliferator-activated receptor alpha transcription by inhibition of ligand-independent activation function region-1 transactivation domain. *J Biol Chem* **274**, 29874–82.

71. Carlsson, L., Linden, D., Jalouli, M. & Oscarsson, J. (2001). Effects of fatty acids and growth hormone on liver fatty acid binding protein and PPARalpha in rat liver. *Am J Physiol Endocrinol Metab* **281**, E772–81.

72. Tata, J. (1970). *Regulation of protein synthesis by growth and developmental hormones.* Biochemical Actions of Hormones (Litwack, G., Ed.), pp. 89–129, Academic Press, New York.

73. Engel, F. & Kostyo, J. (1964). *Metabolic actions of pituitary hormones.* The Hormones (Pincus, G., Thimann, K. V. & Astwood, E., Eds.), V, Academic Press, New York.

74. Kostyo, J. & Nutting, D. (1974). *Growth hormone and protein metabolism.* Handbook of Physiology (Sawyer, W. H. & Knobil, E., Eds.), pp. 187–210, American Physiology Society, Washington DC.

75. Kostyo, J. L. & Isaksson, O. (1977). Growth hormone and the regulation of somatic growth. *Int Rev Physiol* **13**, 255–74.

76. Goodman, H. & Schwartz, J. (1974). *Growth hormone and lipid metabolism.* Handbook of Physiology, (Knobil, E. & Swayer W. H. Eds.), American Physiology Society, pp. 211–231. American Physiology Society, Washington DC.

77. Lagreid, A., Hvidsten, T. R., Midelfart, H., Komorowski, J. & Sandvik, A. K. (2003). Predicting gene ontology biological process from temporal gene expression patterns. *Genome Res* **13**, 965–79.

78. Sjoberg, A., Oscarsson, J., Boren, J., Eden, S. & Olofsson, S. O. (1996). Mode of growth hormone administration influences triacylglycerol synthesis and assembly of apolipoprotein B-containing lipoproteins in cultured rat hepatocytes. *J Lipid Res* **37**, 275–89.

79. Sjoberg, A., Oscarsson, J., Eden, S. & Olofsson, S. O. (1994). Continuous but not intermittent administration of growth hormone to hypophysectomized rats increases apolipoprotein-E secretion from cultured hepatocytes. *Endocrinology* **134**, 790–8.

80. Sjoberg, A., Oscarsson, J., Bostrom, K., Innerarity, T. L., Eden, S. & Olofsson, S. O. (1992). Effects of growth hormone on apolipoprotein-B (apoB) messenger ribonucleic acid editing, and apoB 48 and apoB 100 synthesis and secretion in the rat liver. *Endocrinology* **130**, 3356–64.

81. Horton, J. D., Shah, N. A., Warrington, J. A., Anderson, N. N., Park, S. W., Brown, M. S. & Goldstein, J. L. (2003). Combined analysis of oligonucleotide microarray data from transgenic and knockout mice identifies direct SREBP target genes. *Proc Natl Acad Sci U S A* **100**, 12027–32.

82. Jalouli, M., Carlsson, L., Ameen, C., Linden, D., Ljungberg, A., Michalik, L., Eden, S., Wahli, W. & Oscarsson, J. (2003). Sex difference in hepatic peroxisome proliferator-activated receptor alpha expression: Influence of pituitary and gonadal hormones. *Endocrinology* **144**, 101–9.

83. Zhang, Y. & Dufau, M. L. (2004). Gene silencing by nuclear orphan receptors. *Vitam Horm* **68**, 1–48.

84. Watanabe, M., Houten, S. M., Wang, L., Moschetta, A., Mangelsdorf, D. J., Heyman, R. A., Moore, D. D. & Auwerx, J. (2004). Bile acids lower triglyceride levels via a pathway involving FXR, SHP, and SREBP-1c. *J Clin Invest* **113**, 1408–18.

85. Kullak-Ublick, G. A., Stieger, B. & Meier, P. J. (2004). Enterohepatic bile salt transporters in normal physiology and liver disease. *Gastroenterology* **126**, 322–42.

86. Falany, C. N., Xie, X., Wheeler, J. B., Wang, J., Smith, M., He, D. & Barnes, S. (2002). Molecular cloning and expression of rat liver bile acid CoA ligase. *J Lipid Res* **43**, 2062–71.

87. Gartner, L. M. & Arias, I. M. (1972). Hormonal control of hepatic bilirubin transport and conjugation. *Am J Physiol* **222**, 1091–9.

88. Rudling, M., Parini, P. & Angelin, B. (1997). Growth hormone and bile acid synthesis. Key role for the activity of hepatic microsomal cholesterol 7alpha-hydroxylase in the rat. *J Clin Invest* **99**, 2239–45.

89. Boulias, K., Katrakili, N., Bamberg, K., Underhill, P., Greenfield, A. & Talianidis, I. (2005). Regulation of hepatic metabolic pathways by the orphan nuclear receptor SHP. *EMBO J* **24**, 2624–33.

90. Ameen, C. & Oscarsson, J. (2003). Sex difference in hepatic microsomal triglyceride transfer protein expression is determined by the growth hormone secretory pattern in the rat. *Endocrinology* **144**, 3914–21.

91. Chakravarthy, M. V., Pan, Z., Zhu, Y., Tordjman, K., Schneider, J. G., Coleman, T., Turk, J. & Semenkovich, C. F. (2005). "New" hepatic fat activates PPARalpha to maintain glucose, lipid, and cholesterol homeostasis. *Cell Metab* **1**, 309–22.

92. Abu-Elheiga, L., Matzuk, M. M., Abo-Hashema, K. A. & Wakil, S. J. (2001). Continuous fatty acid oxidation and reduced fat storage in mice lacking acetyl-CoA carboxylase 2. *Science* **291**, 2613–6.

93. Pennacchio, L. A. & Rubin, E. M. (2003). Apolipoprotein A5, a newly identified gene that affects plasma triglyceride levels in humans and mice. *Arterioscler Thromb Vasc Biol* **23**, 529–34.

94. Shah, P. K., Kaul, S., Nilsson, J. & Cercek, B. (2001). Exploiting the vascular protective effects of high-density lipoprotein and its apolipoproteins: An idea whose time for testing is coming, part I. *Circulation* **104**, 2376–83.

95. Terauchi, Y., Tsuji, Y., Satoh, S., Minoura, H., Murakami, K., Okuno, A., Inukai, K., Asano, T., Kaburagi, Y., Ueki, K., Nakajima, H., Hanafusa, T., Matsuzawa, Y., Sekihara, H., Yin, Y., Barrett, J. C., Oda, H., Ishikawa, T., Akanuma, Y., Komuro, I., Suzuki, M., Yamamura, K., Kodama, T., Suzuki, H., Yamamura, K., Kodama, T., Suzuki, H., Koyasu, S., Aizawa, S., Tobe, K., Fukui, Y., Yazaki, Y. & Kadowaki, T. (1999). Increased insulin sensitivity and hypoglycaemia in mice lacking the p85 alpha subunit of phosphoinositide 3-kinase. *Nat Genet* **21**, 230–5.

96. Luo, J., Field, S. J., Lee, J. Y., Engelman, J. A. & Cantley, L. C. (2005). The p85 regulatory subunit of phosphoinositide 3-kinase down-regulates IRS-1 signaling via the formation of a sequestration complex. *J Cell Biol* **170**, 455–64.

97. Mathews, S. T., Singh, G. P., Ranalletta, M., Cintron, V. J., Qiang, X., Goustin, A. S., Jen, K. L., Charron, M. J., Jahnen-Dechent, W. & Grunberger, G. (2002). Improved insulin sensitivity and resistance to weight gain in mice null for the Ahsg gene. *Diabetes* **51**, 2450–8.

98. Liu, J., DeYoung, S. M., Hwang, J. B., O'Leary, E. E. & Saltiel, A. R. (2003). The roles of Cbl-b and c-Cbl in insulin-stimulated glucose transport. *J Biol Chem* **278**, 36754–62.

99. Mitra, P., Zheng, X. & Czech, M. P. (2004). RNAi-based analysis of CAP, Cbl, and CrkII function in the regulation of GLUT4 by insulin. *J Biol Chem* **279**, 37431–5.

100. Cheung, N. W., Liddle, C., Coverdale, S., Lou, J. C. & Boyages, S. C. (1996). Growth hormone treatment increases cytochrome P450-mediated antipyrine clearance in man. *J Clin Endocrinol Metab* **81**, 1999–2001.

101. Rendic, S. & Di Carlo, F. J. (1997). Human cytochrome P450 enzymes: A status report summarizing their reactions, substrates, inducers, and inhibitors. *Drug Metab Rev* **29**, 413–580.

102. MacGeoch, C., Morgan, E. T., Halpert, J. & Gustafsson, J. A. (1984). Purification, characterization, and pituitary regulation of the sex-specific cytochrome P-450 15 beta-hydroxylase from liver microsomes of untreated female rats. *J Biol Chem* **259**, 15433–9.

103. Mode, A., Gustafsson, J. A., Jansson, J. O., Eden, S. & Isaksson, O. (1982). Association between plasma level of growth hormone and sex differentiation of hepatic steroid metabolism in the rat. *Endocrinology* **111**, 1692–7.

104. Westin, S., Strom, A., Gustafsson, J. A. & Zaphiropoulos, P. G. (1990). Growth hormone regulation of the cytochrome P-450IIC subfamily in the rat: Inductive, repressive, and transcriptional effects on P-450f (IIC7) and P-450PB1 (IIC6) gene expression. *Mol Pharmacol* **38**, 192–7.

105. Mode, A., Wiersma-Larsson, E. & Gustafsson, J. A. (1989). Transcriptional and posttranscriptional regulation of sexually differentiated rat liver cytochrome P-450 by growth hormone. *Mol Endocrinol* **3**, 1142–7.

106. Ahluwalia, A., Clodfelter, K. H. & Waxman, D. J. (2004). Sexual dimorphism of rat liver gene expression: Regulatory role of growth hormone revealed by deoxyribonucleic acid microarray analysis. *Mol Endocrinol* **18**, 747–60.

107. Stahlberg, N., Merino, R., Hernández, L. H., Fernández-Pérez, L., Sandelin, A., Engstrom, P., Tollet-Egnell, P., Lenhard, B. & Flores-Morales, A. (2005). Exploring hepatic hormone actions using a compilation of gene expression profiles. *BMC Physiol* **5**, 8.

108. Chang, T. K., Teixeira, J., Gil, G. & Waxman, D. J. (1993). The lithocholic acid 6 beta-hydroxylase cytochrome P-450, CYP 3A10 is an active catalyst of steroid-hormone 6 beta-hydroxylation. *Biochem J* **291 (Pt 2)**, 429–33.

109. Thummel, K. E. & Wilkinson, G. R. (1998). In vitro and in vivo drug interactions involving human CYP3A. *Annu Rev Pharmacol Toxicol* **38**, 389–430.

110. Choi, H. S., Chung, M., Tzameli, I., Simha, D., Lee, Y. K., Seol, W. & Moore, D. D. (1997). Differential transactivation by two isoforms of the orphan nuclear hormone receptor CAR. *J Biol Chem* **272**, 23565–71.

111. Holck, H. G. O., Munir, A.K., Mills, L. & Smith, E.L. (1937). Studies upon the sex difference in rats in tolerance to certain barbiturates and to nicotine. *J Pharmacol Exp Ther* **60**, 323–346.

112. Shapiro, B. H. (1986). Sexually dimorphic response of rat hepatic monooxygenases to low-dose phenobarbital. *Biochem Pharmacol* **35**, 1766–8.

113. Mulder, G. J. (1986). Sex differences in drug conjugation and their consequences for drug toxicity. Sulfation, glucuronidation, and glutathione conjugation. *Chem Biol Interact* **57**, 1–15.

114. Srivastava, P. K. & Waxman, D. J. (1993). Sex-dependent expression and growth hormone regulation of class alpha and class mu glutathione S-transferase mRNAs in adult rat liver. *Biochem J* **294 (Pt 1)**, 159–65.

115. Zhu, B. T., Suchar, L. A., Huang, M. T. & Conney, A. H. (1996). Similarities and differences in the glucuronidation of estradiol and estrone by UDP-glucuronosyltransferase in liver microsomes from male and female rats. *Biochem Pharmacol* **51**, 1195–202.

116. Runge-Morris, M. & Wilusz, J. (1991). Age- and gender-related gene expression of hydroxysteroid sulfotransferase-a in rat liver. *Biochem Biophys Res Commun* **175**, 1051–6.

117. Fletcher, C. V., Acosta, E. P. & Strykowski, J. M. (1994). Gender differences in human pharmacokinetics and pharmacodynamics. *J Adolesc Health* **15**, 619–29.

118. Veldhuis, J. D. (1996). New modalities for understanding dynamic regulation of the somatotropic (GH) axis: Explication of gender differences in GH neuroregulation in the human. *J Pediatr Endocrinol Metab* **9 Suppl 3**, 237–53.

119. Winer, L. M., Shaw, M. A. & Baumann, G. (1990). Basal plasma growth hormone levels in man: New evidence for rhythmicity of growth hormone secretion. *J Clin Endocrinol Metab* **70**, 1678–86.

120. Grofte, T., Wolthers, T., Jensen, S. A., Moller, N., Jorgensen, J. O., Tygstrup, N., Orskov, H. & Vilstrup, H. (1997). Effects of growth hormone and insulin-like growth factor-I singly and in combination on in vivo capacity of urea synthesis, gene expression of urea cycle enzymes, and organ nitrogen contents in rats. *Hepatology* **25**, 964–9.

121. Kurtz, A., Zapf, J., Eckardt, K. U., Clemons, G., Froesch, E. R. & Bauer, C. (1988). Insulin-like growth factor I stimulates erythropoiesis in hypophysectomized rats. *Proc Natl Acad Sci U S A* **85**, 7825–9.

122. Anderson, G. J., Frazer, D. M., McKie, A. T., Vulpe, C. D. & Smith, A. (2005). Mechanisms of haem and non-haem iron absorption: Lessons from inherited disorders of iron metabolism. *Biometals* **18**, 339–48.

123. Bornfeldt, K. E., Arnqvist, H. J., Enberg, B., Mathews, L. S. & Norstedt, G. (1989). Regulation of insulin-like growth factor-I and growth hormone receptor gene expression by diabetes and nutritional state in rat tissues. *J Endocrinol* **122**, 651–6.

124. Straus, D. S. & Takemoto, C. D. (1990). Effect of fasting on insulin-like growth factor-I (IGF-I) and growth hormone receptor mRNA levels and IGF-I gene transcription in rat liver. *Mol Endocrinol* **4**, 91–100.

125. Martínez, V., Balbín, M., Ordóñez, F. A., Rodríguez, J., García, E., Medina, A. & Santos, F. (1999). Hepatic expression of growth hormone receptor/binding protein and insulin-like growth factor I genes in uremic rats. Influence of nutritional deficit. *Growth Horm IGF Res* **9**, 61–8.

126. Brameld, J. M., Gilmour, R. S. & Buttery, P. J. (1999). Glucose and amino acids interact with hormones to control expression of insulin-like growth factor-I and growth hormone receptor mRNA in cultured pig hepatocytes. *J Nutr* **129**, 1298–306.

127. Chung, C. S. & Etherton, T. D. (1986). Characterization of porcine growth hormone (pGH) binding to porcine liver microsomes: Chronic administration of pGH induces pGH binding. *Endocrinology* **119**, 780–6.

128. Gluckman, P. D., Breier, B. H. & Sauerwein, H. (1990). Regulation of the cell surface growth hormone receptor. *Acta Paediatr Scand Suppl* **366**, 73–8.

129. Posner, B. I., Patel, B., Vezinhet, A. & Charrier, J. (1980). Pituitary-dependent growth hormone receptors in rabbit and sheep liver. *Endocrinology* **107**, 1954–8.

130. Maiter, D., Underwood, L. E., Maes, M. & Ketelslegers, J. M. (1988). Acute down-regulation of the somatogenic receptors in rat liver by a single injection of growth hormone. *Endocrinology* **122**, 1291–6.

131. Chen, N. Y., Chen, W. Y. & Kopchick, J. J. (1997). Liver and kidney growth hormone (GH) receptors are regulated differently in diabetic GH and GH antagonist transgenic mice. *Endocrinology* **138**, 1988–94.

132. Camarillo, I. G., Thordarson, G., Ilkbahar, Y. N. & Talamantes, F. (1998). Development of a homologous radioimmunoassay for mouse growth hormone receptor. *Endocrinology* **139**, 3585–9.

133. Bennett, P. A., Levy, A., Carmignac, D. F., Robinson, I. C. & Lightman, S. L. (1996). Differential regulation of the growth hormone receptor gene: Effects of dexamethasone and estradiol. *Endocrinology* **137**, 3891–6.
134. Leung, K. C., Doyle, N., Ballesteros, M., Waters, M. J. & Ho, K. K. (2000). Insulin regulation of human hepatic growth hormone receptors: Divergent effects on biosynthesis and surface translocation. *J Clin Endocrinol Metab* **85**, 4712–20.
135. Ji, S., Guan, R., Frank, S. J. & Messina, J. L. (1999). Insulin inhibits growth hormone signaling via the growth hormone receptor/JAK2/STAT5B pathway. *J Biol Chem* **274**, 13434–42.
136. Mullis, P. E., Eble, A., Marti, U., Burgi, U. & Postel-Vinay, M. C. (1999). Regulation of human growth hormone receptor gene transcription by triiodothyronine (T3). *Mol Cell Endocrinol* **147**, 17–25.
137. Mirpuri, E., García-Trevijano, E. R., Castilla-Cortázar, I., Berasain, C., Quiroga, J., Rodríguez-Ortigosa, C., Mato, J. M., Prieto, J. & Avila, M. A. (2002). Altered liver gene expression in CCl4-cirrhotic rats is partially normalized by insulin-like growth factor-I. *Int J Biochem Cell Biol* **34**, 242–52.

4

Gene Expression Profiling in Leiomyoma in Response to GnRH Therapy and TGF-β

Nasser Chegini, PhD and Xiaoping Luo, MD

CONTENTS

INTRODUCTION
EXPERIMENTAL DESIGN
NOTES
ACKNOWLEDGMENTS
REFERENCES

Abstract

Microarray technology has proven to be a powerful tool for simultaneous profiling of the expression of a large number of genes in various cells and tissues under normal physiological or pathological conditions. The introduction of microarray in profiling the pattern of gene expression in various female reproductive tract tissues and cells resulted in the generation of a large body of new information representing their molecular environments under both normal physiological and pathological conditions. The identification of transcripts for many genes, whose products' functional relevance in reproductive tissues was not previously predicated, has opened new research opportunities in the field of reproductive endocrinology. Here, we shall discuss the application of microarray in studying gene expression profiling in myometrium and leiomyoma and the consequence of GnRHa therapy on their expression, as well as their profiles in smooth muscle cells isolated from these tissues in response to direct action of GnRHa. We shall also discuss the influence of transforming growth factor beta (TGF-β), a profibrotic cytokine, on leiomyoma and myometrial smooth muscle gene expression profile and the consequence of TGF-β receptor type II antisense treatment as an alternative approach to suppress leiomyoma growth. Furthermore, we shall discuss the experimental design with appropriate steps and procedures for RNA preparation, microarray hybridization, image acquisition and data analysis, interpretation, confirmation, and finally the biological significance of these genes.

Key Words: leiomyoma; myometrium; GnRH; TGF-β; tissue; cell culture; microarry; Western blotting; immunohistochemistry; antisense oligomers

INTRODUCTION

Leiomyomas, or fibroids, are benign uterine tumors, which develop during the reproductive years and regress after menopause. It is estimated that approximately 70% of women are affected by leiomyoma, with higher risk of developing clinically significant

From: *Contemporary Endocrinology: Genomics in Endocrinology:*
DNA Microarray Analysis in Endocrine Health and Disease
Edited by S. Handwerger and B. Aronow © Humana Press, Totowa, NJ

tumors in African-American women. Although the transformation of normal myometrial smooth muscle cells/connective tissue fibroblasts leiomyomas is believed to cause leiomyoma, molecule(s) that initiate such cellular transformation and orchestrate their subsequent growth still remains unknown. Clinically, ovarian steroids are leiomyomas' major growth promoting factors and medical management to control their growth center around therapies that suppress the sex steroid production and actions. Gonadotropin releasing hormone agonists (GnRHa) therapy acting at the level of pituitary-ovarian axis has remained the common approach to suppress leiomyoma growth. Recent preclinical and clinical trials with selective estrogen and progesterone receptor modulators (SERM and SPRM), such as Raloxifene, Mifepristone and Asoprisnil, and CDB-2914, have also shown efficacy in regressing leiomyomas' growth *(1–5)*. Despite these advances in medical management of leiomyoma, our knowledge of the molecular environment that distinguishes leiomyoma from myometrium remains limited. Our current understanding of leiomyomas' molecular environment has been based on the results derived from several conventional methods. These investigations revealed the identification of many locally expressed molecules with ability to mediate cellular transformation, growth, differentiation, hypertrophy and apoptosis, as well as extracellular matrix (ECM) accumulation, events central to leiomyoma growth and regression *(5–7)*.

The recent introduction of microarray technology, allowing simultaneous assessment of the expression of a large number of genes in various cells and tissues, has proven to be a powerful tool for obtaining the signature of their molecular environment during normal physiological and/or pathological conditions. This approach has been utilized in studying the pattern of gene expression in several reproductive tract tissues and cells, including leiomyoma and myometrium, as well as their isolated smooth muscle cells *(8–23)*. In this chapter, we shall discuss the application of microarray in studying gene expression profiling in leiomyoma and myometrium and the consequence of GnRHa therapy on their expression. We have presented an experimental approach for examining gene expression profile of the smooth muscle cells isolated from these tissues following treatment with GnRHa and TGF-β. In the following, we present the experimental design with appropriate steps and procedures for RNA preparation, microarray hybridization, image acquisition and data analysis, interpretation, confirmation, and finally assays to define the biological significance of these genes.

EXPERIMENTAL DESIGN

Materials and Methods

Portions of leiomyoma and matched myometrium are collected from premenopausal women scheduled to undergo hysterectomy for indications related to symptomatic leiomyomas. Since leiomyoma develops during reproductive years, consideration is given to the influence of menstrual cycle prior to collection. In addition, if comparative analysis is being made between leiomyoma from patients who received hormonal therapy, the untreated (control) group must include patients who did not receive any medications (including hormonal) during the three months prior to surgery. According to standards and to minimize variability, leiomyoma sampling must be made from similar size tumors, and if large tumors were to be used, samples should be obtained from identical locations.

Following collection, the tissues are cut into several pieces and either immediately snapped frozen and stored in liquid nitrogen for RNA and protein extraction, or fixed and paraffin-embedded for histological evaluation and immunohistochemistry, or used for preparation of isolated leiomyoma and myometrial smooth muscle cells and culturing. Approval from appropriate Institutional Review Board must be obtained prior to the initiation of any study.

Isolation and Culture of Leiomyoma and Myometrial Smooth Muscle Cells

To determine the global gene expression in leiomyoma and myometrial smooth muscle cells (LSMC and MSMC) in response to locally expressed molecules or agents that influence their cellular behavior, these cells must be isolated and cultured in a defined condition. Only untreated tissues must be used for isolation of LSMC and MSMC. There are several well-described procedures for isolation of LSMC and MSMC, including the one from our laboratory *(24, 25)*. Prior to being used in any experiments, the primary culture of LSMC and MSMC must be characterized. This is performed by seeding the cells in 8-well culture slides (Nalge Nunc, Naperville, IL, USA) and culturing in DMEM-supplemented media containing 10% FBS for 24 h. The cells are washed in serum-free media and characterized using immunofluroscence microscopy and antibodies to α smooth muscle actin, desmin, and vimentin *(24, 25)*. The LSMC and MSMC cell preparation selected for further experiments should be 95% pure. The cells are cultured in 6-well plates at a density of approximately 10^6 cells/well in the presence of 10% FBS. After the cells reach visual confluence, they are washed in serum-free media and incubated for 24 h under serum-free, phenol red-free condition. The cells can then be used to study the effect of estrogen, progesterone, SERM, SPRM or, in the case of our study, GnRHa (leuprolide acetate, Sigma Chemical, St Louis, MO, USA), and TGF-β. In either case, the cells are often treated with several doses for one time point, or with one concentration for different time points. Because of the cost associated with microarray experiments, dose- and time-dependent experiments can be designed based on previous experience and/or when adequate data are available with the test agent to support the specific selection of dose and time of treatment. For examining the effect of GnRHa and TGF-β on gene expression profile in LSMC and MSMC, we selected one dose and three treatment time points of 2, 6 and 12 h based on our previous work with the effects of GnRHa and TGF-β on regulation of other genes in these cells *(21, 24, 26)*.

Gene Expression Profiling

Currently, there are several microarray platforms for profiling gene expression in cells and tissues. These platforms are classified into two types, oligonucleotide and cDNA arrays. They utilize synthesized gene-specific oligonucleotides or single-stranded DNA probes of different length, respectively, incorporated on nylon membrane or glass and are exposed to labeled cDNA prepared from cell and tissue RNA samples *(27–29)*. The membrane arrays are hybridized with radioactive- or chemiluminescent-labeled cDNA, whereas oligonucleotide arrays are hybridized with a single fluorescently labeled cDNA. The high-density oligonucleotide arrays use photolithography and solid-phase DNA synthesis to generate synthetic 25-base oligomers (25mers) probes. Each gene on these arrays is represented by 11 to 20 different 25mers, either as "perfect match" or "mismatch" sequence pairs, the latter serving as an internal control for every gene. These probes are generated to represent a unique part of a gene transcript allowing discrimination between closely related genes, or between splice variants of a gene. Longer oligonucleotide (50 to 100mers) microarrays are also available and provide even greater hybridization specificity *(27–29)*. The probes imprinted on these arrays represent both known genes and expressed sequence tags (ESTs) denoting partially sequenced novel genes, some corresponding to a segment of a known gene. The online database at http://www.ncbi.nlm.nih.gov/dbEST/index.html contains all the EST sequences, and http://www.ncbi.nlm.nih.gov/UniGene/index.html contains genes assigns overlapping sequences to a single cluster, which may or may not have a known identity.

Although both platforms have been extensively utilized, results obtained from cDNA microarray platforms are less reproducible as compared to oligonucleotide microarray because of irregularity in probe sizes and target RNA that results in variable binding capacity *(27–29)*. Restriction guidelines must be established for utilizing different commercial oligonucleotide and cDNA microarrays, as well as individually constructed cDNA microarrays *(27–29)*. This is specifically critical when the study design involves comparative analysis of results obtained from different experiments using different or similar platforms, even if similar analysis is employed to obtain a list of significant genes *(30,31)*. Table 1 provides a list of commercially available arrays.

RNA Preparation and Hybridization

In our studies, we used both membrane (http/www.clontech.com; *(10)* and oligonucleotide arrays (Affymetrix at http://www.affymetrix.com, *(20, 21)* for profiling gene expression in leiomyoma, myometrium and their isolated smooth muscle cells. In each case, total RNA was isolated from paired leiomyoma and myometrium, as well as from LSMC and MSMC cultured in a defined condition, using Trizol (Invcrogcn, Carlsbad, CA, USA). The total RNA was treated with DNase I (Roche Applied Science,

Table 1
Commercial Microarrays

Company	Web address
Affymetrix	http://www.affymetrix.com
Agilent Technologies	http://www.home.agilent.com
Applied Biosystems	http://www.appliedbiosystems.com
CombiMatrix Corporation	http://www.combimatrix.com
Illumina	http://www.illumina.com
Mergen	http://www.mergen.com
NimbleGen Systems	http://www.nimblegen.com
Oxford Gene Technologies	http://www.ogt.co.uk
Panomics	http://www.panomics.com
Phalanx Biotech	http://www.phalanxbiotech.com
Plexigen	http://www.plexigen.com
Spectral Genomics	http://www.spectralgenomics.com
SuperArray Bioscience Corporation	http://www.superarray.com
Xeotron Corporation	http://www.xeotron.com

Indianapolis, IN USA.) at 1 unit/10 µg of RNA, incubated at 25°C for 20 min, heat-inactivated at 75°C for 10 min, and subjected to a further purification step using RNeasy Kit (Qiagen, Valencia, CA, USA) according to the protocol. The isolated RNA was subjected to amplification by reverse transcription using SuperScript Choice system (Invitrogen), with final concentrations in the 20-µl first-strand reaction of 100 pmol of T7-(dT)24 primer (high performance liquid chromatography-purified, Genset Corp., La Jolla, CA, USA.), 8 µg of total RNA, 1× first-strand buffer, 10 mM dithiothreitol, 500 µM of each dNTP, and 400 units of Superscript II reverse transcriptase. The second-strand cDNA synthesis was performed in a 150-µl reaction consisting of, in final concentrations, 1× second-strand reaction buffer, 200 µM of each dNTP, 10 units of DNA ligase, 40 units of DNA polymerase I, and 2 units of RNase H (Invitrogen). The double-stranded cDNA was purified with phenol:chloroform extraction using phase lock gels (Eppendorf-5 Prime, Inc. a division of Eppendorf AG, Boulder, CO, USA. Westbury, NY) and an ethanol precipitation (T7 Megascript kit; Ambion, Austin, TX, USA).

The mRNA present in 5 µg of total RNA is reverse-transcribed into cDNA and amplified using an oligo (dT) primer containing a T7 RNA polymerase promoter (T7-primer), followed by transcription into cRNA. In a newly developed protocol, the amplification step using a T7-primer is repeated by allowing only 50 ng total RNA utilization. Although this alternative approach is useful when sample size is small, or total cellular RNA is limited, the correlation between the two different protocols using the same sample is not very high (r2 < 0.7). This is because the T7-primer binds to the 3′ flanking region of mRNA and sometimes fails to amplify sufficient amount of 5′ sequences in the small-sample protocol. Five micrograms of purified cDNA was reverse transcribed using the Enzo BioArray high-yield RNA transcript labeling kit (Affymetrix Inc., Santa Clara, CA, USA), and the product was purified in RNeasy spin columns (Qiagen) according to the manufacture's instructions. Following an overnight ethanol precipitation, cRNA was resuspended in 15 µl of diethyl pyrocarbonate-treated water (Ambion) and quantified using UV-visible spectrophotometer.

Following the quantification of cRNA, to reflect any carryover of unlabeled total RNA according to the equation given by Affymetrix; adjusted cRNA yield = cRNA (mg) measured after in vitro transcription (starting total RNA) (fraction of cDNA reaction used in in vitro transcription), 20 µg of cRNA was fragmented (0.5 µg/µl), according to the instructions of Affymetrix using 5× fragmentation buffer containing 200 mM Tris acetate, pH 8.1, 500 mM potassium acetate, and 150 mM magnesium acetate (Sigma-Aldrich, Inc, St. Louis MS, USA). Twenty micrograms of the adjusted and fragmented cRNA was added to 300 µl of an hybridization mixture containing, at final concentrations, 0.1 mg/ml herring sperm DNA (Promega/Fisher, Madison, WI, USA), 0.5 mg/ml acetylated bovine serum albumin (Invitrogen), and 2× MES hybridization buffer (Sigma). Two hundred microliters of the mixture is used for hybridization to human Affymetrix GeneChip arrays each consisting of a specific number of probe sets representing known and expressed sequence tags (ESTs) according to Affymetrix procedures. Each probe set typically consists of 16 perfectly complementary 25-base long probes as well as 16 mismatch probes that are identical except for an altered central base.

To maintain a standard, all the microarray GeneChips should be purchased at once from the same lot numbers and used shortly after purchase. An aliquot of random samples should be first hybridized to an Affymetrix Test 2 Array to determine sample quality according to manufacturer's criteria. After fulfilling the recommendation criteria for use of expression arrays, hybridization is performed for 14–16 h at 45°C, followed by washing, staining, signal amplification with biotinylated antistrepavidin antibody, and final staining according to manufacture's protocol (Affymetrix).

Data Acquisition

Following the hybridization process, the membrane or GeneChips, are scanned to obtain raw hybridization values (for review see *27–29*). In the case of membrane array, scanning is carried out using a phosphorimager screen. Following image acquisition, the membrane arrays are subjected to specific alignment according to commercially available software to identify spots and quantify signal intensity GeneChips and glass microarray scanners are used for fluorescent image acquisition, and depending on the system utilized, they differ accordingly (http://www.biocompare.com). In the case of Affymetrix GeneChips, Genepix 5000A scanner is used to obtain the level of fluorescence spot intensity of the 25mer oligonucleotides and their mismatches. These intensity values represent the rate of hybridization and are used for analysis of multiple decision matrices to determine the presence or absence of hybridization and an average difference score representing the relative level of expression of a given gene on the array. The qualities of spots and local background for each array are automatically assessed by Genepix software with manual supervision to detect any inaccuracies in automated spot detection. Background and noise corrections are made to account for nonspecific hybridization and minor variations in hybridization conditions to obtain net hybridization values.

Data Analysis

The net hybridization values are first subjected to Affymetrix Analysis Suit V 5.0 to identify alteration in gene expression values. Briefly, probe sets that were flagged as absent on all arrays using default settings must be removed from the datasets. This filtering results in a substantial reduction in the datasets; in our study, the datasets reduced from 12,625 probe sets to 8552 probe sets. After obtaining the dataset values for each array, they are subjected to normalization to reduce experimental variability across different arrays while maintaining biological variability. Several statistical methods have been developed to normalize microarray datasets, as well as further analyses were conducted to address biological variability, some containing useful and simplified, step-by-step procedure to achieve normalization of hybridization values *(32–36)*. We used a global normalization method for the analysis of data obtained from membrane and Affymetrix arrays *(28, 29)*. Briefly, the net hybridization intensity value for each gene is multiplied by a scaling factor so that the average intensity of elements on a given array is equal after scaling, with scaling factors ranging from 2.5 to 10 within each array. The globally normalized values are then used for comparative analysis. To identify changes in the pattern of gene expression, the average and standard deviation (SD) of globally normalized values are calculated, followed by the subtraction of mean value from each observation and division by the SD. The mean transformed expression value of each gene in the transformed dataset was set at 0 and the SD at 1 *(10, 20)*. After filtering, the coefficient of variation is calculated for each probe set across all chips and the probe sets are ranked by the coefficient of variation of the observed single intensities.

The datasets are then analyzed either by grouping or classifying samples and genes according to their expression patterns (unsupervised methods such as hierarchical and K-mean clustering) and dimensionality-reduction methods that map high-dimensional expression datasets into low dimensional space based either on structure of covariance matrix or local similarity *(35–40)*. These methods are useful in describing changes in gene expression and obtaining a list of genes whose expression is considered to change as compared to control. However, because of differences in stringency of different statistical methods, careful attention must be given to the quality of data obtained at this stage.

Currently, there are several statistical methods available to obtain a list of differentially expressed genes *(30–46)*. Many software packages are available for data normalization, statistical analysis and visualization. They include Focus *(44)*, the statistical language "R" *(47, 48)*; http//www.r-project.org), SAM *(49)*, QVALUE *(50)*, Cyber-T *(51)*, and BRB-ArrayTools (http://linus.nci.nih.gov/BRB-ArrayTools). Most of these and other statistical software packages are based on "R" commands, and because of the requirement for a familiar software application, a user-friendly version is made

Table 2
List of Useful Web Sites for Microarray Data Analysis

FatIGO (GEPAS) Fisher Exact test, various multiple testing correction, permutation *(52, 53)*.	http://fatigo.bioinfo.cnio.es
GO-cluster Pearson's correlation coefficient *(54)*	http://mpibpc.mpg.de/go-cluster/
GOget, GOview Visual, search tool *(55)*	http://db.math.macalester.edu/goproject
Gominer Fisher Exact test, Relative enrichment *(56)*	http://discover.nci.nih.gov/gominer/
GOStat Fisher Exact test, multiple testing correction *(57)*	http://gostat.wehi.edu.au/
Gosurfer Pearson's χ^2, q value *(58)*	http://biostat.harvard.edu/complab/gosurfer/
MAPPfinder Standardized difference score (z) from hypergeometric distribution *((59))*	http://genmapp.org
NetAffx GO mining tool *(60)*	http://affymetrix.com/analysis/query/goanalysis.affx
Pathway Miner/Biocarta/GenMAPP/ KEGG *(61)*	http://biorag.org/pathway.html
Gene Ontology project (GO)	http://geneontology.org
The Kyoto Encyclopedia of Genes and Genomes (KEGG) provides searchable pathways	http://genome.jp/kegg/)
GenMAPP provides an image of a pathway	http://genmapp.org
Database for Annotation, Visualization, and Integrated Discovery (DAVID)	http://david.niaid.nih.gov)
Gene Expression Profile Analysis Suite	http://gepas.org
High-Throughput GoMiner	http://discover.nci.nih.gov/gominer/htgm.jsp

available at http://www.bioconductor.org. We used R and SAM statistical packages in our studies (Table 2).

For statistical analysis, we used normalized datasets obtained from all experiments involving tissues and cells and subjected either to SAM or "R" analysis *(47–50)*. The "R" analysis uses multiple test correction and false discovery rate to identify statistically significant gene expression values. For the final list, we selected genes whose expression values displayed a statistical significance at $p < 0.02$ between leiomyoma and myometrium from GnRH-treated and untreated cohorts, and $p < 0.005$ and 0.001 between GnRHa- and TGF-β-treated and untreated cells (control), respectively *(20,21)*. The validity of probe sets identified at these p values in predicting treatment class was established using "leave-one-out" cross validation where the data from one array were left out of the training set and probe sets with differential hybridization signal intensities were identified from the remaining arrays.

Microarray data are also normalized against the expression level of certain housekeeping genes, total mRNA quantity, or other assessments considered universal among samples. However, such relative measurements cannot be applied beyond different sample protocols or microarray platforms. The Toxicogenomic Projects at National Institute of Health Sciences have established a bioinformatics system called "Percellome" allowing normalization of gene expression values on a per cell basis *(62)*. Briefly, the system uses internal standards and a compensation program for each transcriptional level, thus the data once normalized from all samples and studies can be expressed as a copy number per genomic DNA level. This system was primarily for use with the Affymetrix GeneChip but can be expanded to other platforms.

Gene Classification, Ontology Assessment and Biological Pathways

The above analysis provides a list of genes whose expression is considered to change as compared to control without revealing any of their cellular and functional significance. To further extend the information, the data can be subjected to pathway analysis allowing the identification of more subtle changes in expression than the gene lists obtained from statistic p-value and/or fold-change. Although genes with large changes in expression might be interesting, so might those in which there are more subtle changes, such as small, but consistent, changes in expression of a group of genes with related function. Pathway analysis is suited to detecting such trends. There are three main sources of pathway and functional information, which can be either generic or species-specific and describe metabolic and cellular processes and genetic networks. The Gene Ontology project (GO) (http://www.geneontology.org) classifies genes into a hierarchy, placing gene products with similar functional category. Because GO is hierarchical, a gene that is in one category is automatically a part of all its parent classifications. The Kyoto Encyclopedia of Genes and Genomes (KEGG) (http://www.genome.jp/kegg/) provides searchable pathways, and GenMAPP (http://www.genmapp.org) provides an image of a pathway that is annotated with accession numbers. Many gene ontology classifications are available from GenMAPP (Table 2).

Some of the above mentioned software designed for data normalization and classification can also assess functional annotation and visualization of differentially expressed genes. One such useful software is Database for Annotation, Visualization, and Integrated Discovery (DAVID; http://www.david.niaid.nih.gov), which contains integrated GoCharts that assign genes to specific gene ontology functional categories based on selected classifications, KeggCharts that assign genes to KEGG metabolic processes and context of biochemical pathway maps, and Domain Charts that allow genes to be grouped according to PFAM conserved protein domains *(56–63)*. We used this software for assessing functional annotation and visualization of differentially expressed and regulated genes in leiomyoma and myometrium from GnRHa-treated and untreated cohorts as well as LSMC and MSMC treated in vitro with GnRHa and TGF-β. A major drawback of pathway annotation is that, for a large number of genes, no pathway or functional classification exists; this might be the case for more than half the number of genes represented on a microarray. However, most of the above mentioned software allow for other analysis such as promoter elements, chromosome position, protein structure or interaction data, and text enrichment to gain more information about an individual gene or group of genes.

Verification

mRNA EXPRESSION

Although gene microarray provides a substantial amount of information with respect to the molecular environment of cells and tissues at the time of assay, the biological significance of the expression of any gene may be difficult to interpret without verification of the information by quantitative means. Many reasons for such assumptions are recognized and reviewed in numerous articles, which include, but not limited to, differences in array platform, lack of uniform standards for experimental procedures, sample variability, and method of data analysis *(64–67)*.

After obtaining the list of genes based either on statistical significance or fold change, the expression of the gene(s) of interest must be validated and confirmed. Standard RT-PCR, semi-quantitative or quantitative RT-PCR as well as Northern blot analysis or RNase protection assays are the methods of choice for confirming mRNA expression. Although RT-PCR is used to confirm the expression of a gene of interest with a "yes" or "no" answer, the semi-quantitative RT-PCR is more accurate in determining the expression of a gene in relation to a housekeeping gene, i.e., beta action or GAPDH. Quantitative RT-PCR is an accurate and highly sensitive assay and requires a low level of RNA as compared to Northern blot analysis, and is useful for assaying genes with low expression levels, or limited availability of sample for RNA isolation *(68)*. Northern blot and RNase protection assay are not only quantitative but also allow for identification of the number and size of transcripts *(67)*. To further define tissue distribution of mRNA expression of a given gene identified by microarray, in situ hybridization is the method of choice and may provide important information reflecting the potential function of the gene product in that tissue.

We used RT-PCR to confirm the expression of more than 30 genes identified as differently expressed and regulated in leiomyoma and myometrium, as well as in LSMC and MSMC in response to TGF-β and GnRHa treatments. The selection of those genes was based not only on their expression values (up- or downregulation) but also on classification and biological functions important to leiomyoma growth and regression, regulation by ovarian steroids, GnRHa and TGF-β as indicated in the literature, and genes whose expression has not been previously reported in these cells and tissues. For realtime PCR, cDNA was generated with 2 μg of total RNA using Taqman Reverse Transcription (RT) reagent (Applied Biosystems Foster City, CA, USA). The RNA was incubated in 100 ml of a RT reaction mixture (1 × RT buffer, 5.5 mM $MgCl_2$, 2 mM dNTP, 2.5 mM random hexamers, 0.4 U Rnasin (RNase inhibitor), and 1.25 U MultiScribe reverse transcriptase at 25°C for 10 min and then at 48°C for 30 min. The reverse transcriptase was inactivated by heating at 95°C for 5 min. PCR was performed in 96-well optical reaction plates (Applied Biosystems) on cDNA equivalent to 100 ng RNA in a volume of 50 ml, containing 25 ml TaqMan universal master mix and optimized concentrations of FAM-labeled probe, forward and reverse primers for the above genes selected from Assay on Demand and/or Assay by Design (Applied Biosystems). RT-PCR was using ABI-Prism 7700 Sequence System (Applied Biosystems) at the following conditions: 2 min at 50°C, 10 min at 95°C for 1 cycle; and 15 s at 95°C, 1 min at 60°C for 40 cycles. Controls included RNA subjected to RT-PCR without reverse transcriptase and PCR with water replacing cDNA. All the controls gave a Ct value of 40, indicating no detectable PCR product under these cycle conditions. ABI Sequence Detection System 1.6 software (Applied Biosystems) was used to determine the cycle number at which fluorescence emission crossed the automatically determined threshold level (Ct). Results were analyzed using the comparative method; values were normalized to the 18S rRNA expression by subtracting mean Ct of 18S rRNA from mean target Ct for each sample, to obtain the mean DCt. Mean DCt values were then converted into fold change based on a doubling of PCR product in each PCR cycle, according to the manufacturer's guidelines (Applied Biosystems).

PROTEIN EXPRESSION

Confirming the mRNA expression of a differentially expressed gene identified by microarray and correlating that with the gene products at protein level allows for a better assessment of their biological function. Western blotting or ELISA of total protein isolated from the same tissue or cells used in microarray study is the method of choice, allowing for the establishment of whether gene transcription and translation are co-coordinately regulated. However, utilization of these methods depends on the availability of the antibody specific to the gene product. In addition, the specificity of the available antibodies for such applications may limit the detection of any protein in these samples. Immunohistochemical and immunocytochemical techniques can also be utilized to localize tissue distribution and cellular compartment assessment of the protein.

For immunoblotting, total protein was isolated from small portions of the tissues or cells cultured in defined conditions. The tissues and cells were directly lysed in a lysating buffer and the tissue homogenates and cell lysates centrifuged, supernatants collected and their total protein content was determined using a conventional method (Pierce, Rockford, IL, USA). An equal amount of protein sample is subjected to SDS-polyacrylamide gel electrophoresis and transferred to polyvinyldiene difluoride (PVDF) membrane. Subsequently, the blots are processed according to established procedures and then incubated with antibody specific to the target protein and/or anti-b actin (control) antibodies. The membranes are exposed to the corresponding HRP-conjugated IgG, and immunostained proteins are visualized using enhanced chemiluminesence reagents (Amersham-Pharmacia Biotech, Piscataway, NJ, USA). Many ELISA kits are commercially available for various proteins with detailed procedure for their application. For immunohistochemical localization, tissue sections are prepared from formalin-fixed and paraffin-embedded or frozen tissues. We have utilized Western blot analysis and immunohistochemistry to confirm protein production and tissue localization of several newly discovered gene products in leiomyoma and myometrium.

Biological Function

To define the biological function(s) of differentially expressed genes, a well-defined in vitro experiment using primary culture isolated from the tissue of interest, or isolated cell or cell line utilized for gene expression profiling, is recommended. Functional studies may include examining the effect of gene products on cell survival and regulation of other genes whose products are known to be essential for various biological activities of the cells/tissues. Further investigation may include dominant-negative vectors, RNA interference, or antisense morpholino oligonucleotides to determine the function of the gene of interest at cellular level in vitro. Under in vivo condition, experiments may be designed using antisense oligonucleotides, knockout or conditional knockout technologies as part of functional analysis of microarray experimental design stage.

NOTES

Microarray technology has been useful in profiling the expression of a large number of genes in many reproductive tract tissues. In recent years, this technology has been applied in leiomyoma and myometrium to identify their overall molecular environments. Several studies, including that of our own, have utilized different microarray platforms consisting of membrane and GeneChip arrays and a wide range of methods of data analysis identifying the expression of several hundred genes as differentially expressed in these tissues and their isolated smooth muscle cells (Table 3). Perhaps the most critical aspect of a microarray experiment is data analysis *(27–50)*. As such the analysis to identify genes whose expression displays significant difference between leiomyoma and myometrium used simple fold change, ANOVA, SAM and "R" statistical programs (Table 3). However, obtaining the list of genes is not the end-point of microarray investigation; rather the results provide a direction toward selection and confirmation of specific gene expression. With integration of gene expression data combined with proteomic analysis and in vitro regulation, these approaches can become effective tool toward understanding the monocular environment that govern normal physiological and pathological conditions associated with uterine and other reproductive tract tissues.

Using these tools, we have obtained large-scale gene expression profiling of leiomyoma and myometrium as well as their response to GnRHa therapy. We validated the expression of several genes at tissue level and in response to time-dependent action of GnRHa and TGF-β in LSMC and MSMC. Many of these genes and their products are known to play a key role in cellular transformation, growth, differentiation, apoptosis and matrix accumulation. Since these processes are critical in leiomyoma growth and regression, the information should assist us in forming specific hypothesis with therapeutic strategies for leiomyoma medical management.

Table 3
Microarray Platforms and Analysis in Leiomyoma Research

Platform	Method of analysis	#Significant gene	Validation	References
HuGeneFL6800 (6800 genes) and Affymetrix U95A (12,000 genes)	Fold change	67 ↑ 78 ↓ Regulated genes	No	(8)
GeneTrack HSVC 307 chip (17,000 genes)	Fold change	21 ↑ 50 ↓ Regulated genes	RT-PCR	(9)
Clontech membrane array (1200 known genes)	SAM	18 ↑ 82 ↓ Regulated genes	No	(10)
Affymetrix U133 (33,000)	Fold change		RT-PCR/Western blot immunohisto-chemistry	(11)
Affymetrix U95 (12,000 genes and 48,000 EST)	Fold change	245 genes/EST ↑ 358 gene/EST ↓	No	(16)
HuGeneFL6800 microarray chip (6800 genes)	Fold change	23 ↑ 45 ↓ Regulated genes	RT-PCR immuno-histochemistry	(12)
UniGEM V (7800 genes/EST)	GEM tools 2.4	15 ↑↓ Regulated genes	Northern/Western blots immunohisto-chemistry	(15)
In-house microarray (10,500 genes)	ANOVA	14 ↑ 11 ↓ Regulated genes	Quantitative RT-PCR	(13)
GeneTrack HSVC 307 chip (17,000 genes)	Fold change	21 ↑ 50 ↓ Regulated genes	RT-PCR	(14)
Affymetrix U133 (22,500)	Fold change	226 ↑↓ Regulated genes	RT-PCR	(17)
Affymetrix HuFL GeneChips (7000 gene)	ANOVA	146 genes	No	(18)
Affymetrix U95 GeneChip (12,000 genes)	"R"ANOVA	164 ↑↓ Regulated genes	RT-PCR/Western blot immunohisto-chemistry	(20)
Affymetrix U95 GeneChip (12,000 genes)	"R"ANOVA	331 ↑↓ Regulated genes	RT-PCR/Western blot immunohisto-chemistry	(21)
NIEHS 2K ToxChip Version 1.0 (1901 genes)	Z-score/MAD	11 ↑↓ Regulated genes	Northern/Western analysis immunocy-tochemistry	(22)

ACKNOWLEDGMENTS

We would like to thank Drs Li Ding and Jingxia Xu for their contributions toward the array analysis. Supported by the National Institute of Health grant HD 37432.

REFERENCES

1. Wallach, E.E. and Vlahos, N.F. (2004) Uterine myomas: an overview of development, clinical features, and management. Obstet. Gynecol. 104, 393–406.
2. Ohara, N. (2005) Selective estrogen receptor modulator and selective progesterone receptormodulator: therapeutic efficacy in the treatment of uterine leiomyoma. Clin. Exp. Obstet. Gynecol. 32, 9–11.
3. Chabbert-Buffet N., Meduri, G., Bouchard, P., and Spitz, I.M. (2005) Selective progesterone receptor modulators and progesterone antagonists: mechanisms of action and clinical applications. Hum. Reprod. Update. 11, 293–307.
4. Chwalisz, K., Perez, M.C., Demanno, D., Winkel, C., Schubert, G., and Elger, W. (2005) Selective progesterone eceptor modulator development and use in the treatment of leiomyomata and endometriosis. Endocr. Rev. 26, 423–438.
5. Chegini, N. (2000) Implication of growth factor and cytokine networks in leiomyomas. In: Hill J, ed. Cytokines in human reproduction. New York: Wiley & Sons; 133–162.
6. Sandberg, A.A. (2005) Updates on the cytogenetics and molecular genetics of bone and soft tissue tumors: leiomyoma. Cancer Genet. Cytogenet. 158, 1–26.
7. Walker, C.L., and Stewart, E.A. (2005) Uterine fibroids: the elephant in the room. Science 308, 1589–592.
8. Tsibris, J.C., Segars, J., Coppola, D., Mane, S., Wilbanks, G.D., O'Brien, W.F., and Spellacy, W.N. (2002) Insights from gene arrays on the development and growth regulation of uterine leiomyomata. Fertil. Steril. 78, 114–121.
9. Wu, X., Blanck, A., Norstedt, G., Sahlin, L., and Flores-Morales, A. (2002) Identification of genes with higher expression in human uterine leiomyomas than in the corresponding myometrium. Mol. Hum. Reprod. 8, 246–254.
10. Chegini, N., Verala, J., Luo, X., Xu, J, and Williams, R.S. (2003) Gene expression profile of leiomyoma and myometrium and the effect of gonadotropin releasing hormone analogue therapy. J. Soc. Gynecol. Investig. 10, 161–171.
11. Catherino, W.H., Prupas, C., Tsibris, J.C., Leppert, P.C., Payson, M., Nieman, L.K., and Segars, J.H. (2003) Strategy for elucidating differentially expressed genes in leiomyomata identified by microarray technology. Fertil. Steril. 80, 282–290.
12. Wang, H., Mahadevappa, M., Yamamoto, K., Wen, Y., Chen, B., Warrington, J.A., and Polan, M.L. (2003) Distinctive proliferative phase differences in gene expression in human myometrium and leiomyomata. Fertil. Steril. 80, 266–276.
13. Weston, G., Trajstman, A.C., Gargett, C.E., Manuelpillai, U., Vollenhoven, B.J., and Rogers, P.A. (2003) Fibroids display an anti-angiogenic gene expression profile when compared with adjacent myometrium. Mol. Hum. Reprod. 9, 541–549.
14. Ahn, W.S., Kim, K.W., Bae, S.M., Yoon, J.H., Lee, J.M., Namkoong, S.E., Kim, J.H., Kim, C.K., Lee, Y.J., and Kim, Y.W. (2003) Targeted cellular process profiling approach for uterine leiomyoma using cDNA microarray, proteomics and gene ontology analysis. Int. J. Exp. Pathol. 84, 267–279.
15. Kanamori, T., Takakura, K., Mandai, M., Kariya, M., Fukuhara, K., Kusakari, T,, Momma, C., Shime, H., Yagi, H., Konishi, M., Suzuki, A., Matsumura, N., Nanbu, K., Fujita, J., and Fujii, S. (2003) PEP-19 overexpression in human uterine leiomyoma. Mol. Hum. Reprod. 9, 709–717.
16. Skubitz, K.M., and Skubitz, A.P. (2003) Differential gene expression in uterine leiomyoma J. Lab. Clin. Med. 141, 297–308.
17. Hoffman, P.J., Milliken, D.B., Gregg, L.C., Davis, R.R. and Gregg, J.P. (2004) Molecular characterization of uterine fibroids and its implication for underlying mechanisms of pathogenesis. Fertil. Steril. 82, 639–649

18. Quade, B.J., Wang, T.Y., Sornberger, K., DalCin, P., Mutter, G.L., and Morton, C.C. (2004) Molecular pathogenesis of uterine smooth muscle tumors from transcriptional profiling. Genes Chromosomes Cancer 40, 97–108.

19. Lee, E.J., Kong, G., Lee, S.H., Rho, S.B., Park, C.S., Kim, B.G., Bae, D.S., Kavanagh, J.J., and Lee, J.H. (2005) Profiling of differentially expressed genes in human uterine leiomyomas. Int. J. Gynecol. Cancer 15, 146–154.

20. Luo, X., Ding, L., Xu, J., Williams, R.S., and Chegini, N. (2005) Leiomyoma and myometrial gene expression profiles and their responses to gonadotropin-releasing hormone analog therapy. Endocrinology 146, 1074–1096.

21. Luo, X., Ding, L., Xu, J., and Chegini, N. (2005) Gene expression profiling of leiomyoma and myometrial smooth muscle cells in response to transforming growth factor-beta. Endocrinology 146, 1097–1118.

22. Swartz, C.D., Afshari, C.A., Yu, L., Hall, K.E., and Dixon, D. (2005) Estrogen-induced changes in IGF-I, Myb family and MAP kinase pathway genes in human uterine leiomyoma and normal uterine smooth muscle cell lines. Mol. Hum. Reprod. 11, 441–450.

23. Arslan, A.A., Gold, L.I., Mittal, K., Suen, T.C., Belitskaya-Levy, I., Tang, M.S., and Toniolo, P. (2005) Gene expression studies provide clues to the pathogenesis of uterine leiomyoma: new evidence and a systematic review. Hum. Reprod. 20, 852–863.

24. Chegini, N., Ma, C., Tang, X.M., and Williams, R.S. (2002) Effects of GnRH analogues, 'add-back' steroid therapy, antiestrogen and antiprogestins on leiomyoma and myometrial smooth muscle cell growth and transforming growth factor-beta expression. Mol. Hum. Reprod. 8, 1071–1078.

25. Rossi, M.J., Chegini, N., and Masterson, B.J. (1992) Presence of epidermal growth factor, platelet-derived growth factor, and their receptors in human myometrial tissue and smooth muscle cells: their action in smooth muscle cells in vitro. Endocrinology 130, 1716–1727.

26. Ding, L., Xu, J., Luo, X., and Chegini, N. (2004) Gonadotropin releasing hormone and transforming growth factor beta activate mitogen-activated protein kinase/extracellularly regulated kinase and differentially regulate fibronectin, type I collagen, and plasminogen activator inhibitor-1 expression in leiomyoma and myometrial smooth muscle cells. J. Clin. Endocrinol. Metab. 89, 5549–5557.

27. Stoughton, R.B. (2005) Applications of DNA microarrays in biology. Annu. Rev. Biochem. 74, 53–82.

28. Auburn, R.P., Kreil, D.P., Meadows, L.A., Fischer, B., Matilla, S.S., and Russell, S. (2005) Robotic spotting of cDNA and oligonucleotide microarrays. Trends Biotechnol. 23, 374–379.

29. Barrett, J.C., and Kawasaki, E.S. (2003) Microarrays: the use of oligonucleotides and cDNA for the analysis of gene expression. Drug Discov. Today 8, 134–141.

30. Larkin, J.E., Frank, B.C., Gavras, H., Sultana, R., and Quackenbush, J. (2005) Independence and reproducibility across microarray platforms. Nat. Methods 2, 337–344.

31. Irizarry, R.A., Warren, D., Spencer, F., Kim, I.F., Biswal, S., Frank, B.C., Gabrielson, E., Garcia, J.G., Geoghegan, J., Germino, G., Griffin, C., Hilmer, S.C., Hoffman, E., Jedlicka, A.E., Kawasaki, E., Martinez-Murillo, F., Morsberger, L., Lee, H., Petersen, D., Quackenbush, J., Scott, A., Wilson, M., Yang, Y., Ye, S.Q., and Yu, W. (2005) Multiple-laboratory comparison of microarray platforms. Nat. Methods 2, 345–350.

32. Engelen, K., Coessens, B., Marchal, K., and DeMoor, B. (2003) MARAN: normalizing micro-array data. Bioinformatics 19, 893–894.

33. Quackenbush, J. (2002) Microarray data normalization and transformation. Nat. Genet. 32 (Suppl), 496–501.

34. Smyth, G.K, and Speed, T. (2003) Normalization of cDNA microarray data. Methods 31, 265–273.

35. Park, T., Yi, S.G., Kang, S.H., Lee, S.Y., Lee, Y.S., and Simon, R. (2003) Evaluation of normalization methods for microarray data. BMC Bioinformatics 4, 33.

36. Butte, A. (2002) The use and analysis of microarray data. Nat. Rev. Drug. Discov. 1, 951–960.

37. Curtis, R.K., Oresic, M., and Vidal-Puig, A. (2005) Pathways to the analysis of microarray data. Trends Biotechnol. 23, 429–435.

38. Raychaudhuri, S., Sutphin, P.D., Chang, J.T., and Altman, R.B. (2001) Basic microarray analysis: grouping and feature reduction. Trends Biotechnol. 19, 189–193.

39. Toronen, P., Kolehmainen, M., Wong, G., and Castren, E. (1999) Analysis of gene expression data using self-organizing maps. FEBS. Lett. 451, 142–146.

40. Eisen, M.B., Spellman, P.T., Brown, P.O., and Botstein, D. (1998) Cluster analysis and display of genome-wide expression patterns. Proc. Natl. Acad. Sci. USA 95, 14863–14868.

41. Chen, J.J., Delongchamp, R.R., Tsai, C.A., Hsueh, H.M., Sistare, F., Thompson, K.L., Desai, V.G., and Fuscoe, J.C. (2004) Analysis of variance components in gene expression data. Bioinformatics 20, 1436–1446.

42. Chuaqui, R.F., Bonner, R.F., Best, C.J., Gillespie, J.W., Flaig, M.J., Hewitt, S.M., Phillips, J.L., Krizman, D.B., Tangrea, M.A., Ahram, M., Linehan, W.M., Knezevic, V., and Emmert-Buck, M.R. (2002) Post-analysis follow-up and validation of microarray experiments. Nat. Genet. 32 (Suppl), 509–514.

43. Churchill, G.A. (2002) Fundamentals of experimental design for cDNA microarrays. Nature Genet 32 (Suppl), 490–495.

44. Cole, S.W., Galic, Z., and Zack, J.A. (2003) Controlling false negative errors in microarray differential expression analysis: a PRIM approach. Bioinformatics 19, 1808–1816.

45. Cui, X., and Churchill, G.A. (2003) Statistical tests for differential expression in cDNA microarray experiments. Genome Biol. 4, 210.

46. Dudoit, S., Yang, Y.H., Callow, M.J., and Speed, T.P. (2002) Statistical methods for identifying differentially expressed genes in replicated cDNA microarray experiments. Statistica Sin. 12, 111–139.

47. Dudoit, S., Yang, Y.H., and Bolstad, B. (2002) Using R for the analysis of DNA microarray data. R News 2, 24–32.

48. Ihaka, R., and Gentleman, R. (1996) R: a language for data analysis and graphics. J. Comput. Graph. Stat. 5, 299–314.

49. Tusher, V.G., Tibshirani, R., and Chu, G. (2001) Significance analysis of microarrays applied to the ionizing radiation response. Proc. Natl. Acad. Sci. USA 98, 5116–5121.

50. Storey, J.D., and Tibshirani, R. (2003) Statistical significance for genome-wide studies. Proc. Natl. Acad. Sci. USA 100 9440–9445.

51. Baldi, P., and Long, A. (2001) A Bayesian framework for the analysis of microarray expression data: regularized t-test and statistical inferences of gene changes. Bioinformatics 17, 509–519.

52. Herrero, J., Al-Shahrour, F., Diaz-Uriarte, R., Mateos, A., Vaquerizas, J.M., Santoyo, J., and Dopazo, J. (2003), GEPAS: A web-based resource for microarray gene expression data analysis. Nucleic Acids Res. 31, 3461–3467.

53. Al-Shahrour, F., Diaz-Uriarte, R., and Dopazo, J. (2004) FatiGO: a web tool for finding significant associations of Gene Ontology terms with groups of genes. Bioinformatics 20, 578–580.

54. Adryan, B., and Schuh R. (2004) Gene ontology-based clustering of gene expression data. Bioinformatics 20, 2851–2852.

55. Shoop, E., Casaes, P., Onsongo, G., Lesnett, L., Petursdottir, E.O., Donkor, E.K., Tkach, D., and Cosimini, M. (2004) Data exploration tools for the Gene Ontology database, Bioinformatics 20, 3442–3454.

56. Zeeberg, B.R., Qin, H., Narasimhan, S., Sunshine, M., Cao, H., Kane, D.W., Reimers, M., Stephens, R.M., Bryant, D., Burt, S.K.., Elnekave, E., Hari, D.M., Wynn, T.A., Cunningham-Rundles, C., Stewart, D.M., Nelson, D., and Weinstein, J.N. (2005) High-Throughput GoMiner, an 'industrial-strength' integrative gene ontology tool for interpretation of multiple-microarray experiments, with application to studies of Common Variable Immune Deficiency (CVID). BMC Bioinformatics 6, 168.

57. Beissbarth T., and Speed T.P. (2004) GOstat: find statistically overrepresented Gene Ontologies within a group of genes. Bioinformatics 20, 1464–1465.

58. Zhong, S., Li, C., and Wong, W.H. (2003) ChipInfo: software for extracting gene annotation and gene ontology information for microarray analysis. Nucleic Acids Res. 31 3483–3486.

59. Doniger, S.W., Salomonis, N., Dahlquist, K.D., Vranizan, K., Lawlor, S.C., and Conklin, B.R. (2003) MAPPFinder: using Gene Ontology and GenMAPP to create a global gene-expression profile from microarray data. Genome Biol. 4, R7.

60. Cheng, J., Sun, S., Tracy, A., Hubbell, E., Morris, J., Valmeekam, V., Kimbrough, A., Cline, M.S., Liu, G., Shigeta, R., Kulp, D., and Siani-Rose, M.A. (2004) NetAffx Gene Ontology Mining Tool: a visual approach for microarray data analysis. Bioinformatics 20, 1462–1463.

61. Khatri, P., and Draghici, S. (2005) Ontological analysis of gene expression data: current tools, limitations, and open problems. Bioinformatics 21, 3587–3595.

62. Kanno J., Aisaki, K-I., Igarashi, K., Nakatsu, N., Ono, A., Kodama, Y., and Nagao T. (2006) Per cell" normalization method for mRNA measurement by quantitative PCR and microarrays. BMC Genomics. 7, 64.

63. Pandey, R., Guru, R.K., and Mount, D.W. (2004) Pathway Miner: extracting gene association networks from molecular pathways for predicting the biological significance of gene expression microarray data. Bioinformatics 20, 2156–2158.

64. Knight, J. (2001) When the chips are down. Nature 410, 860–861.

65. Kothapalli, R., Yoder, S.J., Mane, S, and Loughran, T.P. Jr. (2002) Microarray results: how accurate are they? BMC Bioinformatics 3, 22.

66. Stears, R.L., Martinsky, T., and Schena, M. (2003) Trends in microarray analysis. Nat. Med. 9, 140–145.

67. Taniguchi, M., Miura. K., Iwao, H., and Yamanaka, S. (2001) Quantitative assessment of DNA microarrays – comparison with Northern blot analyses. Genomics 71, 34–39.

68. Rajeevan, M.S., Vernon, S.D., Taysavang, N., and Unger, E.R. (2001) Validation of array-based gene expression profiles by real-time (kinetic) RT-PCR. J. Mol. Diagn. 3, 26–31.

5 Gene Profiling Analysis of Androgen Receptor Mediated Function

Clay E. S. Comstock, PhD,
Craig J. Burd, PhD, Walter J. Jessen, PhD,
and Karen E. Knudsen, PhD

Contents

Introduction
Development and Reproductive Function
Androgen Insensitivity Syndrome
Kennedy's Disease
Prostate Cancer
Potential Androgen Receptor Targets
Potential AR Target Genes from The LNCaP Model
    in Prostate Cancer
Conclusions
References

Abstract

Androgens, particularly testosterone and its potent metabolite 5-α-dihydrotestosterone, serve as critical mediators in the development and maintenance of the male reproductive and non-reproductive systems. Androgen-dependent signaling is conveyed by the androgen receptor (AR), which is a member of the nuclear receptor superfamily. AR binding of androgen stimulates its ability to bind DNA and regulate gene transcription. The importance of the AR in human physiology is exemplified by the fact that disruption of this key receptor is causative for androgen insensitivity syndrome (AIS) and spinal and bulbar muscular atrophy (SBMA), also known as Kennedy's disease. In contrast, AR activity is essential for benign prostatic hyperplasia (BPH) and the development and progression of prostate cancer. The advent of microarray technology has provided significant contributions toward the understanding of the underlying mechanisms that govern the AR function in these physiological and pathophysiological conditions. The aim of this chapter is to summarize the recent advances using genomics to study the role of androgens in developmental and reproductive processes, AIS, Kennedy's disease, and prostate cancer. As the body of work relative to prostate cancer is large, a gene list comparison was performed with prostate cancer studies that utilized a well-characterized human prostatic adenocarcinoma cell line treated with physiological concentrations of androgen. The consensus targets were then evaluated on their physiological relevance as well as on their potential as direct AR targets. In addition, alternate models of AR action in prostate cancer were examined. These studies provide evidence for AR-dependent regulation of other identified genes

From: *Contemporary Endocrinology: Genomics in Endocrinology:*
DNA Microarray Analysis in Endocrine Health and Disease
Edited by S. Handwerger and B. Aronow © Humana Press, Totowa, NJ

involved in metabolism, transcription, and signaling pathways. Together, these collective observations reveal insightful information concerning androgen action in human health and disease.

Key Words: Microarray; androgen; androgen receptor (AR); androgen response element (ARE); androgen responsive genes (ARG); androgen insensitivity syndrome (AIS); spinal and bulbar muscular atrophy (SBMA); Kennedy's disease; prostate cancer; development; spermatogenesis; LNCaP cells; animal models; T877A mutation; alternate ligands; metabolism; transcription; proliferation; signaling

INTRODUCTION

Androgens are essential to a variety of developmental and physiological responses. The most freely circulating androgen is testosterone, which is produced primarily in the testes. Additional androgen synthesis occurs within the adrenal gland. However, in distinct androgen-responsive organs (e.g., prostate, penis, scrotum), testosterone is converted to the more potent 5-α-dihydrotestosterone (DHT) by the enzyme 5-α reductase *(1)*. Alternatively, the enzymatic activity of aromatase can convert testosterone to estradiol, resulting in estrogen receptor (ER) activation. Testosterone exerts its biological effect through the androgen receptor (AR) (Fig. 1).

The AR is a member of the steroid subclass of the ligand-activated nuclear receptor superfamily of transcription factors *(2)*. Receptors in this transcription factor family share a conserved structure *(3)* consisting of an amino terminal transactivation domain containing transactivation functions (AF1), a conserved DNA-binding domain (DBD), and a carboxyl terminus containing the ligand-binding domain (LBD) and a second transactivation domain (AF2). In contrast to most nuclear receptors that rely on the AF2 domain for activity, the majority of AR activity is based within the amino-terminal domain (NTD), which contains two alternate transactivation domains (AF1 and AF5). In the absence of ligand, cytoplasmic AR is sequestered and held inactive through its association with heat shock proteins (HSP). Subsequent to stimulation with androgen, the receptor disassociates from the HSPs, followed by rapid dimer-

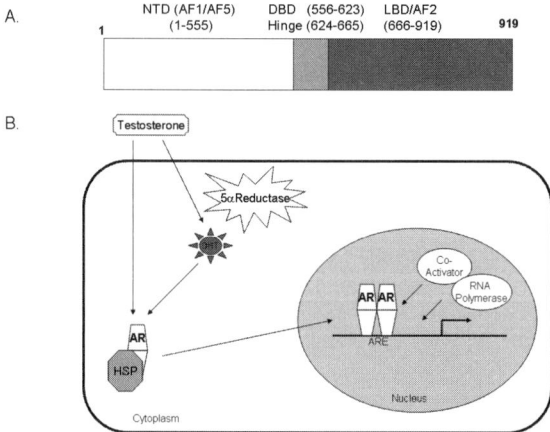

Fig. 1. Androgen receptor domains and androgen signaling. (a) The functional domains of the androgen receptor depicting the amino-terminal domain (NTD), DNA-binding domain (DBD), hinge region, and the ligand-binding domain (LBD). Amino acid designations depicted in parenthesis. (b) Testosterone or DHT act through the AR to stimulate expression of target genes. Refer to the text for more detail.

ization and translocation to the nucleus. These events confer the proper configuration of the receptor necessary for activity and recruitment of accessory proteins. Within the nucleus, the receptor occupies distinct promoter regions that contain androgen response elements (ARE). The consensus ARE is GG(A/T)ACAnnnTGTTCT and plays an essential role in transcription factor initiation (4). Additionally, many gene regulatory regions contain multiple AREs that work in a cooperative fashion with each other or additional regulatory elements to promote AR binding or activity (5–9). Once the receptor is bound to its cognate DNA sequence, subsequent recruitment of co-regulatory proteins to the AREs facilitates gene regulation through chromatin modification and complex stabilization. Thus, androgens act through the AR to induce a specific gene expression program that can elicit a wide spectrum of biological outcomes, including cellular differentiation and proliferation. In addition to this classical paradigm of AR action, recent evidence suggests that androgens can exert rapid effects on cellular processes through modulation of discrete signal transduction pathways (e.g., MAPK). These immediate responses to androgen occur with kinetics that exclude the AR function at AREs and are proposed to occur via membrane-associated and/or cytoplasmic receptors (10). The relevance of these "non-genomic" effects for the biological functions of androgens has yet to be discerned.

Androgen and AR activity are critical for a myriad of processes, including development of the male reproductive tract, muscle formation, body composition, bone mineralization, fat metabolism, and nervous system development (11). The importance of androgen is illustrated by the alterations in AR function, which are tightly associated with a number of diseases in these tissue types. Specifically, disruption of AR function is causative in spinal and bulbar muscular atrophy (SBMA), also termed Kennedy's disease, and androgen insensitivity syndrome (AIS). By contrast, the AR activity plays an essential role in the development of benign prostatic hyperplasia (BPH), prostatic intraepithelial neoplasia (PIN), and prostatic adenocarcinoma. For these hyperplastic and neoplastic diseases of the prostate, AR is the major therapeutic target for patient management. For each condition, delineation of AR function is essential for clarifying the mechanisms by which these disease states arise. Recent advances in genomics have been paramount to the study of genome-wide changes initiated by androgen and have yielded preliminary insight into AR function. This chapter is focused on the use of DNA microarray technology to reveal AR function under physiological and pathophysiological conditions.

DEVELOPMENT AND REPRODUCTIVE FUNCTION

The role of the AR in development is prominent in male sexual dimorphism and function (12). Androgen action is most notable at the onset of puberty, wherein males become sexually competent. However, the requirement for androgen begins much earlier, wherein androgen is necessary for the development of Wolffian ducts, vas deferens, seminal vesicle, prostate, penis, and scrotum. Androgen signaling is also pivotal in the differentiation of bone between males and females during puberty, allowing for greater bone volume and length that is associated with the increased size of males after puberty (13). Despite the established importance of androgen in these systems, few studies have investigated the AR-dependent gene targets that govern its influence over development. By contrast, several studies have addressed the action of androgen and AR in spermatogenesis.

Spermatogenesis occurs initially in the testes and is critical for male reproduction. It is well established that initiation and progression of spermatogenesis requires androgen action, and the testicular action of AR has been addressed using murine model systems. To assess androgen action, hypogonadal mice were utilized, which lack gonadal releasing hormone (GnRH), a positive regulator of testosterone production (14). These mice exhibit infantile testes and halted spermatogenesis, which can be qualitatively rescued through androgen supplementation. Using this model, testes were analyzed to identify genes that respond to testosterone after either acute (up to 24 hours) or chronic (up to 21 days) stimulation. Interestingly, short-term testosterone exposure at short times (<12 hours) led to a uniform decrease in transcripts. These changes were attributed to testosterone-induced stimulation of meiosis in the immature testes. However, at 24 hours of treatment, changes resulted in a net increase of androgen-dependent gene transcription (139 upregulated genes at $t = 24$ hours, compared to 19 at $t = 4$ hours). From these studies, placentae and embryos oncofetal (Pem) gene expression was induced by 8 hours and was sustained during acute exposure. Although the mechanism of altered Pem expression was not determined, Pem has been previously identified as an androgen-responsive target gene. Long-term exposure to androgen allowed for study in postmeiotic germ cells and revealed induction of 78 genes, of which only three were shared with the 24-hour treatment.

A second study examining the effects of androgen on testicular action utilized the effects of androgens on neonatal mice as well as the AR antagonist flutamide (Chimax® and Drogenil®) on adult mouse testes (15). As neonatal mice lack a definitive hypothalamus-pituitary-testis axis, they provide a model by which androgen supplementation can be examined in the absence of global signaling effects. Pem was utilized in this study as a control and was upregulated 3.8-fold at 16 hours by treatment with testosterone. As expected for an AR target gene, Pem expression was significantly repressed in the adult testes of mice treated with flutamide. Interestingly, androgen treatment did not induce transcripts of numerous cell cycle markers such as PCNA or Myc. AR, a putative target of itself, was also unchanged. Among the notable regulated genes were steroidogenic proteins such as CYP17a1, also identified in the GnRH model, which converts C(21) to C(19) steroids. Combined, these studies identify Pem as a putative marker of AR action in the murine testes, although it remains to be determined whether Pem is a direct AR target gene.

Spermatozoa maturation in the epididymis is also dependent on androgen action, and murine model systems have proved useful in assessing AR action in this tissue (16). For these studies, intact or castrated mice were treated with oil (vehicle) or DHT for 48 hours (10 mg total). Subsequently, the epididymis was dissected into the cauda, caput, and corpus, regions for RNA extraction and subjected to microarray analysis. As such, each functional region was separately analyzed to identify the androgen-responsive genes. Of the regions examined, the cauda was identified as the most androgen-responsive, with 133 uniquely regulated genes. The caput and corpus demonstrated 62 and 55 gene-specific effects, respectively. Among the notable genes proposed to be androgen regulated was myo-inositol 1-phosphate synthase A1 (Isyna1) corresponding to a loss of myo-inositol, which regulates osmolarity, seen in the epididymis lumen after castration. In addition, within the cauda region, members of the kallikrein family of serine proteases and serpine-2, a thrombin and urokinase inhibitor previously identified as critical for spermatogenesis, were also upregulated by androgen. Again, these gene targets await further validation and analyses to be assigned as direct or indirect targets of AR action. However, these studies provide further evidence that the action of androgen and its effect on gene regulation are tissue-specific. Moreover, these studies indicate that murine model systems may prove useful in the analysis of tissue-specific AR activity.

ANDROGEN INSENSITIVITY SYNDROME

Given the importance of androgen in mediating development of multiple tissues, aberrations in AR are known to give rise to multiple disease states *(17)*. Loss or reduction of AR function results in AIS, which is subclassified into categories of complete (C) or partial (P) AIS based on severity *(18)*. Severe CAIS results from plenary loss of AR function. Consequently, CAIS patients with 46XY genotypes display external female genitalia, gynecomastia, and present with absence of all male sexual organs, pubic and axillary hair. Patients with PAIS exhibit a range of symptoms from a predominantly female phenotype to an ambiguous or male phenotype. However, all AIS results in infertility. The predominant cause of AIS is mutations within the AR gene, resulting in a functionally impaired protein product. A list of AR mutations known to result in AIS is banked in the *McGill's Androgen Receptor Gene Mutations Database (19)*.

Although there is a clear link between loss or reduction of AR function and AIS, the use of microarray technology to reveal the basis of disease has been very limited. One study utilized genital fibroblasts from a CAIS patient to delineate the effects of androgen *(20)*. The AR expression was retained within these cells and two comparisons were made. First, baseline alterations in gene expression were compared between AIS-derived and normal genital fibroblasts. Second, gene expression in the presence and absence of androgen was analyzed, so as to specifically identify defects in AR signaling. While differences in homeobox A13, T-box, cell adhesion, and matrix genes were detected between AIS and normal cells, no effects of androgen were significant in either of the genital fibroblast backgrounds. These findings may indicate that the effects of androgen are cell type and temporally specific, thus creating difficulty for identification of critical AR target genes during development. Generation of additional models that recapitulate AIS will assist in identifying AR targets that are critical for disease development.

KENNEDY'S DISEASE

SBMA is another X-chromosome-linked disease that is attributed to aberrations in AR function. The disease is characterized by progressive muscular weakness and atrophy that usually presents at the beginning of the third decade of life in males. Muscular weakness often leads to trembling and muscle cramps, concurrent with loss of motor neurons in the brain stem and spinal cord *(21)*. Patients also demonstrate some signs of androgen insensitivity but display increased serum testosterone levels *(22)*. The underlying genetic basis of SBMA is attributed to expansion of a trinucleotide (CAG) repeat region within the AR gene *(23)*. While the modal number of repeats is below 35, in SBMA individuals these repeats can be present in excess of 60 copies. These CAG or polyglutamine (polyQ) repeats are known to result in protein misfolding, aggregation, and general reduction in AR activity. However, since AIS patients do not present symptoms of SBMA, it is hypothesized that the polyQ repeats render a "gain of function" effect *(21)*. The increase in repeats confers a new activity to AR in neurons generated by protein aggregation and results in cell death. Moreover, ligand stimulation results in an increase in disease progression theorized to be generated by AR dissociation from HSPs *(24)*. As HSPs tend to stabilize protein conformation, they may also inhibit protein aggregation. Therefore, dissociation of HSPs from AR may increase disease progression. As such, the role of AR in the development and progression of SBMA is not associated with its canonical function, which is diminished, but with novel activity associated with the polyglutamine extension.

Due to the known etiology of SBMA, characteristic symptoms, and cell-specific background, it provides a tantalizing model to study using genomic approaches. To address the causes of the disease, mouse motor neuron neuroblastoma cell lines have been generated, wherein AR variants expressing either 24 or 65 polyglutamine repeats were integrated into the genome *(25)*. Expression of the expanded polyglutamine AR generated a basal change in transcription of genes involved in ubiquitin-mediated degradation as compared to cells expressing a wild-type AR. Among these genes are components of the 20S proteasome, Lmp7 and MECL1, as well as the proteasome activator PA28. In addition, the molecular chaperone HSP22 was also upregulated. The effect of ligand in this system caused changes in 57 genes for the Q24 receptor and 17 genes for the Q65 receptor. In the Q65 receptor group, three genes were unique: retinol-binding protein, adrenomedullin precursor, and PP2A-β (reduced expression). Many of the genes activated by the wild-type receptor were unchanged by the expanded polyQ receptor, supporting the theory of diminished ligand activity associated with the mutant. Thus, microarray analyses have yielded early insight into the effect of the polyQ region on AR function in vivo. Interestingly, a transgenic mouse model of Kennedy's disease has been developed and further investigation using microarray analysis should reveal novel insights into the development and progression of the disease *(24)*.

PROSTATE CANCER

Advances in our knowledge of AR function have largely emerged from the study of AR function in the prostate. It is known that AR is required for the development, growth, and secretory function of the prostate. The action of the AR during development is complex and is mediated by stromal–epithelial interactions *(26)*. During development, it is actually the stromal cells that are AR positive and it is thought that activation of the AR in this cell type promotes induction and secretion of paracrine factors that promote epithelial cell growth and differentiation. This hypothesis was borne out using models of tissue recombination, wherein proliferation of the epithelia requires AR stimulation in the stromal compartment *(27)*. However, the gene targets of stromal cell action have yet to be clearly defined. Postdevelopment, the differentiated epithelia gain AR expression and AR activation in this cell type induces expression of the well-characterized AR target gene, PSA (prostate-specific antigen). PSA expression is generally confined to the secretory epithelia and the PSA protein encodes a protease that contributes to the overall function of the prostate. Most importantly, this marker has been proven to be invaluable in the diagnosis and monitoring of therapeutic efficacy in prostate cancer *(28)*.

AR Function in Prostate Cancer

Prostate cancer is a major health concern as it is the most diagnosed malignancy and second leading cause of cancer death among American men. For organ-confined disease, radical prostatectomy or radiation therapy results in a high cure rate. However, for metastatic disease, androgen deprivation therapy is the mainstay of treatment, as prostate cancer cells require AR activity for growth and survival *(29)*. Androgen ablation therapies target AR action through removal of endogenous androgen production and/or through the use of direct AR antagonists. This treatment is initially effective, in that tumor cells either undergo cell cycle arrest *(30)* or programmed cell death *(31)*, resulting in disease remission. Efficacy of therapy is monitored biochemically, wherein tumor regression is accompanied by dramatic reductions in serum PSA *(28)*. However, recurrent tumors ultimately arise (within a median time of 2–3 years), wherein AR has been re-activated *(32)*. Tumor recurrence is almost invariably preceded by a rise in serum PSA levels, also referred to as "biochemical failure" *(33)*. Unfortunately, there is no effective treatment for recurrent tumors that fail androgen deprivation therapy, leading to significant patient morbidity. Thus, substantial efforts have been directed at delineating the multiple mechanisms by which AR is regulated and re-activated in recurrent disease. A large body of evidence has demonstrated that AR can be re-activated in therapy-resistant tumors by at least four mechanisms *(34)*, including: (a) AR amplification (thus sensitizing to environments of low ligand); (b) mutation of the AR LBD (thus broadening ligand specificity); (c) overexpression of AR co-activators; and (d) ligand-independent AR activation (typically mediated by growth factor pathways). These examples illustrate common mechanisms by which tumors evade therapeutic intervention and underscore the substantial selective pressure for AR activation that is induced by hormone therapy. However, the precise mechanisms by which AR regulates prostate cancer cell proliferation remain elusive. It is hoped that, through DNA microarray analyses, critical targets of AR action in early and recurrent tumors will reveal novel mechanisms to control AR function in prostate cancer development and progression.

Consideration of Parameters in Cell-based Assays: The LNCaP Model

The study of androgen action in prostate cancer has been significantly aided by the use of many prostate cancer cell lines, arguably the most common and well characterized being the LNCaP cell line *(35)*. These cells, derived from a metastatic lymph node deposit of prostate cancer, exhibit many of the properties of androgen dependence seen in clinical disease. These cells retain AR and PSA expression and, therefore, provide good biochemical markers for the analysis of AR activity in response to ligand stimulation. LNCaP cells are also dependent on physiological levels of androgen for cellular proliferation both in vitro and in vivo, thus providing analyses of androgen-dependent tumor growth when utilized as a xenograft. Lastly, LNCaP cells undergo growth arrest *(36)* and exhibit reduced PSA expression in response to bicalutamide (Casodex®), which is an AR antagonist that is frequently utilized in prostate cancer therapy. Bicalutamide both prevents DHT binding to the receptor *(37)* and also facilitates the recruitment of co-repressors to the AR complex *(38)*, thus utilizing both active and passive mechanisms to block AR function. Thus, LNCaP cells respond appropriately to both androgen deprivation therapy and to therapeutic AR antagonists.

Despite the obvious utility of this model system in analyzing AR action, several critical parameters must be considered in experimental design. First, it is essential to note that these cells express a relatively common tumor-derived mutant of the AR (T877A), which broadens its ligand specificity (39). Study of androgen-dependent gene regulation in this model system, therefore, requires careful manipulation of cultured cells under steroid-depleted conditions, wherein androgen is re-supplemented as the only steroid source. Second, androgen dosage is a critical factor for data interpretation. As PSA is considered the gold standard for AR activity and its expression is directly dependent on the dose of ligand, many investigators have performed assays with supraphysiological concentrations of androgen (1 to 10 nM) (40–43). These supraphysiological androgen levels preclude the chance of identifying physiologic targets that are important for prostate cancer proliferation. As with most hormone-dependent cancer cells, there is a bell-shaped response of LNCaP cells to androgen stimulation, wherein physiological androgen levels (0.1 to 1 nM) stimulate cell cycle progression, while doses above 10 nM and below 0.01 nM potently inhibit cellular proliferation. The mechanism of "high androgen"-mediated growth inhibition has been attributed to increased p27kip1 (44) and TGF β (45) levels and holds limited biological relevance. Therefore, delineating the significance of AR target genes identified using high-dose androgen requires subsequent stringent analyses using physiologic conditions. In sum, identification of AR target genes using the LNCaP model system necessitates a priori consideration of culture conditions and appropriate androgen dosage.

AR Activity in the LNCaP Model

To identify AR target genes critical to cancer growth and survival, a multitude of studies have utilized the LNCaP model. However, disparities in experimental design and platform have resulted in abundant putative target genes, thus confounding comparison between datasets. To circumvent these issues and bolster the potential for identifying AR target genes, we refined the datasets for a baseline comparison of nine LNCaP studies (46–53) that utilized androgen at concentrations less than 10 nM and/or utilized bicalutamide to confirm the AR association. Selection for analysis also required that gene lists were available. In the majority of studies, an initial period of steroid deprivation (1–5 days) was utilized to achieve residual androgen activity, followed by a 24 to 72- hour stimulation of androgen in the presence or absence of bicalutamide. To complement the LNCaP studies, two additional genome-wide comparisons were incorporated. The first compared androgen sensitive- (LNCaP, 22Rv1, LAPC4, MDA PCa 2a and 2b) and androgen insensitive- (PC-3, PPC-1, DU 145) prostate cancer cell lines (15), and the second analyzed therapy-resistant human tumors as compared to changes in LNCaP cells under the conditions of androgen ablation (54). Each study was represented as a sample within the expression data analysis software package GeneSpring GX 7.3 (Agilent Technologies) using a binary score to indicate the presence or absence of genes. Genes were filtered for those appearing in a minimum of three samples and clustered using Pearson correlation around zero (complete linkage) for genes and Euclidean distance (complete linkage) for conditions. The cross-comparison revealed 104 common targets, with PSA (gene designation KLK3) being the most frequently identified target (>90%) (Fig. 2). This observation validates the established utility of PSA as a universal marker of AR action in prostate cancer cells and lends confidence

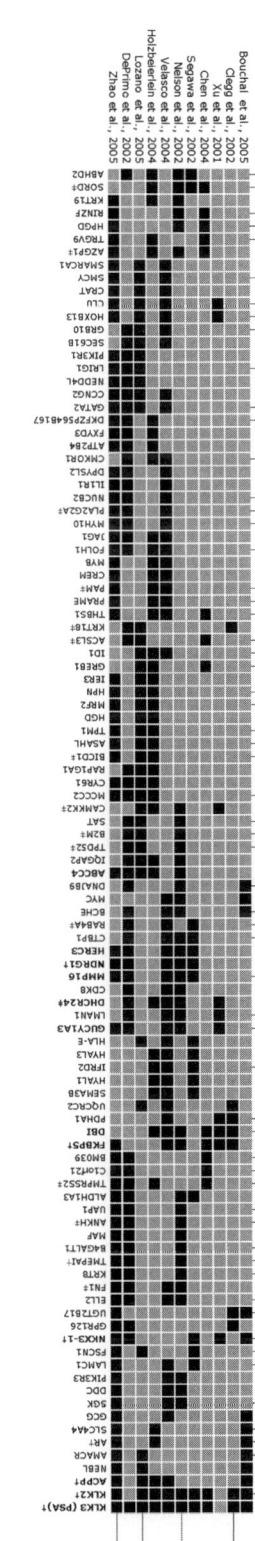

Fig. 2. Hierarchical cluster analysis of genes identified in LNCaP cells. Gene comparisons were performed as described in the text. Prior to clustering, genes were filtered for those appearing in at least 3 of 11 samples ((45%, in **bold type**). Genes appearing frequently ((≥ 27%) tend to cluster towards the top of the heat map. A number of these genes are known to be expressed in prostate tissue, including KLK3, KLK2, ACPP, NKX3-1, DHCR24, NDRG1, and ABCC4. Genes containing functional androgen response elements (†, Table 2) or putative androgen response elements (‡, Table 3) are indicated to the right of each gene symbol.

to subsequent gene comparisons. Henceforth, the genes mentioned in the text will be referred to by their abbreviations as a full description of each gene; and the frequency within the dataset is summarized in Table 1. Although these genes are scored as androgen-responsive, most are yet to be validated as direct AR target genes.

By comparison to the large number of identified androgen-responsive genes, few loci have been established as direct AR targets. For the purposes of this chapter, the existing literature was scrutinized to identify androgen-induced genes with functional AREs, as defined by the ability of AR to bind directly to oligonucleotide AREs in vitro or associate with ARE-containing promoter regions as determined by chromatin immunoprecipitation analysis (ChIP). At least 30 genes were identified, which meet the criteria stated above, and summarized in Table 2. Of the genes commonly identified in the LNCaP microarray analysis (Fig. 2), 8% were validated direct AR targets, while an additional 14% of the genes have been shown to harbor putative (but not yet validated) AREs, shown in Table 3. Other microarray studies using androgen or anti-AR therapies have documented similar androgen-responsive targets. For instance, the genes KLK2, TMPRSS2, and DBI were among six common targets similarly regulated at 48 hours by antisense AR treatment or androgen deprivation in combination with bicalutamide in LNCaP cells *(55)*. Likewise, serial analysis of gene expression after 24 hours of androgen treatment in LNCaP cells identified ACPP, CLU, and KRT19, while 72 hours of treatment identified only KRT18 using a proteomic approach *(56)*. Combined, these analyses lend further confidence in the identification of these genes as potential AR targets. The known and potential AR targets identified represent a myriad of cellular processes, thus revealing the diversity of the androgen-dependent signaling pathways. Of significant concern, however, is that few of the identified or putative AR targets have clear implications for prostate cancer growth or survival. This is surprising as AR activation is clearly linked to these biological outcomes in prostate cancer cells. This disparity in AR function and biological outcome indicates the need for parallel study of gene expression with analyses of cellular proliferation in cancer model systems.

Comparison to Animal Models of AR Action

An alternate model for studying androgen action can be achieved in rodents through removal of endogenous androgen by orchiectomy. Generally, young adult male Sprague-Dawley rats are castrated and, following a period of 7 to 10 days, are injected daily with 5 to 7 mg/kg androgen. One particular study identified 234 differentially expressed transcripts in the ventral prostate by 6 hours of androgen treatment, of which 36% of the genes were cell signaling and response related *(57)*. Unfortunately, the data was not available for comparison against the LNCaP model. A separate study, wherein rodents were supplemented with androgen for 48 hours, identified 162 upregulated and 143 downregulated genes in the prostate *(58)*. Of these, selected genes were also identified in the LNCaP cluster analysis (Fig. 2), including ALDH, CLU, DBI, and FKBP5. Similar targets were identified using subtractive hybridization of the rat ventral prostate *(59)*. The Dunning Rat model has also been employed to assess AR function, which utilizes transplantable prostate-derived tumors that progress after castration to a therapy-resistant state. A comprehensive castration study using this model revealed a number of differentially expressed transcripts *(60)*; however, 73 transcripts identified with human homologs had no commonality with the 104 human genes in the LNCaP dataset.

Table 1
Description and Frequency of Genes Appearing in Fig. 2

Gene	Accessionƒ	Frequency	Description
ABCC4	**NM_005845**	**45%**	**ATP-binding cassette, sub-family C (CFTR/MRP), member 4**
ABHD2	NM_007011	36%	Abhydrolase domain containing 2
ACPP†	**NM_001099**	**45%**	**Acid phosphatase, prostate**
ACSL3‡	NM_004457	27%	Acyl-CoA synthetase long chain family member 3 (Fatty-acid-Coenzyme A ligase, long-chain 3)
ALDH1A3	NM_000693	36%	Aldehyde dehydrogenase 1 family, member A3
AMACR	NM_014324	27%	Alpha-methylacyl-CoA racemase
ANKH‡	NM_054027	27%	Ankylosis, progressive homolog (mouse)
AR†	NM_000044	27%	Androgen receptor (Dihydrotestosterone receptor)
ASAHL	NM_014435	27%	N-acylsphingosine amidohydrolase (acid ceramidase)-like
ATP2B4	NM_001684	27%	ATPase, Ca++ transporting, plasma membrane 4
AZGP1‡	NM_001185	36%	Alpha-2-glycoprotein 1, zinc
B2M‡	NM_004048	27%	Beta-2-microglobulin
B4GALT1	NM_001497	27%	UDP-Gal:betaGlcNAc beta 1,4-galactosyltransferase, polypeptide 1
BCHE	NM_000055	36%	Butyrylcholinesterase
BICD1‡	NM_001714	27%	Bicaudal D homolog 1 (Drosophila)
BM039	NM_018455	27%	Uncharacterized bone marrow protein BM039
C1orf21	NM_030806	27%	Chromosome 1 open reading frame 21
CAMKK2‡	NM_006549	36%	Calcium/calmodulin-dependent protein kinase kinase 2, beta
CCNG2	NM_004354	36%	Cyclin G2
CDK8	NM_001260	27%	Cyclin-dependent kinase 8
CLU	NM_001831	36%	Clusterin (Testosterone-repressed prostate message 2, Apolipoprotein J)
CMKOR1	NM_020311	27%	Chemokine orphan receptor 1
CRAT	NM_000755	27%	Carnitine acetyltransferase
CREM	NM_183013	27%	cAMP responsive element modulator
CTBP1	NM_001328	36%	C-terminal binding protein 1
CYR61	NM_001554	36%	Cysteine-rich, angiogenic inducer, 61 (Insulin-like growth factor-binding protein 10)
DBI	**NM_020548**	**55%**	**Diazepam binding inhibitor (GABA receptor modulator, Acyl-Coenzyme A binding protein)**
DDC	AK057400	27%	Homo sapiens cDNA FLJ32838 fis, clone TESTI2003299.
DHCR24‡	**NM_014762**	**45%**	**24-dehydrocholesterol reductase**
DKFZP564B167	NM_015415	27%	Brain protein 44
DNAJB9	NM_012328	27%	DnaJ (Hsp40) homolog, subfamily B, member 9 (Microvascular endothelial differentiation gene 1)
DPYSL2	NM_001386	27%	Dihydropyrimidinase-like 2
ELL2	NM_012081	36%	Elongation factor, RNA polymerase II, 2
FKBP5†	**NM_004117**	**55%**	**FK506 binding protein 5 (Androgen-regulated protein 6)**

(Continued)

Table 1
(Continued)

Gene	Accession*f*	Frequency	Description
FN1‡	NM_002026	36%	Fibronectin 1
FOLH1	NM_004476	36%	Folate hydrolase 1 (Prostate-specific membrane antigen 1, Glutamate carboxypeptidase II)
FSCN1	NM_003088	27%	Fascin homolog 1, actin-bundling protein (Strongylocentrotus purpuratus)
FXYD3	NM_021910	27%	FXYD domain containing ion transport regulator 3 (Mammary tumor 8 kDa protein)
GATA2	NM_002050	36%	Endothelial transcription factor GATA-2 (GATA binding protein 2, NFE1B)
GCG	NM_002054	27%	Glucagon
GPR126	NM_020455	27%	G-protein coupled receptor 126
GRB10	NM_005311	27%	Growth factor receptor-bound protein 10
GREB1	NM_014668	27%	GREB1 protein
GUCY1A3	**NM_000856**	**45%**	**Guanylate cyclase 1, soluble, alpha 3**
HERC3	**NM_014606**	**45%**	**Hect domain and RLD 3**
HGD	NM_000187	27%	Homogentisate 1,2-dioxygenase (Homogentisate oxidase)
HLA-E	NM_005516	27%	Major histocompatibility complex, class I, E
HOXB13	NM_006361	36%	Homeo box B13
HPGD	NM_000860	27%	Hydroxyprostaglandin dehydrogenase 15-(NAD)
HPN	NM_182983	27%	Hepsin (Transmembrane protease, serine 1)
HYAL1	NM_007312	27%	Hyaluronoglucosaminidase 1
HYAL3	NM_003549	27%	Hyaluronoglucosaminidase 3
ID1	NM_181353	27%	Inhibitor of DNA binding 1, dominant negative helix-loop-helix protein
IER3	NM_052815	27%	Radiation-inducible immediate-early gene IEX-1 (Immediate early response 3)
IFRD2	NM_006764	27%	Interferon-related developmental regulator 2
IL1R1	NM_000877	27%	Interleukin 1 receptor, type I
IQGAP2	NM_006633	36%	IQ motif containing GTPase activating protein 2
JAG1	NM_000214	36%	Jagged 1 (Alagille syndrome)
KLK2†	**NM_005551**	**82%**	**Kallikrein 2, prostatic**
KLK3 (PSA)†	**NM_001648**	**91%**	**Kallikrein 3, (Prostate specific antigen)**
KRT18‡	NM_000224	27%	Keratin 18
KRT19	NM_002276	27%	Keratin 19
KRT8	NM_002273	27%	Keratin 8
LAMC1	NM_002293	27%	Laminin, gamma 1 (formerly LAMB2)
LMAN1	NM_005570	36%	Lectin, mannose-binding, 1
LRIG1	NM_015541	27%	Leucine-rich repeats and immunoglobulin-like domains 1
MAF	NM_005360	27%	v-maf musculoaponeurotic fibrosarcoma oncogene homolog (avian)
MCCC2	NM_022132	36%	Methylcrotonoyl-Coenzyme A carboxylase 2 (beta)
MMP16	**NM_005941**	**45%**	**Matrix metalloproteinase 16 (membrane-inserted)**
MRF2	BU616749	27%	Modulator recognition factor 2 (AT-rich interactive domain-containing protein 5B)

MYB	NM_005375	27%	v-myb myeloblastosis viral oncogene homolog (avian)
MYC	NM_002467	27%	v-myc myelocytomatosis viral oncogene homolog (avian)
MYH10	NM_005964	27%	Myosin heavy chain, nonmuscle type B
NDRG1†	**NM_006096**	**45%**	**N-myc downstream regulated gene 1**
NEBL	NM_006393	27%	Nebulette
NEDD4L	NM_015277	27%	Neural precursor cell expressed, developmentally down-regulated 4-like
NKX3-1†	**NM_006167**	**45%**	**NK3 transcription factor related, locus 1 (Drosophila)**
NUCB2	NM_005013	27%	Nucleobindin 2
PAM‡	NM_000919	27%	Peptidylglycine alpha-amidating monooxygenase
PDHA1	NM_000284	27%	Pyruvate dehydrogenase (lipoamide) alpha 1
PIK3R1	NM_181523	27%	Phosphoinositide-3-kinase, regulatory subunit, polypeptide 1 (p85 alpha)
PIK3R3	NM_003629	27%	Phosphoinositide-3-kinase, regulatory subunit, polypeptide 3 (p55, gamma)
PLA2G2A‡	NM_000300	27%	Phospholipase A2, group IIA (platelets, synovial fluid)
PRAME	NM_006115	27%	Melanoma antigen preferentially expressed in tumors (MAPE)
RAB4A‡	NM_004578	27%	RAB4A, member RAS oncogene family
RAP1GA1	NM_002885	27%	RAP1, GTPase activating protein 1
RINZF	NM_023929	27%	Zinc finger and BTB domain containing protein 10 (ZBTB10)
SAT	NM_002970	27%	Spermidine/spermine N1-acetyltransferase
SEC61B	NM_006808	27%	Sec61 beta subunit
SEMA3B	NM_004636	27%	Sema domain, immunoglobulin domain (Ig), short basic domain, secreted, (semaphorin) 3B
SGK	NM_005627	27%	Serum/glucocorticoid regulated kinase
SLC4A4	NM_003759	27%	Solute carrier family 4, sodium bicarbonate cotransporter, member 4
SMARCA1	NM_003069	27%	SWI/SNF related, matrix associated, actin dependent regulator of chromatin, subfamily a, member 1
SMCY	NM_004653	27%	Selected cDNA homolog, Y-linked (mouse) (Jumonji/ARID domain-containing protein 1D)
SORD‡	NM_003104	36%	Sorbitol dehydrogenase
THBS1	NM_003246	36%	Thrombospondin 1
TMEPAI†	NM_020182	27%	Transmembrane, prostate androgen induced RNA
TMPRSS2‡	NM_005656	36%	Transmembrane protease, serine 2
TPD52‡	NM_005079	27%	Tumor protein D52
TPM1	NM_000366	27%	Tropomyosin 1 (alpha)
TRGV9	BC030554	27%	T cell receptor gamma variable 9
UAP1	NM_003115	27%	UDP-N-acteylglucosamine pyrophosphorylase 1 (Sperm-associated antigen 2)
UGT2B17	NM_001077	27%	UDP glycosyltransferase 2 family, polypeptide B17
UQCRC2	NM_003366	27%	Ubiquinol-cytochrome c reductase core protein II

Bold type genes with frequency ≥45%.
†Genes common to Fig. 2 and Table 2.
‡Genes common to Fig. 2 and Table 3.
ƒ Human Ref. Seq

While rodents provide a unique perspective about AR action, a few considerations must be noted about the use of rodents for studies involving the prostate. First, rodents are not especially prone to prostate cancer without genetic manipulation (e.g., transgenic expression of SV40 T-antigen or loss of either p27 or PTEN). Second, androgen-action in a normal prostate involves complex stromal–epithelial interactions, whereas isolated prostate cancer cells are often directly dependent on androgen. Third, cross-comparison of datasets is complicated by the observation that rodents do not express the same complement of prostate-specific genes. For example, rodents do not express common human prostatic markers such as KLK2 or KLK3 *(61)*. Finally, elimination of androgen

Table 2
Functional Androgen Response Elements

Gene	Accessionf	Sequence	Reference
ACPP†	NM_001099h	Multiple	*(130)*
ANDPRO	NM_012718r	Multiple	*(131)*
AR†	NM_000044h	Multiple	*(8, 47)*
CASP2	NM_032982h	Multiple	*(132)*
CFLAR	NM_003879h	Multiple	*(133)*
CRISP1	NM_009638h	Multiple	*(134)*
CRISP3	NM_006061h	Multiple	*(134)*
FGF8	NM_033165h	Multiple	*(135)*
FKBP5†	NM_004117h	Multiple	*(136)*
F2R	NM_001992h	Multiple	*(137)*
KLK2†	NM_005551h	Multiple	*(138, 139)*
KLK3†	NM_001648h	Multiple	*(7, 9, 50, 140, 141)*
MAK	NM_005906h	Multiple	*(142, 143)*
NKX3-1†	NM_006167h	Multiple	*(47, 50)*
PBSN	NM_017471m	Multiple	*(9, 144, 145, 148)*
PIGR	NM_002644h	Multiple	*(146, 147)*
SARG	NM_023938h	Multiple	*(148)*
SLP	NM_011413m	Multiple	*(6)*
F9	NM_000133h	AGCTCAgctTGTACT	*(47)*
GPX5	NM_001509h	TGTTCTctcTCAACA	*(149, 150)*
GUSB	NM_000181h	AGTACTtgtTGTTCT	*(151)*
P21	NM_000389h	AGCACGcgaGGTTCC	*(47)*
MME	NM_000902h	CTCACAaagAGTTCT	*(152)*
NDRG1†	NM_006096h	TGATTAaccTGTTCT	*(47)*
ODC	NM_002539h	AGTCCCactTGTTCT	*(153)*
P22K12	NM_199266r	AGAAGAaaaTGTACA	*(154)*
SCAP	NM_012235h	TGCACAtgtTGTTCT	*(50, 87)*
SCGB2A2(PBP)	NM_002567h	AGTACGtgaTGTTCT	*(155)*
TAT	NM_000353h	TGTACAggaTGTTCT	*(156)*
TMEPAI†	NM_020182h	TGAAGAatgTGTTCT	*(47, 54, 157)*

†Genes common to Fig. 2.
fRef.Seq. and speices (h = human, r = rat, m = mouse) with the identified ARE.
Consensus ARE excluding multiple AREs was generated using WebLogo.

Table 3
Putative Androgen Response Elements

Gene‡	Accessionƒ	Sequence	Reference
DHCR24	NM_014762h	Multiple	(47, 54)
TMPRSS2	NM_005656h	Multiple	(47, 88)
ACSL3	NM_004457h	GAAAGAtaaTGTTCT	(47)
ANKH	NM_054027h	AGAACAcacTTTCCT	(?, 47)
AZGP1	NM_001185h	AGAATAgcaTGTTGC	(47)
BICD1	NM_001714h	AGAATAttcTGTTGT	(54)
B2M	NM_004048h	AGATCTtaaTCTTCT	(47)
CAMKK2	NM_006549h	GGCACActtTGTTAT	(47)
FN1	NM_002026h	AGAACAcagAGTTCT	(157)
KRT18	NM_000224h	AAATCAcagTGTTCC	(47)
PAM	NM_000919h	AAGACCtttTGTTCT	(157)
PLA2G2A	NM_000300h	GAGGTAaatGGTATTCTC	(76)
RAB4A	NM_004578h	ACAAAAgtaTGTACT	(47)
SORD	NM_003104h	AAATCAccgTCTTCT	(47)
TPD52	NM_005079h	AAAACAgatTTTTAT	(47)

‡Genes common to Fig. 2.
ƒRef.Seq. and species (h = human) with the identified ARE.
Consensus ARE excluding multiple AREs and PLA2G2A was generated using WebLogo.

in vivo by castration does not eliminate androgen production of the adrenal glands. Combined, these mitigating factors may explain why few common targets have been identified using rodent models. However, additional microarray studies and a more detailed comparison between rodents and human data may increase concordance.

AR Alterations in Prostate Cancer Progression

As described earlier, androgen deprivation therapy selects for AR mutations that alter ligand specificity. These mutants assist in restoring AR activity and are directly related to therapeutic relapse. A detailed listing of tumor-derived mutants has been catalogued (see *McGill's Androgen Receptor Gene Mutations Database*) *(19)*, and these analyses reveal that the majority of AR mutations occur within the LBD. These LBD mutants are capable of using androgen as an agonist, but gain responsiveness to a broad spectrum of endogenous hormones (e.g., progesterone, estrogen, or cortisol). Selected mutants have also adapted to utilize therapeutic AR antagonists (e.g., flutamide) as agonists *(62)*. The clinical relevance of this finding was documented in the "flutamide withdrawal syndrome," wherein tumor regression was observed in a subset of patients when flutamide treatment was halted *(63)*. Examination of these tumors revealed that T877A mutation in AR altered the receptor confirmation, allowing flutamide to be perceived as an agonist rather than an antagonist *(64)*. Thus, a plethora of clinical data demonstrates that AR mutation and alternate AR ligands can play a major role in therapeutic relapse. In addition, there is some evidence that the wild-type AR can be activated by other steroids (e.g., estrogen) in the presence of selected co-activators known to be overexpressed in cancer. For example, the ARA70 co-activator sensitizes AR to activation by estrogen, and its expression is increased as a function of prostate cancer grade *(65)*. Lastly, recent evidence suggests that even low-dose exposure to estrogens or diethylstilbestrol (DES) can also have profound effects on prostate *(66,67)*. In sum, the response of AR to alternate ligands may contribute to its effect on prostate cancer progression.

Although it is known that AR target gene selection is heavily influenced by differential ligand binding, only a handful of studies have addressed the impact of these ligands on AR target gene expression. Microarray analysis has been used to determine the genes involved in primary prostatic stromal cells treated with 100 nM estradiol *(68)*. Ninety-six genes were shown from a total of 933 estrogen-regulated genes in at least two out of three time points. However, MYC was the only gene common with the LNCaP list (Table 1). This disparity is likely attributed to either the differential ligand utilized or the cellular context, and more study will be required to assess estrogen action directly in prostate cancer cells. The effects of DES have also been examined, as this estrogenic compound has been utilized clinically to block testosterone synthesis in prostate cancer patients. Gene expression profiles were collected from DES-treated androgen-dependent (LNCaP) and androgen-independent, AR negative (PC-3) cells *(69)*. Six genes were regulated exclusively in LNCaP cells, while 18 genes were common to both cell lines, thus indicating that DES elicits AR-independent effects in prostate cancer cells. Two DES-regulated genes, CCNG2 and GRB10, were identified as frequent AR target genes (Fig. 2). The signal transduction adaptor molecule GRB10 was specifically regulated in the LNCaP cells, suggesting that DES and androgen may regulate numerous metabolic and mitogenic signaling responses *(70)*. It is important to note that the AR mutation present in LNCaP (AR-T877A) is responsive to DES as a ligand *(71)*. Moreover, the study of estrogenic compounds in the prostate is further complicated by the fact that ER has been shown to be present and functionally important in the prostatic epithelia *(72)*. Thus, discerning the effects of these substances acting aberrantly through the AR or more canonically through ER will be difficult. Continued study of AR action in the presence of these alternate ligands will facilitate our understanding of AR action in cancer progression.

Distinct from the alternate ligands, growth factors and cytokines such as EGF, IGF-I, and IL-6 have been implicated in "ligand-independent" activation of AR in the absence of androgen *(73)*. The mechanisms by which ligand-independent AR activation occurs are poorly understood, but likely involve the MAPK and/or AKT signal transduction pathways. A recent article attempted to profile the outcome of EGF treatment on LNCaP cells and a derivative of the same cell line that has been passaged to select for therapy-resistant sublines *(74)*. Results from these studies suggested a very high percentage of overlap (75%) between genes regulated by androgen and EGF. Of the limited number of genes presented, KLK2, KLK3, and BCHE were in common with the LNCaP comparison (Table 1). Further study of these factors will most definitely enhance our knowledge concerning the role they play in modulating prostate cancer.

POTENTIAL ANDROGEN RECEPTOR TARGETS

Based on the LNCaP comparisons shown in Table 1, commonly identified androgen-responsive genes can be classified into functional categories of genes affecting metabolism, transcriptional regulation, cellular proliferation, and signal transduction pathways.

Metabolism and Protein Metabolism Targets

Consistent with androgens being critical for growth and maintenance of the prostate, and/or its malignant transformation, a number of potential androgen-regulated genes can be unified as participating in specific metabolic processes. Four putative ARE-containing metabolic genes, DHCR24, SORD, ASCL3, and PLA2G2A, were identified as part of the androgen-responsive pathway in LNCaP cells. The most frequently identified of these metabolic genes is DHCR24 (45%), with SORD being the second most common (36%). While the androgen-mediated events of SORD have not been defined, DHCR24 may influence androgen biosynthesis through its involvement in cholesterol production. Some evidence exists that mRNA of ASCL3, a fatty acid synthetase important for regulating acyl-CoA levels, is modestly increased in the presence of androgen (proliferative), but is synergistic with vitamin D_3 (antiproliferative) in prostate cells (75). Similarly, PLA2G2A is another enzyme involved in fatty acid regulation and its mRNA is increased in response to androgens in LNCaP cells (76). The homeo-static signals that modulate these two enzymes may be pivotal for their many fatty acid-mediated functions such as energy production and ability to influence apoptosis.

Several other metabolic genes were identified as androgen-responsive in the LNCaP comparison, which include CRAT, SAT, and UGT2B17 (27%), but no ARE has yet been identified in the regulatory regions. Early data have shown that the activity of CRAT, an important enzyme in the regulation of carnitine levels, is under andro-genic control in rat testes and epididymis, with the highest activity occurring in the spermatozoa (77, 78), although the androgen dependence of CRAT in prostate has not been tested. SAT is also induced by androgen (79); the encoded protein is the rate-limiting enzyme in the catabolism of spermidine/spermine, and is implicated in altering proliferation (80). Others have documented a similar upregulation in the kidney of androgen-treated female mice that express low levels of SAT (81). Interestingly, the offsprings from SAT over-expressing mice crossed with mice genetically predis-posed for prostate cancer (TRAMP mice) have reduced prostate growth (82), clearly indicating that SAT has physiological implications for prostate function.

Genes associated with steroid regulation were also revealed by the analysis. For example, UGT2B17, a detoxification-related enzyme important for steroid inactivation, is decreased in androgen-treated LNCaP cells (83). Downregulation of UGT2B17 may reveal a mechanism by which androgen levels are auto-regulated. Protein analysis also demonstrated a potential increase in PDHA1 in androgen-treated rat ventral prostate (84), which is thought to increase acetyl-CoA for energy production and sterol biosynthesis. Either androgen or estrogen treatment of LNCaP cells resulted in increased HPGD (85), a key enzyme for the destruction of prostaglandins. Moreover, immunomodulatory hormones such as IL-6 can synergize with androgens to enhance HPGD activity in LNCaP cells (86). These data may reveal a potential link between immune response and prostate cancer progression. Finally, a key component in the regulation of lipid metabolism is DBI, an acyl-CoA binding protein, which is associated with the LNCaP gene list (54%). It appears that androgen-dependent regulation of DBI likely occurs through a transcriptional mechanism involving sterol regulatory-element binding protein (SREBP) (160), which in turn is presumably regulated by an ARE-containing gene coding for an escort protein known as SREBP cleavage-activating protein (SCAP) (87).

Functions relating to androgen-mediated protein metabolism can also be generated from the list of LNCaP-associated genes. The most notable gene with an ARE is TMPRSS2 (36%); other genes include ANKH, BICD1, and PAM (27%). TMPRSS2 is a transmembrane serine protease that is elevated in neoplastic prostate epithelium *(88, 89)*. The protease domain is auto-catalytically cleaved and secreted into the extracellular environment, and has recently been suggested to function in prostate cancer metastasis through proteolysis of the protease-activated receptor-2 (PAR-2) *(90)*. PAM is an important enzyme for the amidation and subsequent activation of many peptide hormones and growth factors. However, conflicting evidence concerning the androgen-dependence of PAM has been documented. A series of reports using xenograft models suggests that PAM is decreased after castration *(91–93)*. However, a second group has shown a modest increase in PAM after castration in another xenograft model *(94)* or androgen-withdrawal in cell lines *(95)*. Both groups demonstrate an increase in PAM from human prostate; therefore, the discrepancy may be attributed to differences in the xenograft models or cell lines used.

HPN is one of the few genes identified in the protein metabolism group, which lacks a consensus, putative ARE. HPN is a serine protease thought to be increased during prostate cancer progression *(96–101)*, although its function and dependence on androgen has not been clearly discerned. Lastly, genes involved in the ubiquitin-mediated protein metabolism pathway have been identified as putative AR targets. One of the potential androgen-regulated genes identified is NEDD4L, a ubiquitin ligase, which has been shown to be highly expressed in prostate and under the control of androgen in LNCaP cells *(102)*. Further study of androgens in protease and proteasome-mediated events will possibly reveal important ramifications for prostate physiology.

Transcription and Proliferation Targets

In addition to metabolic targets, androgen-dependent regulation of transcriptional factors is likely to be important for the cancer survival and proliferation. Commonly identified targets in the LNCaP studies include the ARE-containing gene FN1 (36%) and other genes such as ID1, MYC, and MAF (27%). FN1 is a component of the seminal coagulum and a biologically relevant substrate for PSA during the semen liquefaction process *(103)*. While the function of prostate-derived FN1 has not been studied, it is hypothesized to affect formation of the extracellular matrix *(104)* and may be involved in protection from protease activity. Likewise, another probable action for FN1 is its ability to regulate proliferation or apoptosis in prostate cells *(105)*. A second potential androgen-regulated transcriptional target is ID1, a dominant negative helix-loop-helix (HLH) protein. ID1 binds and alters the activity of HLH containing proteins and non-HLH proteins such as the retinoblastoma protein to assist in the control of cellular proliferation *(106)*. Gene expression profiling of the prostate from androgen-treated 20-day-old rats identified a minimal decrease in ID1 transcripts *(107)*; however, in this model system, ID1 downregulation is not consistent with the proliferative action of androgen, but may be related to the development/differentiation process. The MYC family of transcription factors is known to control proliferation and tumorogenic potential *(108)*, and MYC expression in prostatic cell lines is reported to be induced by mitogenic doses of androgen *(109, 110)*, which is inconsistent with its proliferative functions. Interestingly, the MYC expression was inhibited at higher doses of androgen, consistent with its role in stimulating proliferation (111). MAF is a member of the basic-leucine-zipper (bZIP) transcription factor family and has been shown to have oncogenic potential as well as important roles in differentiation (112). Evidence of MAF as an androgen-induced gene is limited, but a recent report has shown MAF to be induced in calf prostate after androgen stimulation (113).

Finally, two androgen-repressed genes were identified, which may impact prostate cancer. Several genes of the keratin family were found within the LNCaP comparison: KRT8, KRT18, and KRT19 (27%). Although keratins are generally considered useful as epithelial markers, some evidence suggests that keratin regulation may be involved in cancer progression (114). In agreement with the LNCaP studies, transcript levels of KRT8 in the rat prostate increase after castration and partially rebound with androgen re-stimulation (115). Second, CLU was identified as an androgen-repressed gene (36%). CLU is a stress-activated survival factor and appears to be regulated in a similar fashion as KRT8 in rat testes and prostate (115–117). Moreover, recent antisense strategies against CLU have indicated that it might serve as an efficacious target for combinatorial therapy in the treatment of prostate cancer (118).

Signaling and Other Targets

The last class of AR target genes identified in the LNCaP comparison is associated with signal transduction. Two potential androgen-regulated genes with putative AREs are CAMKK2 (36%) and RAB4A (27%). CAMKK2, a calcium/calmodulin-activated kinase, has the potential to regulate multiple processes through activation of downstream kinases such as CAMKI and CAMK4 as well as other pathways, including AKT (119). The small-GTPase RAB4A has frequently been associated with its ability to regulate endocytosis (120). While neither gene has been shown to be regulated by androgen, they could potentially regulate important functions in the prostate such as proliferation. TPD52 is an adaptor protein known for its regulation of vesicular trafficking and secretion, and is frequently overexpressed in many human malignancies (121) including prostate cancer (122). Immunoblot analyses of LNCaP cells have confirmed that androgen induces a modest increase in TPD52 (122,123).

A number of potential androgen-responsive genes involved in signaling were identified by the LNCaP model. In particular, two genes had a higher association, GUCY1A3 (45%) and HOXB13 (36%), while most others had a lower association (27%). Furthermore, the evidence for their androgen dependence is somewhat circumstantial. For instance, SGK, a serum/glucocorticoid-induced kinase with high homology to protein kinase B (PKB), may be linked to androgen regulation in granulose cells differentiated for 24 to 48 hours with androgen in combination with follicle-stimulating hormone (124). Likewise, GUCY1A3, a guanylate cyclase, has been reported to have no significant correlation in transcript levels in androgen- and estrogen-treated prostate from calves (113). In both studies, the individual contributions of each differentiation component were not evaluated, making it difficult to solidify conclusions. Lastly, a few potential androgen targets identified have been investigated for their direct androgen dependence and appear not to be regulated or minimally regulated by androgen; these genes include AMACR (125), CREM (126), IER3 (127), HOXB13 (128), and SMCY (129).

POTENTIAL AR TARGET GENES FROM THE LNCaP MODEL IN PROSTATE CANCER

A goal for understanding the contributions of androgen in the LNCaP model is to apply this paradigm to prostate cancer pathogenesis. To this end, a supervised clustering analysis was performed using the 104 potential AR-target genes (Fig. 2) and 84 prostate samples (5 BPH, 79 adenocarcinoma) *(161)* within GeneSpring GX 7.3 using Pearson correlation around zero (complete linkage) for genes and Euclidean distance (complete linkage) for conditions. Two principle clusters were identified (Fig. 3). Significantly over-represented biological processes and pathways in each of the clusters were classified using DAVID 2.1 (Database for Annotation, Visualization and Integrated Discovery). Approximately, 45% of the genes identified by the LNCaP model were also associated within major functional pathways in the prostate samples.

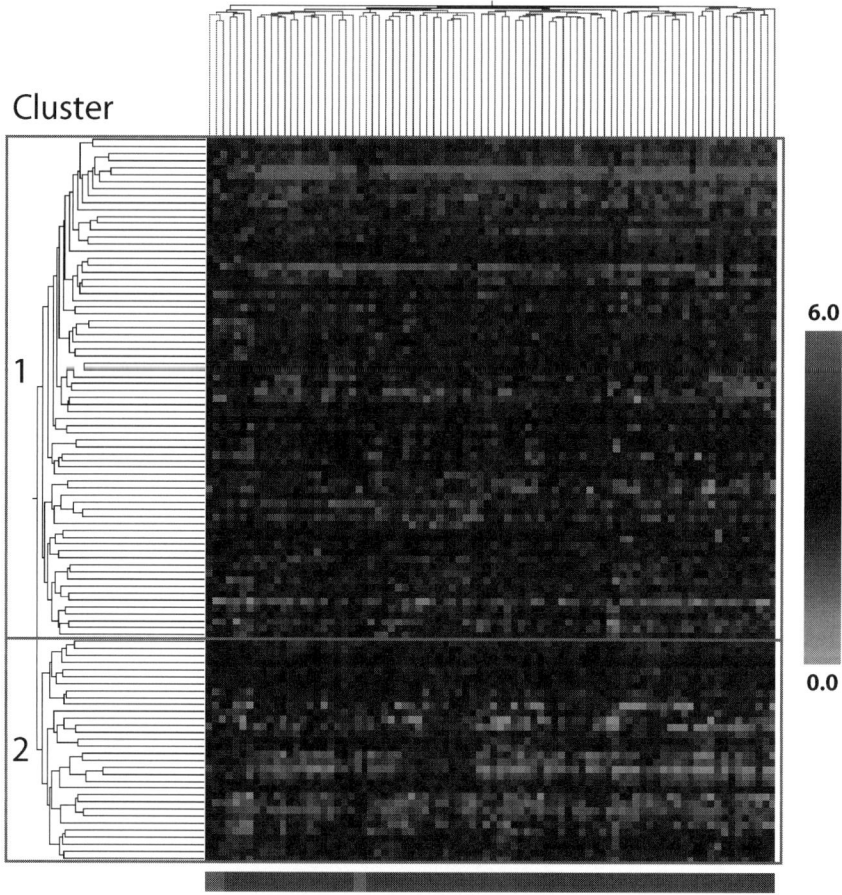

Fig. 3. Cluster analysis of potential AR-regulated genes and human prostate tissue samples. The 104 potential AR-regulated genes identified by comparison of 11 microarray studies were clustered across 84 human prostate tissue samples, consisting of 5 BPH (red) and 79 adenocarcinoma (blue). Genes in cluster one are significantly associated with response to wounding, cell migration, carboxylic acid metabolism and amine metabolism, development, and the insulin receptor signaling pathway. Genes in cluster two are associated with antigen processing and presentation.

Interestingly, 20% of the annotated genes contained a functional- (KLK3, ACPP, AR, NKX3-1, NDRG1) or a putative-ARE (FN1, TPD52, B2M, ACSL3, KRT18).

Cluster one contained 72 genes. A large number of genes within this cluster were significantly associated with development (AR, CYR61, HOXB13, IER3, IFRD2, JAG1, KLK3, KRT18, LAMC1, MYH10, NDRG1, NKX3-1, THBS1, TPD52) and response to wounding (FN1, HOXB13, IL1R1, LMAN1, MYH10, THBS1, TRGV9). Additional significant functional annotations included cell migration (FN1, JAG1, MYH10), carboxylic acid metabolism (ACSL3, CRAT, CTBP1, DDC, HPGD, MCCC2), amine metabolism (CTBP1, DDC, MCCC2, SAT, UAP1), and the insulin receptor signaling pathway (GRB10, PIK3R3).

Eight genes (11%) in cluster one were consistently upregulated (80% of samples with normalized data ≤ 1.5) across the 79 adenocarcinoma samples (AMACR, DKFZP564B167, FOLH1, GUCY1A3, HPN, LRIG1, MYC, UAP1). AMACR has been shown to have increased expression in the androgen-sensitive cell lines MDA PCa 2a and MDA PCa 2b *(158)* as well as in tumors and LNCaP cells *(53)*. AMACR was repressed in LNCaP cells after treatment with bicalutamide *(46)* and was downregulated in androgen-independent cell lines (DU 145, PPC-1, PC-3) *(158)*. DKFZP564B167 was induced with synthetic androgen in LNCaP cells *(51)*, but was downregulated in androgen-insensitive cell lines *(158)* and in prostate tumors after 3 months of androgen ablation therapy *(54)*. FOLH1 had increased expression in the androgen-sensitive cell lines LAPC4, LNCaP, MDA PCa 2a, and MDA PCa 2b *(158)*. In contrast, FOLH1 was repressed with synthetic androgen in LNCaP cells *(51)* and was downregulated in the androgen-insensitive cell lines PPC-1 and PC-3, as well as in prostate tumors after 3 months of androgen ablation therapy *(54)*. GUCY1A3 was induced in LNCaP cells by treatment with synthetic androgen *(47,48,51)* or DHT *(52)*, but was downregulated in androgen-insensitive cell lines *(158)*. HPN has been shown to be upregulated in tumors and LNCaP cells *(53,158)*, but was downregulated in androgen-insensitive cell lines *(158)* and in prostate tumors after 3 months of androgen ablation therapy *(54)*. LRIG1 had increased expression in LNCaP cells *(158)* and was induced with synthetic androgen *(51)*. Conversely, LRIG1 was repressed in the androgen-insensitive cell lines DU 145 and PPC-1 *(158)*. MYC was upregulated in androgen-independent DU145 cells after treatment with bicalutamide *(46)*. However, in LNCaP cells, MYC was repressed with synthetic androgen *(47)* and bicalutamide treatment *(46)*. UAP1 was induced with synthetic androgen in LNCaP cells *(46,51)*, but downregulated in the androgen-sensitive cell line 22Rv1 as well as the androgen-insensitive cell line PC-3 *(158)*. Thus, with the exception of MYC, genes displaying increased expression in androgen-sensitive cell lines or after treatment with DHT or synthetic androgen were consistently upregulated in human prostate tumors.

Cluster two is composed of 32 genes. The only significant functional annotations of cluster two were antigen processing and antigen presentation, associated with the genes B2M and HLA-E. Ten genes (31%) within cluster two were consistently downregulated across the 79 adenocarcinoma samples (ACPP, C1orf21, CLU, CMKOR1, DBI, DPYSL2, MAF, PIK3R1, TPM1, UGT2B17). ACPP has been shown to be repressed after bicalutamide treatment in LNCaP cells *(46)*, repressed in LNCaP cells with DHT treatment *(52)*, downregulated in PC-3, 22Rv1, LAPC4, and LNCaP cell lines *(158)*, and downregulated in prostate tumors after 3 months of androgen ablation therapy

(54). C1orf21 was downregulated in androgen-insensitive cell lines *(158)*. In contrast, C1orf21 was induced with synthetic androgen *(51)* and bicalutamide treatment *(49)* in LNCaP cells. CLU was downregulated in PPC-1, PC-3, LAPC4, and LNCaP cell lines *(158)*. In LNCaP cells, CLU was repressed by treatment with DHT *(52)* or synthetic androgen *(48)*. However, CLU had increased expression in DU 145, 22Rv1, MDA PCa 2a, and MDA PCa 2b cell lines. In prostate tumors, CLU had varied expression *(53)*. CMKOR1 was repressed in LNCaP cells by treatment with DHT *(52)* or synthetic androgen *(51)*, but was upregulated in prostate tumors after 3 months of androgen ablation therapy *(54)*. DBI was downregulated in prostate tumors after 3 months of androgen ablation therapy *(54)*. Conversely, DBI was induced with synthetic androgen *(47, 48, 159)* or DHT *(52)* in LNCaP cells. DBI was also induced in LNCaP cells after treatment with bicalutamide *(49)*. DPYSL2 had decreased expression in LAPC4, LNCaP, MDA PCa 2a, and MDA PCa 2b cells, but was upregulated in androgen-insensitive cell lines *(158)*. In LNCaP cells, DPYSL2 was repressed with synthetic androgen *(51)*. MAF was downregulated in DU 145, PC-3, 22Rv1, and MDA PCa 2a cells but had increased expression in LAPC4, LNCaP, and MDA PCa 2b cell lines *(158)*. MAF was also induced in LNCaP cells with synthetic androgen *(47, 51)*. PIK3R1 was repressed in LNCaP cells with synthetic androgen *(51)* and had decreased expression in DU 145, PC-3, 22Rv1, and MDA PCa 2b cell lines, but was upregulated in PPC-1, LAPC4, LNCaP, and MDA PCa 2a cells *(158)*. TPM1 was downregulated in tumors and LNCaP cells *(53)*, had decreased expression in androgen-independent cell lines *(158)*, and was downregulated in prostate tumors after 3 months of androgen ablation therapy *(54)*. In LNCaP cells, UGT2B17 was repressed with synthetic androgen *(159)* and bicalutamide *(46)*. UGT2B17 had decreased expression in androgen-insensitive cell lines as well as MDA PCa 2b cells, but had increased expression in 22Rv1, LAPC4, LNCaP, and MDA PCa 2a cells *(158)*. Surprisingly, a number of genes induced in LNCaP cells with DHT or synthetic androgen treatment were consistently downregulated in human prostate tumors. Collectively, the biological processes identified from our clustering analysis using potential AR-regulated genes from the LNCaP comparison suggests that androgen directly or indirectly regulates a diverse set of pathways to control growth and development of the prostate.

CONCLUSIONS

While the effects of androgens encompass a broad range of physiological and pathophysiological conditions, the specific targets of the AR are only beginning to be fully understood. These effects have been best characterized in the prostate in relation to the development and progression of prostatic adenocarcinoma. Interestingly, while these cells are dependent on androgens for growth and survival, few validated targets of the AR have been attributed to this function. The use of microarray technology in both cell line and animal models in this effort has had mixed success. While new targets have been identified and potential targets validated, the critical mediators of growth driving prostate cancer remain elusive. Alternate study into the mechanisms of cancer progression has also been investigated through microarray analysis. Specifically, activation of AR through alternate ligands and signaling cascades imitate the restored activity seen in therapy-resistant tumors. These studies have demonstrated similarities between these mechanisms and canonical activation of the receptor, further implicating these processes in re-activation of AR in advanced disease.

While each individual study has provided insight into androgen action, a consensus of the whole is difficult to ascertain. Further studies using microarray analysis need to be in agreement with physiological constraints of the experimental model, specifically the use of biologically relevant concentrations of ligand. Furthermore, the distinction of primary versus secondary targets can also be addressed through time-point selection and validation through alternate biochemical assays. The understanding of androgen action through microarray analysis is still in the future, but the growing body of knowledge continues to elucidate and excite.

REFERENCES

1. Lee, H. J. & Chang, C. (2003). Recent advances in androgen receptor action. *Cell Mol Life Sci* **60**, 1622.
2. Gelmann, E. P. (2002). Molecular biology of the androgen receptor. *J Clin Oncol* **20**, 3015.
3. Heinlein, C. A. & Chang, C. (2004). Androgen receptor in prostate cancer. *Endocr Rev* **25**, 276–308.
4. Roche, P. J., Hoare, S. A. & Parker, M. G. (1992). A consensus DNA-binding site for the androgen receptor. *Mol Endocrinol* **6**, 2235.
5. Huang, W., Shostak, Y., Tarr, P., Sawyers, C. & Carey, M. (1999). Cooperative assembly of androgen receptor into a nucleoprotein complex that regulates the prostate-specific antigen enhancer. *J Biol Chem* **274**, 25768.
6. Adler, A. J., Scheller, A., Hoffman, Y. & Robins, D. M. (1991). Multiple components of a complex androgen-dependent enhancer. *Mol Endocrinol* **5**, 1596.
7. Cleutjens, K. B., van Eekelen, C. C., van der Korput, H. A., Brinkmann, A. O. & Trapman, J. (1996). Two androgen response regions cooperate in steroid hormone regulated activity of the prostate-specific antigen promoter. *J Biol Chem* **271**, 6388.
8. Grad, J. M., Dai, J. L., Wu, S. & Burnstein, K. L. (1999). Multiple androgen response elements and a Myc consensus site in the androgen receptor (AR) coding region are involved in androgen-mediated up-regulation of AR messenger RNA. *Mol Endocrinol* **13**, 1911.
9. Reid, K. J., Hendy, S. C., Saito, J., Sorensen, P. & Nelson, C. C. (2001). Two classes of androgen receptor elements mediate cooperativity through allosteric interactions. *J Biol Chem* **276**, 2952.
10. Lutz, L. B., Jamnongjit, M., Yang, W. H., Jahani, D., Gill, A. & Hammes, S. R. (2003). Selective modulation of genomic and nongenomic androgen responses by androgen receptor ligands. *Mol Endocrinol* **17**, 1116.
11. Keller, E. T., Ershler, W. B. & Chang, C. (1996). The androgen receptor: a mediator of diverse responses. *Front Biosci.* **1**, d71.
12. Holdcraft, R. W. & Braun, R. E. (2004). Hormonal regulation of spermatogenesis. *Int J Androl* **27**, 342.
13. Vanderschueren, D., Vandenput, L., Boonen, S., Lindberg, M. K., Bouillon, R. & Ohlsson, C. (2004). Androgens and bone. *Endocr Rev.* **25**, 425.
14. Sadate-Ngatchou, P. I., Pouchnik, D. J. & Griswold, M. D. (2004). Identification of testosterone-regulated genes in testes of hypogonadal mice using oligonucleotide microarray. *Mol Endocrinol* **18**, 433.
15. Zhou, Q., Shima, J. E., Nie, R., Friel, P. J. & Griswold, M. D. (2005). Androgen-regulated transcripts in the neonatal mouse testis as determined through microarray analysis. *Biol Reprod.* **72**, 1019.
16. Chauvin, T. R. & Griswold, M. D. (2004). Androgen-regulated genes in the murine epididymis. *Biol Reprod* **71**, 569.
17. Dohle, G. R., Smit, M. & Weber, R. F. (2003). Androgens and male fertility. *World J Urol* **21**, 345.
18. Hiort, O. & Holterhus, P. M. (2003). Androgen insensitivity and male infertility. *Int J Androl* **26**, 20.
19. Gottlieb, B., Beitel, L. K., Wu, J. H. & Trifiro, M. (2004). The androgen receptor gene mutations database (ARDB): 2004 update. *Hum.Mutat* **23**, 533.

20. Holterhus, P. M., Hiort, O., Demeter, J., Brown, P. O. & Brooks, J. D. (2003). Differential gene-expression patterns in genital fibroblasts of normal males and 46, XY females with androgen insensitivity syndrome: evidence for early programming involving the androgen receptor. *Genome Biol* **4**.

21. Greenland, K. J. & Zajac, J. D. (2004). Kennedy's disease: pathogenesis and clinical approaches. *Intern Med J* **34**, 279–86.

22. Dejager, S., Bry-Gauillard, H., Bruckert, E., Eymard, B., Salachas, F., LeGuern, E., Tardieu, S., Chadarevian, R., Giral, P. & Turpin, G. (2002). A comprehensive endocrine description of Kennedy's disease revealing androgen insensitivity linked to CAG repeat length. *J Clin Endocrinol Metab* **87**, 3893–901.

23. Piccioni, F., Simeoni, S., Andriola, I., Armatura, E., Bassanini, S., Pozzi, P. & Poletti, A. (2001). Polyglutamine tract expansion of the androgen receptor in a motoneuronal model of spinal and bulbar muscular atrophy. *Brain Res Bull* **56**, 220.

24. Sinnreich, M. & Klein, C. J. (2004). Bulbospinal muscular atrophy: Kennedy's disease. *Arch Neurol* **61**, 1324–6.

25. Lieberman, A. P., Harmison, G., Strand, A. D., Olson, J. M. & Fischbeck, K. H. (2002). Altered transcriptional regulation in cells expressing the expanded polyglutamine androgen receptor. *Hum.Mol Genet* **11**, 1976.

26. Wong, Y. C. & Wang, Y. Z. (2000). Growth factors and epithelial-stromal interactions in prostate cancer development. *Int RevCytol* **199**, 116.

27. Cunha, G. R., Cooke, P. S. & Kurita, T. (2004). Role of stromal-epithelial interactions in hormonal responses. *Arch Histol Cytol* **67**, 434.

28. Roscigno, M., Scattoni, V., Bertini, R., Pasta, A., Montorsi, F. & Rigatti, P. (2004). Diagnosis of prostate cancer. State of the art. *Minerva Urol Nefrol* **56**, 145.

29. Balk, S. P. (2002). Androgen receptor as a target in androgen-independent prostate cancer. *Urology* **60**, 138.

30. Agus, D. B., Cordon-Cardo, C., Fox, W., Drobnjak, M., Koff, A., Golde, D. W. & Scher, H. I. (1999). Prostate cancer cell cycle regulators: response to androgen withdrawal and development of androgen independence. *J Natl Cancer Inst* **91**, 1876.

31. Chatterjee, B. (2003). The role of the androgen receptor in the development of prostatic hyperplasia and prostate cancer. *Mol Cell Biochem* **253**, 101.

32. Burnstein, K. L. (2005). Regulation of androgen receptor levels: implications for prostate cancer progression and therapy. *J Cell Biochem* **95**, 669.

33. Loberg, R. D., Fielhauer, J. R., Pienta, B. A., Dresden, S., Christmas, P., Kalikin, L. M., Olson, K. B. & Pienta, K. J. (2003). Prostate-specific antigen doubling time and survival in patients with advanced metastatic prostate cancer. *Urology* **62 Suppl 1**, 133.

34. Culig, Z., Steiner, H., Bartsch, G. & Hobisch, A. (2005). Mechanisms of endocrine therapy-responsive and -unresponsive prostate tumours. *Endocr Relat Cancer* **12**, 244.

35. van Bokhoven, A., Varella-Garcia, M., Korch, C., Johannes, W. U., Smith, E. E., Miller, H. L., Nordeen, S. K., Miller, G. J. & Lucia, M. S. (2003). Molecular characterization of human prostate carcinoma cell lines. *Prostate* **57**, 225.

36. Knudsen, K. E., Arden, K. C. & Cavenee, W. K. (1998). Multiple G1 regulatory elements control the androgen-dependent proliferation of prostatic carcinoma cells. *J Biol Chem.* **273**, 20222.

37. Farla, P., Hersmus, R., Trapman, J. & Houtsmuller, A. B. (2005). Antiandrogens prevent stable DNA-binding of the androgen receptor. *J Cell Sci* **118**, 4198.

38. Shang, Y., Myers, M. & Brown, M. (2002). Formation of the androgen receptor transcription complex. *Mol Cell* **9**, 610.

39. Steketee, K., Timmerman, L., Ziel-van der Made, A. C., Doesburg, P., Brinkmann, A. O. & Trapman, J. (2002). Broadened ligand responsiveness of androgen receptor mutants obtained by random amino acid substitution of H874 and mutation hot spot T877 in prostate cancer. *Int J Cancer* **100**, 317.

40. Calabrese, E. J. (2001). Androgens: biphasic dose responses. *Crit RevToxicol* **31**, 522.

41. de Launoit, Y., Veilleux, R., Dufour, M., Simard, J. & Labrie, F. (1991). Characteristics of the biphasic action of androgens and of the potent antiproliferative effects of the new pure antiestrogen

EM-139 on cell cycle kinetic parameters in LNCaP human prostatic cancer cells. *Cancer Res* **51**, 5170.

42. Hofman, K., Swinnen, J. V., Verhoeven, G. & Heyns, W. (2001). E2F activity is biphasically regulated by androgens in LNCaP cells. *Biochem Biophys Res Commun* **283**, 101.

43. Nunlist, E. H., Dozmorov, I., Tang, Y., Cowan, R., Centola, M. & Lin, H. K. (2004). Partitioning of 5alpha-dihydrotestosterone and 5alpha-androstane-3alpha, 17beta-diol activated pathways for stimulating human prostate cancer LNCaP cell proliferation. *J Steroid Biochem Mol Biol* **91**, 170.

44. Lloyd, R. V., Erickson, L. A., Jin, L., Kulig, E., Qian, X., Cheville, J. C. & Scheithauer, B. W. (1999). p27kip1: a multifunctional cyclin-dependent kinase inhibitor with prognostic significance in human cancers. *Am.J Pathol* **154**, 323.

45. Ling, M. T., Wang, X., Tsao, S. W. & Wong, Y. C. (2002). Down-regulation of Id-1 expression is associated with TGF beta 1-induced growth arrest in prostate epithelial cells. *Biochim Biophys Acta* **1570**, 152.

46. Bouchal, J., Baumforth, K. R., Svachova, M., Murray, P. G., von Angerer, E. & Kolar, Z. (2005). Microarray analysis of bicalutamide action on telomerase activity, p53 pathway and viability of prostate carcinoma cell lines. *J Pharm Pharmacol* **57**, 92.

47. Nelson, P. S., Clegg, N., Arnold, H., Ferguson, C., Bonham, M., White, J., Hood, L. & Lin, B. (2002). The program of androgen-responsive genes in neoplastic prostate epithelium. *Proc Natl Acad Sci U S A* **99**, 11895.

48. Xu, L. L., Su, Y. P., Labiche, R., Segawa, T., Shanmugam, N., McLeod, D. G., Moul, J. W. & Srivastava, S. (2001). Quantitative expression profile of androgen-regulated genes in prostate cancer cells and identification of prostate-specific genes. *Int J Cancer* **92**, 328.

49. Chen, C. D., Welsbie, D. S., Tran, C., Baek, S. H., Chen, R., Vessella, R., Rosenfeld, M. G. & Sawyers, C. L. (2004). Molecular determinants of resistance to antiandrogen therapy. *Nat Med* **10**, 39.

50. Segawa, T., Nau, M. E., Xu, L. L., Chilukuri, R. N., Makarem, M., Zhang, W., Petrovics, G., Sesterhenn, I. A., McLeod, D. G., Moul, J. W., Vahey, M. & Srivastava, S. (2002). Androgen-induced expression of endoplasmic reticulum (ER) stress response genes in prostate cancer cells. *Oncogene* **21**, 8758.

51. DePrimo, S. E., Diehn, M., Nelson, J. B., Reiter, R. E., Matese, J., Fero, M., Tibshirani, R., Brown, P. O. & Brooks, J. D. (2002). Transcriptional programs activated by exposure of human prostate cancer cells to androgen. *Genome Biol* **3**.

52. Velasco, A. M., Gillis, K. A., Li, Y., Brown, E. L., Sadler, T. M., Achilleos, M., Greenberger, L. M., Frost, P., Bai, W. & Zhang, Y. (2004). Identification and validation of novel androgen-regulated genes in prostate cancer. *Endocrinology* **145**, 3924.

53. Lozano, J. J., Soler, M., Bermudo, R., Abia, D., Fernandez, P. L., Thomson, T. M. & Ortiz, A. R. (2005). Dual activation of pathways regulated by steroid receptors and peptide growth factors in primary prostate cancer revealed by Factor Analysis of microarray data. *BMC Genomics* **6**.

54. Holzbeierlein, J., Lal, P., LaTulippe, E., Smith, A., Satagopan, J., Zhang, L., Ryan, C., Smith, S., Scher, H., Scardino, P., Reuter, V. & Gerald, W. L. (2004). Gene expression analysis of human prostate carcinoma during hormonal therapy identifies androgen-responsive genes and mechanisms of therapy resistance. *Am J Pathol* **164**, 227.

55. Eder, I. E., Haag, P., Basik, M., Mousses, S., Bektic, J., Bartsch, G. & Klocker, H. (2003). Gene expression changes following androgen receptor elimination in LNCaP prostate cancer cells. *Mol Carcinog* **37**, 191.

56. Waghray, A., Feroze, F., Schober, M. S., Yao, F., Wood, C., Puravs, E., Krause, M., Hanash, S. & Chen, Y. Q. (2001). Identification of androgen-regulated genes in the prostate cancer cell line LNCaP by serial analysis of gene expression and proteomic analysis. *Proteomics* **1**, 1338.

57. Nantermet, P. V., Xu, J., Yu, Y., Hodor, P., Holder, D., Adamski, S., Gentile, M. A., Kimmel, D. B., Harada, S., Gerhold, D., Freedman, L. P. & Ray, W. J. (2004). Identification of genetic pathways activated by the androgen receptor during the induction of proliferation in the ventral prostate gland. *J Biol Chem* **279**, 1322.

58. Jiang, F. & Wang, Z. (2003). Identification of androgen-responsive genes in the rat ventral prostate by complementary deoxyribonucleic acid subtraction and microarray. *Endocrinology* **144**, 1265.

59. Wang, Z., Tufts, R., Haleem, R. & Cai, X. (1997). Genes regulated by androgen in the rat ventral prostate. *Proc Natl Acad SciU S A* **94**, 13004.

60. Pfundt, R., Smit, F., Jansen, C., Aalders, T., Straatman, H., van, D. V., Isaacs, J., van Kessel, A. G. & Schalken, J. (2005). Identification of androgen-responsive genes that are alternatively regulated in androgen-dependent and androgen-independent rat prostate tumors. *Genes ChromosomesCancer* **43**, 283.

61. Olsson, A. Y., Valtonen-Andre, C., Lilja, H. & Lundwall, A. (2004). The evolution of the glandular kallikrein locus: identification of orthologs and pseudogenes in the cotton-top tamarin. *Gene* **343**, 355.

62. Hara, T., Miyazaki, J., Araki, H., Yamaoka, M., Kanzaki, N., Kusaka, M. & Miyamoto, M. (2003). Novel mutations of androgen receptor: a possible mechanism of bicalutamide withdrawal syndrome. *Cancer Res* **63**, 153.

63. Kelly, W. K. & Scher, H. I. (1993). Prostate specific antigen decline after antiandrogen withdrawal: the flutamide withdrawal syndrome. *J Urol* **149**, 609.

64. Bohl, C. E., Miller, D. D., Chen, J., Bell, C. E. & Dalton, J. T. (2005). Structural basis for accommodation of nonsteroidal ligands in the androgen receptor. *J Biol Chem* **280**, 37754.

65. Yeh, S., Miyamoto, H., Shima, H. & Chang, C. (1998). From estrogen to androgen receptor: a new pathway for sex hormones in prostate. *Proc Natl Acad SciU S A* **95**, 5532.

66. vom Saal, F. S., Timms, B. G., Montano, M. M., Palanza, P., Thayer, K. A., Nagel, S. C., Dhar, M. D., Ganjam, V. K., Parmigiani, S. & Welshons, W. V. (1997). Prostate enlargement in mice due to fetal exposure to low doses of estradiol or diethylstilbestrol and opposite effects at high doses. *Proc Natl Acad Sci U S A* **94**, 2061.

67. Timms, B. G., Howdeshell, K. L., Barton, L., Bradley, S., Richter, C. A. & vom Saal, F. S. (2005). Estrogenic chemicals in plastic and oral contraceptives disrupt development of the fetal mouse prostate and urethra. *ProcNatl Acad Sci U S A* **102**, 7019.

68. Bektic, J., Wrulich, O. A., Dobler, G., Kofler, K., Ueberall, F., Culig, Z., Bartsch, G. & Klocker, H. (2004). Identification of genes involved in estrogenic action in the human prostate using microarray analysis. *Genomics* **83**, 44.

69. Koike, H., Ito, K., Takezawa, Y., Oyama, T., Yamanaka, H. & Suzuki, K. (2005). Insulin-like growth factor binding protein-6 inhibits prostate cancer cell proliferation: implication for anticancer effect of diethylstilbestrol in hormone refractory prostate cancer. *Br J Cancer* **92**, 1544.

70. Riedel, H. (2004). Grb10 exceeding the boundaries of a common signaling adapter. *Front Biosci* **9**, 618.

71. Kalach, J. J., Joly-Pharaboz, M. O., Chantepie, J., Nicolas, B., Descotes, F., Mauduit, C., Benahmed, M. & Andre, J. (2005). Divergent biological effects of estradiol and diethylstilbestrol in the prostate cancer cell line MOP. *J Steroid Biochem Mol Biol* **96**, 129.

72. Taylor, A. H. & Al Azzawi, F. (2000). Immunolocalisation of oestrogen receptor beta in human tissues. *J Mol Endocrinol* **24**, 155.

73. Culig, Z. (2004). Androgen receptor cross-talk with cell signalling pathways. *Growth Factors* **22**, 184.

74. Oosterhoff, J. K., Grootegoed, J. A. & Blok, L. J. (2005). Expression profiling of androgen-dependent and -independent LNCaP cells: EGF versus androgen signalling. *Endocr Relat Cancer* **12**, 148.

75. Qiao, S. & Tuohimaa, P. (2004). The role of long-chain fatty-acid-CoA ligase 3 in vitamin D3 and androgen control of prostate cancer LNCaP cell growth. *Biochem Biophys Res Commun* **319**, 368.

76. Sved, P., Scott, K. F., McLeod, D., King, N. J., Singh, J., Tsatralis, T., Nikolov, B., Boulas, J., Nallan, L., Gelb, M. H., Sajinovic, M., Graham, G. G., Russell, P. J. & Dong, Q. (2004). Oncogenic action of secreted phospholipase A2 in prostate cancer. *Cancer Res* **64**, 6940.

77. Marquis, N. R. & Fritz, I. B. (1965). Effects of testosterone on the distribution of carnitine, acetyle-carnitine, and carnitine acetyltransferase in tissues of the reproductive system of the male rat. *J Biol Chem* **240**, 2200.

78. Brooks, D. E. (1978). Activity and androgenic control of enzymes associated with the tricarboxylic acid cycle, lipid oxidation and mitochondrial shuttles in the epididymis and epididymal spermatozoa of the rat. *Biochem J* **174**, 752.

79. Stachurska, A., Dudkowska, M., Czopek, A., Manteuffel-Cymborowska, M. & Grzelakowska-Sztabert, B. (2004). Cisplatin up-regulates the in vivo biosynthesis and degradation of renal polyamines and c-Myc expression. *Biochim Biophys Acta* **1689**, 266.

80. Dudkowska, M., Stachurska, A., Grzelakowska-Sztabert, B. & Manteuffel-Cymborowska, M. (2002). Up-regulation of spermidine/spermine N1-acetyltransferase (SSAT) expression is a part of proliferative but not anabolic response of mouse kidney. *Acta BiochimPol* **49**, 977.

81. Levillain, O., Greco, A., Diaz, J. J., Augier, R., Didier, A., Kindbeiter, K., Catez, F. & Cayre, M. (2003). Influence of testosterone on regulation of ODC, antizyme, and N1-SSAT gene expression in mouse kidney. *Am J Physiol Renal Physiol* **285**, F506.

82. Kee, K., Foster, B. A., Merali, S., Kramer, D. L., Hensen, M. L., Diegelman, P., Kisiel, N., Vujcic, S., Mazurchuk, R. V. & Porter, C. W. (2004). Activated polyamine catabolism depletes acetyl-CoA pools and suppresses prostate tumor growth in TRAMP mice. *J Biol Chem* **279**, 40083.

83. Guillemette, C., Levesque, E., Beaulieu, M., Turgeon, D., Hum, D. W. & Belanger, A. (1997). Differential regulation of two uridine diphospho-glucuronosyltransferases, UGT2B15 and UGT2B17, in human prostate LNCaP cells. *Endocrinology* **138**, 3005.

84. Costello, L. C., Liu, Y. & Franklin, R. B. (1995). Prolactin specifically increases pyruvate dehydrogenase E1 alpha in rat lateral prostate epithelial cells. *Prostate* **26**, 193.

85. Tong, M. & Tai, H. H. (2000). Induction of NAD(+)-linked 15-hydroxyprostaglandin dehydrogenase expression by androgens in human prostate cancer cells. *Biochem Biophys Res Commun* **276**, 81.

86. Tong, M. & Tai, H. H. (2004). Synergistic induction of the nicotinamide adenine dinucleotide-linked 15-hydroxyprostaglandin dehydrogenase by an androgen and interleukin-6 or forskolin in human prostate cancer cells. *Endocrinology* **145**, 2147.

87. Heemers, H., Verrijdt, G., Organe, S., Claessens, F., Heyns, W., Verhoeven, G. & Swinnen, J. V. (2004). Identification of an androgen response element in intron 8 of the sterol regulatory element-binding protein cleavage-activating protein gene allowing direct regulation by the androgen receptor. *J Biol Chem* **279**, 30887.

88. Lin, B., Ferguson, C., White, J. T., Wang, S., Vessella, R., True, L. D., Hood, L. & Nelson, P. S. (1999). Prostate-localized and androgen-regulated expression of the membrane-bound serine protease TMPRSS2. *Cancer Res* **59**, 4184.

89. Afar, D. E., Vivanco, I., Hubert, R. S., Kuo, J., Chen, E., Saffran, D. C., Raitano, A. B. & Jakobovits, A. (2001). Catalytic cleavage of the androgen-regulated TMPRSS2 protease results in its secretion by prostate and prostate cancer epithelia. *Cancer Res* **61**, 1692.

90. Wilson, S., Greer, B., Hooper, J., Zijlstra, A., Walker, B., Quigley, J. & Hawthorne, S. (2005). The membrane-anchored serine protease, TMPRSS2, activates PAR-2 in prostate cancer cells. *Biochem J* **388**, 972.

91. Fina, F., Muracciole, X., Rocchi, P., Nanni-Metellus, I., Delfino, C., Daniel, L., Dussert, C., Ouafik, L. H. & Martin, P. M. (2005). Molecular profile of androgen-independent prostate cancer xenograft LuCaP 23.1. *J Steroid Biochem.Mol Biol* **96**, 365.

92. Rocchi, P., Boudouresque, F., Zamora, A. J., Muracciole, X., Lechevallier, E., Martin, P. M. & Ouafik, L. (2001). Expression of adrenomedullin and peptide amidation activity in human prostate cancer and in human prostate cancer cell lines. *Cancer Res* **61**, 1206.

93. Rocchi, P., Muracciole, X., Fina, F., Mulholland, D. J., Karsenty, G., Palmari, J., Ouafik, L., Bladou, F. & Martin, P. M. (2004). Molecular analysis integrating different pathways associated with androgen-independent progression in LuCaP 23.1 xenograft. *Oncogene* **23**, 9119.

94. Jimenez, N., Jongsma, J., Calvo, A., van der Kwast, T. H., Treston, A. M., Cuttitta, F., Schroder, F. H., Montuenga, L. M. & van Steenbrugge, G. J. (2001). Peptidylglycine alpha-amidating monooxygenasc- and proadrenomedullin-derived peptide-associated neuroendocrine differentiation are induced by androgen deprivation in the neoplastic prostate. *Int J Cancer* **94**, 34.

95. Jimenez, N., Abasolo, I., Jongsma, J., Calvo, A., Garayoa, M., van der Kwast, T. H., van Steenbrugge, G. J. & Montuenga, L. M. (2003). Androgen-independent expression of adrenomedullin and peptidylglycine alpha-amidating monooxygenase in human prostatic carcinoma. *Mol Carcinog.* **38**, 24.

96. Brooks, J. D. (2002). Microarray analysis in prostate cancer research. *Curr.Opin.Urol.* **12**, 399.

97. De Marzo, A. M., DeWeese, T. L., Platz, E. A., Meeker, A. K., Nakayama, M., Epstein, J. I., Isaacs, W. B. & Nelson, W. G. (2004). Pathological and molecular mechanisms of prostate carcinogenesis: implications for diagnosis, detection, prevention, and treatment. *J Cell Biochem.* **91**, 477.

98. Luo, J. H. (2002). Gene expression alterations in human prostate cancer. *Drugs Today (Barc)* **38**, 719.

99. Luo, J. H., Yu, Y. P., Cieply, K., Lin, F., Deflavia, P., Dhir, R., Finkelstein, S., Michalopoulos, G. & Becich, M. (2002). Gene expression analysis of prostate cancers. *Mol Carcinog.* **33**, 35.

100. Fromont, G., Chene, L., Vidaud, M., Vallancien, G., Mangin, P., Fournier, G., Validire, P., Latil, A. & Cussenot, O. (2005). Differential expression of 37 selected genes in hormone-refractory prostate cancer using quantitative taqman real-time RT-PCR. *Int J Cancer* **114**, 181.

101. Foley, R., Hollywood, D. & Lawler, M. (2004). Molecular pathology of prostate cancer: the key to identifying new biomarkers of disease. *Endocr Relat Cancer* **11**, 488.

102. Qi, H., Grenier, J., Fournier, A. & Labrie, C. (2003). Androgens differentially regulate the expression of NEDD4L transcripts in LNCaP human prostate cancer cells. *Mol Cell Endocrinol* **210**, 62.

103. Seregni, E., Botti, C., Ballabio, G. & Bombardieri, E. (1996). Biochemical characteristics and recent biological knowledge on prostate-specific antigen. *Tumori* **82**, 77.

104. Albrecht, M., Janssen, M., Konrad, L., Renneberg, H. & Aumuller, G. (2002). Effects of dexamethasone on proliferation of and fibronectin synthesis by human primary prostatic stromal cells in vitro. *Andrologia* **34**, 21.

105. Morgan, M., Saba, S. & Gower, W. (2000). Fibronectin influences cellular proliferation and apoptosis similarly in LNCaP and PC-3 prostate cancer cell lines. *Urol Oncol* **5**, 159.

106. Chaudhary, J., Schmidt, M. & Sadler-Riggleman, I. (2005). Negative acting HLH proteins Id 1, Id 2, Id 3, and Id 4 are expressed in prostate epithelial cells. *Prostate* **64**, 264.

107. Asirvatham, A. J., Schmidt, M., Gao, B. & Chaudhary, J. (2005). Androgens regulate the immune/ inflammatory response and cell survival pathways in rat ventral prostate epithelial cells. *Endocrinology.*

108. Hurlin, P. J. & Dezfouli, S. (2004). Functions of myc:max in the control of cell proliferation and tumorigenesis. *Int Rev Cytol* **238**, 226.

109. Masiello, D., Chen, S. Y., Xu, Y., Verhoeven, M. C., Choi, E., Hollenberg, A. N. & Balk, S. P. (2004). Recruitment of beta-catenin by wild-type or mutant androgen receptors correlates with ligand stimulated growth of prostate cancer cells. *Mol Endocrinol* **18**, 2401.

110. Angelucci, C., Iacopino, F., Lama, G., Capucci, S., Zelano, G., Boca, M., Pistilli, A. & Sica, G. (2004). Apoptosis-related gene expression affected by a GnRH analogue without induction of programmed cell death in LNCaP cells. *Anticancer Res* **24**, 2738.

111. Foury, O., Nicolas, B., Joly-Pharaboz, M. O. & Andre, J. (1998). Control of the proliferation of prostate cancer cells by an androgen and two antiandrogens. Cell specific sets of responses. *J Steroid Biochem Mol Biol* **66**, 240.

112. Blank, V. & Andrews, N. C. (1997). The Maf transcription factors: regulators of differentiation. *Trends Biochem Sci* **22**, 441.

113. Toffolatti, L., Rosa, G. L., Patarnello, T., Romualdi, C., Merlanti, R., Montesissa, C., Poppi, L., Castagnaro, M. & Bargelloni, L. (2005). Expression analysis of androgen-responsive genes in the prostate of veal calves treated with anabolic hormones. *Domest Anim Endocrinol.*

114. Woelfle, U., Sauter, G., Santjer, S., Brakenhoff, R. & Pantel, K. (2004). Down-regulated expression of cytokeratin 18 promotes progression of human breast cancer. *Clin Cancer Res* **10**, 2674.

115. Hsieh, J. T., Zhau, H. E., Wang, X. H., Liew, C. C. & Chung, L. W. (1992). Regulation of basal and luminal cell-specific cytokeratin expression in rat accessory sex organs: evidence for a new class of androgen-repressed genes and insight into their pairwise control. *J Biol Chem* **267**, 2310.

116. Turner, K. J., Morley, M., MacPherson, S., Millar, M. R., Wilson, J. A., Sharpe, R. M. & Saunders, P. T. (2001). Modulation of gene expression by androgen and oestrogens in the testis and prostate of the adult rat following androgen withdrawal. *Mol Cell Endocrinol* **178**, 87.

117. Nellemann, C., Dalgaard, M., Lam, H. R. & Vinggaard, A. M. (2003). The combined effects of vinclozolin and procymidone do not deviate from expected additivity in vitro and in vivo. *Toxicol Sci* **71**, 262.

118. Gleave, M. & Miyake, H. (2005). Use of antisense oligonucleotides targeting the cytoprotective gene, clusterin, to enhance androgen- and chemo-sensitivity in prostate cancer. *World J Urol* **23**, 46.

119. Soderling, T. R. (1999). The Ca-calmodulin-dependent protein kinase cascade. *Trends Biochem Sci* **24**, 236.

120. Nagelkerken, B., Van Anken, E., Van Raak, M., Gerez, L., Mohrmann, K., Van Uden, N., Holthuizen, J., Pelkmans, L. & Van Der, S. P. (2000). Rabaptin4, a novel effector of the small GTPase rab4a, is recruited to perinuclear recycling vesicles. *Biochem J* **346 Pt 3**, 601.

121. Boutros, R., Fanayan, S., Shehata, M. & Byrne, J. A. (2004). The tumor protein D52 family: many pieces, many puzzles. *Biochem Biophys Res Commun* **325**, 1121.

122. Rubin, M. A., Varambally, S., Beroukhim, R., Tomlins, S. A., Rhodes, D. R., Paris, P. L., Hofer, M. D., Storz-Schweizer, M., Kuefer, R., Fletcher, J. A., Ilsi, B. L., Byrne, J. A., Pienta, K. J., Collins, C., Sellers, W. R. & Chinnaiyan, A. M. (2004). Overexpression, amplification, and androgen regulation of TPD52 in prostate cancer. *Cancer Res* **64**, 3822.

123. Wang, R., Xu, J., Saramaki, O., Visakorpi, T., Sutherland, W. M., Zhou, J., Sen, B., Lim, S. D., Mabjeesh, N., Amin, M., Dong, J. T., Petros, J. A., Nelson, P. S., Marshall, F. F., Zhau, H. E. & Chung, L. W. (2004). PrLZ, a novel prostate-specific and androgen-responsive gene of the TPD52 family, amplified in chromosome 8q21.1 and overexpressed in human prostate cancer. *Cancer Res* **64**, 1594.

124. Gonzalez-Robayna, I. J., Falender, A. E., Ochsner, S., Firestone, G. L. & Richards, J. S. (2000). Follicle-Stimulating hormone (FSH) stimulates phosphorylation and activation of protein kinase B (PKB/Akt) and serum and glucocorticoid-Induced kinase (Sgk): evidence for a kinase-independent signaling by FSH in granulosa cells. *Mol Endocrinol* **14**, 1300.

125. Kuefer, R., Varambally, S., Zhou, M., Lucas, P. C., Loeffler, M., Wolter, H., Mattfeldt, T., Hautmann, R. E., Gschwend, J. E., Barrette, T. R., Dunn, R. L., Chinnaiyan, A. M. & Rubin, M. A. (2002). Alpha-methylacyl-CoA racemase: expression levels of this novel cancer biomarker depend on tumor differentiation. *Am J Pathol* **161**, 848.

126. West, A. P., Sharpe, R. M. & Saunders, P. T. (1994). Differential regulation of cyclic adenosine 3', 5'-monophosphate (cAMP) response element-binding protein and cAMP response element modulator messenger ribonucleic acid transcripts by follicle-stimulating hormone and androgen in the adult rat testis. *Biol Reprod* **50**, 881.

127. Segev, D. L., Hoshiya, Y., Hoshiya, M., Tran, T. T., Carey, J. L., Stephen, A. E., MacLaughlin, D. T., Donahoe, P. K. & Maheswaran, S. (2002). Mullerian-inhibiting substance regulates NF-kappa B signaling in the prostate in vitro and in vivo. *Proc Natl Acad Sci U S A* **99**, 244.

128. Jung, C., Kim, R. S., Zhang, H. J., Lee, S. J. & Jeng, M. H. (2004). HOXB13 induces growth suppression of prostate cancer cells as a repressor of hormone-activated androgen receptor signaling. *Cancer Res* **64**, 9192.

129. Lau, Y. F. & Zhang, J. (2000). Expression analysis of thirty one Y chromosome genes in human prostate cancer. *Mol Carcinog* **27**, 321.

130. Virkkunen, P., Hedberg, P., Palvimo, J. J., Birr, E., Porvari, K., Ruokonen, M., Taavitsainen, P., Janne, O. A. & Vihko, P. (1994). Structural comparison of human and rat prostate-specific acid phosphatase genes and their promoters: identification of putative androgen response elements. *Biochem Biophys Res Commun* **202**, 57.

131. Ho, K. C., Marschke, K. B., Tan, J., Power, S. G., Wilson, E. M. & French, F. S. (1993). A complex response element in intron 1 of the androgen-regulated 20-kDa protein gene displays cell type-dependent androgen receptor specificity. *J Biol Chem* **268**, 27235.

132. Rokhlin, O. W., Taghiyev, A. F., Guseva, N. V., Glover, R. A., Chumakov, P. M., Kravchenko, J. E. & Cohen, M. B. (2005). Androgen regulates apoptosis induced by TNFR family ligands via multiple signaling pathways in LNCaP. *Oncogene*.

133. Gao, S., Lee, P., Wang, H., Gerald, W., Adler, M., Zhang, L., Wang, Y. F. & Wang, Z. (2005). The androgen receptor directly targets the cellular Fas/FasL-associated death domain protein-like inhibitory protein gene to promote the androgen-independent growth of prostate cancer cells. *Mol Endocrinol* **19**, 1802.

134. Schwidetzky, U., Schleuning, W. D. & Haendler, B. (1997). Isolation and characterization of the androgen-dependent mouse cysteine-rich secretory protein-1 (CRISP-1) gene. *Biochem J* **321 (Pt 2)**, 332.

135. Gnanapragasam, V. J., Robson, C. N., Neal, D. E. & Leung, H. Y. (2002). Regulation of FGF8 expression by the androgen receptor in human prostate cancer. *Oncogene* **21**, 5080.

136. Magee, J. A., Chang, L. W., Stormo, G. D. & Milbrandt, J. (2005). Direct, androgen receptor mediated regulation of the fkbp5 gene via a distal enhancer element. *Endocrinology*.

137. Salah, Z., Maoz, M., Cohen, I., Pizov, G., Pode, D., Runge, M. S. & Bar-Shavit, R. (2005). Identification of a novel functional androgen response element within hPar1 promoter: implications to prostate cancer progression. *FASEB J* **19**, 72.

138. Murtha, P., Tindall, D. J. & Young, C. Y. (1993). Androgen induction of a human prostate-specific kallikrein, hKLK2: characterization of an androgen response element in the 5′ promoter region of the gene. *Biochemistry* **32**, 6464.

139. Yu, D. C., Sakamoto, G. T. & Henderson, D. R. (1999). Identification of the transcriptional regulatory sequences of human kallikrein 2 and their use in the construction of calydon virus 764, an attenuated replication competent adenovirus for prostate cancer therapy. *Cancer Res.* **59**, 1504.

140. Riegman, P. H., Vlietstra, R. J., van der Korput, J. A., Brinkmann, A. O. & Trapman, J. (1991). The promoter of the prostate-specific antigen gene contains a functional androgen responsive element. *Mol Endocrinol* **5**, 1930.

141. Schuur, E. R., Henderson, G. A., Kmetec, L. A., Miller, J. D., Lamparski, H. G. & Henderson, D. R. (1996). Prostate-specific antigen expression is regulated by an upstream enhancer. *J Biol Chem* **271**, 7051.

142. Jia, L. & Coetzee, G. A. (2005). Androgen receptor-dependent PSA expression in androgen-independent prostate cancer cells does not involve androgen receptor occupancy of the PSA locus. *Cancer Res* **65**, 8008.

143. Xia, L., Robinson, D., Ma, A. H., Chen, H. C., Wu, F., Qiu, Y. & Kung, H. J. (2002). Identification of human male germ cell-associated kinase, a kinase transcriptionally activated by androgen in prostate cancer cells. *J Biol Chem* **277**, 35433.

144. Rennie, P. S., Bruchovsky, N., Leco, K. J., Sheppard, P. C., McQueen, S. A., Cheng, H., Snoek, R., Hamel, A., Bock, M. E. & MacDonald, B. S. (1993). Characterization of two cis-acting DNA elements involved in the androgen regulation of the probasin gene. *Mol Endocrinol* **7**, 36.

145. Zhang, J., Gao, N., Kasper, S., Reid, K., Nelson, C. & Matusik, R. J. (2004). An androgen-dependent upstream enhancer is essential for high levels of probasin gene expression. *Endocrinology* **145**, 148.

146. Verrijdt, G., Schoenmakers, E., Alen, P., Haelens, A., Peeters, B., Rombauts, W. & Claessens, F. (1999). Androgen specificity of a response unit upstream of the human secretory component gene is mediated by differential receptor binding to an essential androgen response element. *Mol Endocrinol* **13**, 1570.

147. Haelens, A., Verrijdt, G., Schoenmakers, E., Alen, P., Peeters, B., Rombauts, W. & Claessens, F. (1999). The first exon of the human sc gene contains an androgen responsive unit and an interferon regulatory factor element. *Mol Cell Endocrinol* **153**, 102.

148. Steketee, K., Ziel-van der Made, A. C., van der Korput, H. A., Houtsmuller, A. B. & Trapman, J. (2004). A bioinformatics-based functional analysis shows that the specifically androgen-regulated gene SARG contains an active direct repeat androgen response element in the first intron. *J Mol Endocrinol* **33**, 491.

149. Lareyre, J. J., Claessens, F., Rombauts, W., Dufaure, J. P. & Drevet, J. R. (1997). Characterization of an androgen response element within the promoter of the epididymis-specific murine glutathione peroxidase 5 gene. *Mol Cell Endocrinol* **129**, 46.

150. Ghyselinck, N. B., Dufaure, I., Lareyre, J. J., Rigaudiere, N., Mattei, M. G. & Dufaure, J. P. (1993). Structural organization and regulation of the gene for the androgen-dependent glutathione peroxidase-like protein specific to the mouse epididymis. *Mol Endocrinol* **7**, 272.

151. Lund, S. D., Gallagher, P. M., Wang, B., Porter, S. C. & Ganschow, R. E. (1991). Androgen responsiveness of the murine beta-glucuronidase gene is associated with nuclease hypersensitivity, protein binding, and haplotype-specific sequence diversity within intron 9. *Mol Cell Biol* **11**, 5434.

152. Shen, R., Sumitomo, M., Dai, J., Hardy, D. O., Navarro, D., Usmani, B., Papandreou, C. N., Hersh, L. B., Shipp, M. A., Freedman, L. P. & Nanus, D. M. (2000). Identification and characterization of two androgen response regions in the human neutral endopeptidase gene. *Mol Cell Endocrinol.* **170**, 142.

153. Crozat, A., Palvimo, J. J., Julkunen, M. & Janne, O. A. (1992). Comparison of androgen regulation of ornithine decarboxylase and S-adenosylmethionine decarboxylase gene expression in rodent kidney and accessory sex organs. *Endocrinology* **130**, 1144.

154. Devos, A., Claessens, F., Alen, P., Winderickx, J., Heyns, W., Rombauts, W. & Peeters, B. (1997). Identification of a functional androgen-response element in the exon 1-coding sequence of the cystatin-related protein gene crp2. *Mol Endocrinol* **11**, 1043.

155. Claessens, F., Celis, L., Peeters, B., Heyns, W., Verhoeven, G. & Rombauts, W. (1989). Functional characterization of an androgen response element in the first intron of the C3(1) gene of prostatic binding protein. *Biochem Biophys Res Commun* **164**, 840.

156. Strahle, U., Klock, G. & Schutz, G. (1987). A DNA sequence of 15 base pairs is sufficient to mediate both glucocorticoid and progesterone induction of gene expression. *Proc Natl Acad Sci U S A* **84**, 7875.

157. Dhanasekaran, S. M., Dash, A., Yu, J., Maine, I. P., Laxman, B., Tomlins, S. A., Creighton, C. J., Menon, A., Rubin, M. A. & Chinnaiyan, A. M. (2005). Molecular profiling of human prostate tissues: insights into gene expression patterns of prostate development during puberty. *FASEB J* **19**, 245.

158. Zhao, H., Young, K., Wang, P., Lapointe, J., Tibshirani, R., Pollack, J. R. & Brooks, J. D. (2005). Genome-wide characterization of gene expression variations and DNA copy number changes in prostate cancer cell lines. *Prostate* **63**, 187.

159. Clegg, N., Eroglu, B., Ferguson, C., Arnold, H., Moorman, A. & Nelson, P. S. (2002). Digital expression profiles of the prostate androgen-response program. *J Steroid Biochem.Mol Biol* **80**, 13.

160. Swinnen, J.V., Alen, P., Heyns, W., & Verhoeven, G. (1998). Identification of diazepam-binding Inhibitor/Acyl-CoA-binding protein as a sterol regulatory element-binding protein-responsive gene. *J Biol Chem* **273**, 19938.

161. Stephenson, A.J., Smith, A., Kattan, M.W., Satagopan, J., Reuter,V.E., Scardino,P.T., & Gerald, W.L. (2005). Integration of gene expression profiling and clinical variables to predict prostate carcinoma recurrence after radical prostatectomy. *Cancer* **104**, 290.

6

Interrogating Estrogen Receptor α Signaling in Breast Cancer by Chromatin Immunoprecipitation Microarrays

Alfred S. L. Cheng, PhD,
Huey-Jen L. Lin, PhD,
and Tim H. -M. Huang, PhD

CONTENTS

INTRODUCTION
CHIP
IDENTIFICATION AND CHARACTERIZATION OF ERα TARGET
  GENES USING CHIP-CHIP
COMPUTATIONAL MODELING REVEALS cis-REGULATORY MODULES
ERα SIGNALING NETWORK IN BREAST CANCER
POTENTIAL OF CHIP-CHIP TO IMPROVE UNDERSTANDING
  OF ENDOCRINE SYSTEMS
CONCLUSIONS
ACKNOWLEDGEMENTS
REFERENCES

Abstract

Breast cancer growth is regulated by the coordinated action of estrogen receptor α and multiple signaling pathways. Profiling of estrogen-regulated gene expression provides mechanisms on estrogenic control of proliferation in breast cancer cells. However, little is known of the hierarchy and the *trans*-acting factors in ERα signaling network. Chromatin immunoprecipitation microarray (or ChIP-chip) has emerged as a powerful technique to identify the binding sites of transcription factors and profile the histone modifications in a genome-wide scale. This technology permits, for the first time, comprehensive identification and characterization of direct ERα target genes. Recruitment of ERα to the genome does not only occur at proximal promoter regions, but also involves distal enhancer elements. By using integrated ChIP-chip and bioinformatics analyses, new transcription factor partners (FoxA1 and c-MYC) are being uncovered that play critical roles in ERα-mediated transcription. Based on these findings, a model is constructed to show the complex network of ERα signaling. These new insights will support the development of more precise therapeutic regimen for breast cancer treatment.

From: *Contemporary Endocrinology: Genomics in Endocrinology:*
DNA Microarray Analysis in Endocrine Health and Disease
Edited by S. Handwerger and B. Aronow © Humana Press, Totowa, NJ

115

Key Words: Estrogen receptor α signaling; breast cancer; ChIP-chip; FoxA1; c-MYC; bioinformatics

INTRODUCTION

The steroid hormone estrogen plays a key role in many physiological processes, including reproduction, cardiovascular and nervous functions. In addition to normal homeostatic functions, estrogen is implicated in the development of breast cancer. Data from clinical and animal studies suggested breast cancer is positively correlated with cumulative exposure of the breast epithelium to estrogen *(1)*. The action modes of estrogen are mediated through binding to two genetically distinct estrogen receptors, ERα and ERβ, which belong to the nuclear receptor superfamily of transcription factors.

Here, we further discuss the regulatory role of ERα in target gene expression. In general, ERα-mediated transcription is through its direct binding to specific estrogen response element (ERE) in the promoters of target genes or to other promoter-bound transcription factors via protein–protein interaction. Co-activators or -repressors are further recruited to form a functional complex that remodels chromatin structure specifying transcriptional activities of target genes *(2, 3)*. In addition to the nuclear actions, estrogen regulates target genes in a "nongenomic" manner through the interactions of ERα located in or adjacent to the plasma membrane. Activation of the membrane ERα confers a rapid change in cellular signaling and kinase stimulation resulting in transcription activation *(4–6)*. The ERα-regulated target genes, regardless via a genomic and nongenomic manner, would further signal a cascade of downstream targets, which subsequently lead to cytoarchitectural and phenotypic alterations.

Key insight into the ERα signaling has come from microarray studies on the expression of responsive genes in breast cancer cells *(7–9)*. These studies suggested that growth and development of breast cancer is regulated by coordinated actions of ERα in conjunction with signaling pathways *(10, 11)*. Nevertheless, little is known with regard to the hierarchy and the *trans*-acting factors in ERα signaling network.

In this chapter, we outline the recent advances in chromatin immunoprecipitation microarray (or ChIP-chip) followed by detailed discussion on new insights into ERα signaling in breast cancer revealed by this technology. ChIP-chip contributes, for the first time, a comprehensive understanding of direct ERα target genes. More importantly, new transcription factor partners are being uncovered that play critical roles in ERα-mediated transcription. With this knowledge, more precise therapeutic regimen could be developed to tackle the regulatory targets in ERα signaling for breast cancer treatment.

ChIP

DNA–protein interactions play a crucial role in many cellular processes including regulation of gene expression. Transcription is shaped by the binding of transcription factors and cofactors to *cis*-regulatory elements on DNA that influence the chromatin structure of target promoters. The basic building block of chromatin is the nucleosome, which comprises a histone octamer wrapped with 146-bp DNA in two turns. Nucleosomes can restrict or enhance the access of transcription factors to DNA through diverse histone modifications and thereby modulate gene expression *(12)*.

ChIP is an important technique for determining the genomic locations of endogenous proteins including transcription factors and histones *(13)*. Procedures involved in a standard ChIP are briefly outlined in Fig. 1. First, DNA and proteins within the cells are cross-linked by chemicals (e.g., formaldehyde) or UV light. The chromatin is then isolated from the cells and subjected to fragmentation. The size of smeared chromatin, spanning from ~0.2 to 1 kb, determines the detection limit by subsequent PCR-based assays. Fragmented chromatin is then immunoprecipitated with an antibody, which binds specifically to the nucleosomal protein of interest. The quality of antibody is crucial for the recovery of DNA fragments. Upon the removal of unbound DNA, the

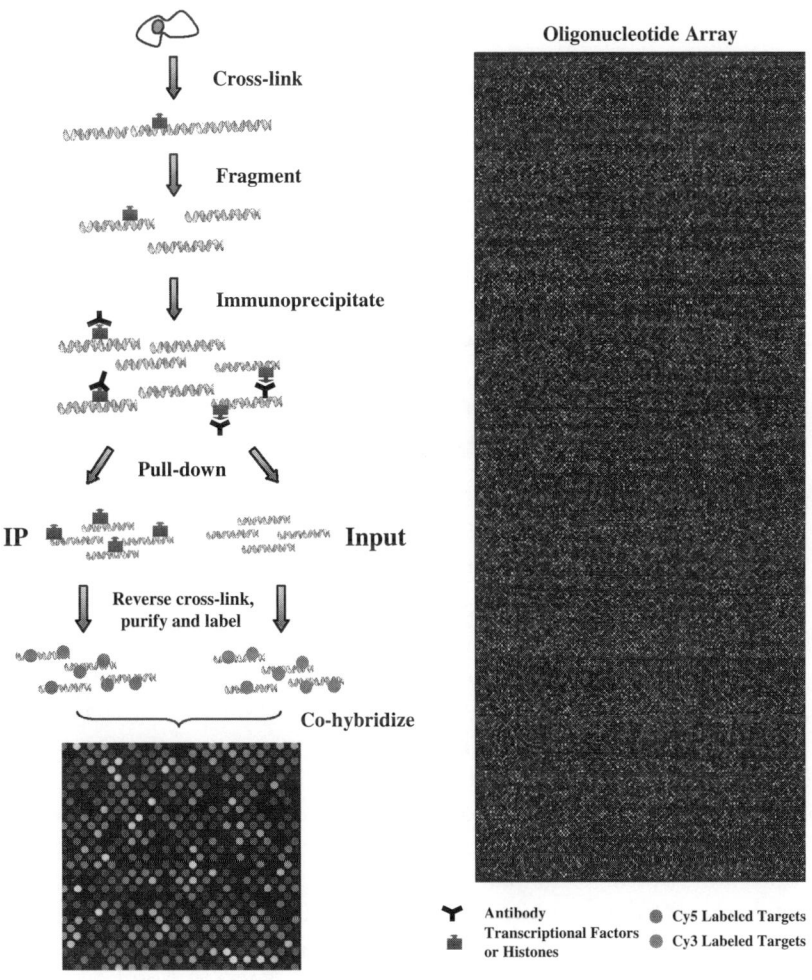

Fig. 1. Schematic diagram for ChIP-chip. The live cells are subjected to cross-linking, followed by shearing. The derived fragmented chromatin DNA is immunoprecipitated with specific antibody. The chromatin DNA of interest is pulled down (IP) while the genomic DNA is collected (Input). After reversal of cross-links and purification, the immunoprecipitated and input DNA is labeled with Cy5 and Cy3 fluorescent dyes, respectively, and co-hybridized onto DNA microarray. An example of scanned image of an oligonucleotide array containing 44,000 60-mer oligonucleotides is shown on the right.

cross-linkage is reversed and the immunoprecipitated DNA is purified. Although ChIP comprises highly versatile procedures, it requires optimization of conditions to achieve a successful DNA extraction. The critical parameters and the variants of ChIP have been discussed in detail in Das et al. *(14)*.

Many standard detection and quantification methods (e.g., real-time PCR) can be applied to study the isolated gene fragments *(15)*. However, they are only useful for investigating DNA–protein interactions of a few selected targets. ChIP combined with microarray (ChIP-chip) has emerged as a powerful technique to interrogate DNA–protein interactions in a genome-wide scale (Fig. 1) *(16, 17)*. This new high-throughput strategy enables the discovery of novel transcription factor targets and provides detailed information regarding specific chromatin architecture along the genome *(18)*.

ChIP-chip and Its Applications

Pioneered in yeast, the ChIP-chip technique was first used to identify binding sites of various transcription factors in a genome-wide manner *(19–21)*. These experiments were conducted on microarrays derived from PCR products of yeast open reading frames, intergenic regions, or a combination of both. The sizes of the mammalian genomes are vastly greater and contain a large number of repetitive elements and, therefore, it is a challenge to construct mammalian DNA microarrays.

Several strategies of microarray construction have been adopted for the human studies. Using a CpG island (CGI) microarray, Weinmann and colleagues have isolated and characterized novel E2F4 targets in HeLa cells *(22)*. This DNA microarray contains 7776 human genomic fragments, which were isolated by methyl-binding MeCP2 protein due to their high CpG content *(23)*. CGIs often correspond to promoter regions *(24)* and, in fact, provide a reliable measure for promoter prediction *(25, 26)*. This approach was later applied to identify target promoters of other transcription factors, namely Rb *(27)*, c-MYC *(28)*, and ERα *(29)*. In addition, genes silenced by specific histone modification, histone 3-lysine 9 dimethylation, were identified in cancer cells using the same approach *(30)*. An extended version of CGI microarray, which contains 20782 genomic fragments, has been developed and characterized recently *(31)*. Coupled with the online CpG Island Library browser (http://data.microarrays.ca/cpg/), this new microarray platform will facilitate ChIP-chip analysis of transcription factor binding and histone modifications in human cells.

A second strategy is to construct arrays from PCR fragments of target promoters. A DNA microarray, which contains approximately 900-bp PCR products spanning the proximal promoters of 4839 human genes, has been utilized to identify the genomic binding sites of c-MYC in Burkitt's lymphoma cells *(32)*. A more elaborate promoter microarray containing approximately 13 000 genomic regions [−750 to +250 bp relative to the transcription start site (TSS)] was constructed and used to discover targets of HNF transcription factors in human hepatocytes and pancreatic island cells *(33)*. This microarray platform was also used to map the gene promoters bound by all five NF-κB proteins before and after stimulation of monocytic cells with bacterial lipopolysaccharide *(34)*. This custom-made microarray excludes repetitive sequences that would interfere with the hybridization, but interrogation of binding sites is biased by the selected 5′-end regions of known genes. Although a significant percentage of transcription factor binding sites (TFBSs) are located within proximal promoters, this promoter array cannot detect binding site that is further upstream of TSS.

The estimates of genomic binding sites based on promoter arrays probably represent only the tip of the iceberg. Intergenic regions and introns in higher eukaryotes comprise a large part of the genome and augment enormous protein-binding potential. Oligonucleotide arrays containing probes spaced on average every 35-bp along all nonrepetitive sequences on human chromosomes 21 and 22 were developed (35). This high-density oligonucleotide arrays have been employed to map binding sites for three transcription factors, Sp1, c-MYC, and p53 (36). Surprisingly, results obtained from these unbiased mappings suggested that the human genome contains roughly comparable numbers of protein-coding and noncoding genes that are bound by common transcription factors and regulated by common environmental signals (36). These microarrays were also used to construct chromosome-wide maps of histone 3-lysine 4 di- and tri-methylation, as well as lysine 9/14 acetylation (37). More recently, a series of DNA microarrays containing ~ 14.5 million of 50-mer oligonucleotides have been designed to represent all the nonrepetitive sequences throughout the human genome at 100-bp resolution (38). Genome-wide mapping of active promoters using these microarrays has suggested extensive usage of multiple promoters by the human genes and widespread clustering of active promoters in the genome (38). Whole-genome location analysis of OCT4, SOX2, and NANOG in human embryonic stem (ES) cells using oligonucleotide arrays have revealed that these transcription factors co-occupy a substantial portion of their target genes, indicating the presence of a complex network of autoregulatory and feedforward loops (39). Subsequently, the binding sites of the Polycomb Repressive Complex 2 (PRC2) subunit SUZ12 were mapped across the entire nonrepeat portion of the genome in human ES cells (40). The results indicated that PRC2 occupies a special set of developmental regulator genes in ES cells that must be repressed to maintain pluripotency and that are poised for activation during ES cell differentiation (40). The utility of human genome tiling arrays will deliver an unbiased view of the entire genome, enabling scientists to look beyond the known protein-coding gene sequences and thoroughly study gene function, structure, and regulation.

BIOINFORMATICS

The massive datasets derived from these genome-wide mapping experiments pose great challenges for data analysis. A growing number of bioinformatics approaches have been developed to detect enriched sites of transcription factor binding and histone modification from the ChIP-chip experiments performed on different DNA microarrays (41–43). While classical DNA binding motif prediction was based on a limited number of experimentally validated binding sites (44), unbiased motif screen from all the authentic sites mapped by ChIP-chip reveals unanticipated biological phenomenon. For example, many in vivo target sites of c-MYC, Sp1, and p53 lack the respective consensus binding motifs, suggesting that these transcription factors are often recruited to their target loci by other proteins rather than through direct protein–DNA interactions (36).

Upon identification of binding sites, valuable information can be gained by characterizing the target genes based on their functions (Gene Ontology, http://www.godatabase.org/cgi-bin/amigo/go.cgi). Online bioinformatics resources are also available to organize the target genes according to different metabolic or signaling pathways, e.g., KEGG database (http://www.genome.ad.jp/kegg/kegg2.html).

Whole-genome binding maps can help to unravel the interactions between transcription factors (and other regulatory proteins). Combined ChIP-chip and expression array analysis has indicated that transcription factor binding is not necessarily associated with transcriptional regulation *(45)*. Perhaps these "inactive" transcription factor sites are conditional *cis*-acting elements whose regulatory activity depends on the presence or absence of other factors *(46)*. Accumulating evidences have postulated that sets of transcription factors can operate in functional *cis*-regulatory modules to achieve specific regulatory properties *(44)*. Thus, new algorithms have been recently developed to detect and select patterns of TFBSs in the regulatory regions that can distinguish activated targets from repressed ones *(47, 48)*. These integrated ChIP-chip and bioinformatics analyses have led to comprehensive views of genetic and epigenetic regulatory networks *(46)*.

IDENTIFICATION AND CHARACTERIZATION OF ERα TARGET GENES USING CHIP-CHIP

Aberrant responses to the mitogenic actions of estrogen occur in the majority of malignant breast tumors. It has been postulated that binding of estrogens to ERα stimulates proliferation of mammary cells which further leads to an increase in the number of target cells within the breast. Enhanced cell division along with elevated DNA synthesis may promote the risk for replication errors, which would render the acquisition of detrimental mutations that disrupt normal cellular processes, such as apoptosis, cellular proliferation, or DNA repair *(49)*. Microarray profiling of gene expression have shed light on estrogenic control of proliferation and cell phenotype in breast cancer cells *(7,50,51)*. However, this experimental approach failed to distinguish direct ERα target genes from downstream secondary genes. To further understand the hierarchy of ERα signaling network, we, and others, have utilized ChIP-chip for the comprehensive identification and characterization of direct ERα target genes *(29,47,52,53)*.

Genome-wide screening of ERα target genes in MCF7 breast cancer cells was first performed using a CGI microarray containing approximately 9000 genomic fragments *(29)*. Within the list of 70 putative targets, a number of enriched DNA-binding motifs were identified, including ERE half site, SP1, and AP2 *(54)*. Using a more recently constructed CGI microarray containing 12 288 genomic fragments, we further identified 92 ERα-responsive promoters *(47)*. When these promoter sequences were subjected to a database for mammalian ERα target promoters, *ERTargetDB (55)*, 74% of the targets were found to have EREs. This result was complimentary with the previous finding that around one-third of known target genes indirectly associate with ERα through intermediary transcription factors *(56)*. We further demonstrated that estrogen stimu-lation led to the recruitment of co-activators, CBP and SRC-3, to specific promoters and enhanced the association of RNA polymerase II. These findings support the notion that different ERE sequences modulate the interaction of ligand-bound ERα with co-activators *(57)*.

The CGI microarrays mentioned earlier contain a total of 8842 single-copy CGI promoters *(31)*, and ERα was found to bind to approximately 1.8% of the promoters in breast cancer cells *(29, 47)*. In a promoter array study, Laganiere and colleagues demonstrated that ERα binds nearly 0.8% of all tested promoters in MCF7 cells *(52)*. This density of promoter binding by ERα resembles some human transcription factors, including E2F4 *(22)*, HNF1α, and HNF6 *(33)*. Other transcription factors seem to associate with a much larger proportion of promoters in the genome *(28, 32, 34, 36)*. The binding patterns of many transcription factors are highly dynamic, depending on different stages of cell cycle *(27)*, cellular differentiation *(58)*, or different cell types *(33)*. Thus, it is anticipated that ERα binding patterns in normal breast cells are different from those in cancer cells.

The ERα target genes are involved in a variety of biological processes, including cellular metabolism, transport, cytoskeleton, and defense response (Fig. 2). These findings highlight the diverse gene networks and pathways through which estrogen operates to achieve its widespread effects on breast cancer cells. Given the well-known property of estrogen in stimulating cell cycle progression of MCF7 and other cancer cell lines *(59)*, it was surprising to find that cell cycle-related genes constitute a relatively small portion of ERα targets identified by ChIP-chip. Profiling of estrogen-regulated gene expression in MCF7 cells has demonstrated a general upregulation of growth promoting molecules, including survivin, multiple growth factors, and genes involved in cell cycle progression *(7)*. Altogether, these results suggest that ERα indirectly governs proliferation regulators via specific transcriptional regulators, which were well-represented among direct ERα target genes (Table 1).

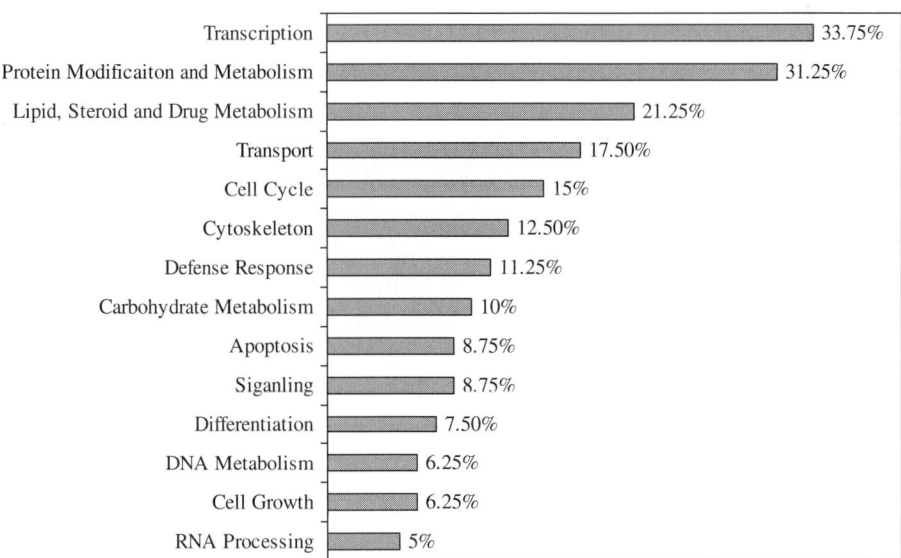

Fig. 2. Biological functions and processes of target genes bound by ERα in estrogen-treated MCF7 cells. Functional categories of 162 targets identified from two independent ChIP-chip studies *(29, 47)* were assigned by Gene Ontology (www.fatigo.org).

Table 1
A Partial List of Direct ERα Target Genes with Transcriptional Regulatory Function

ERα target gene	Symbol	GeneID	Function
B-cell CLL/lymphoma 11A	BCL11A	53335	Transcriptional repressor
Basic helix-loop-helix domain containing, class B, 5	BHLHB5	27319	Transcription factor
Basic leucine zipper nuclear factor 1	BLZF1	8548	Transcription factor
Chromosome 10 open reading frame 48	C10orf48	283078	Transcription factor
Cold shock domain protein A	CSDA	8531	Transcription co-repressor
Friend leukemia virus integration 1	FLI1	2313	Transcriptional activator
Forkhead box D3	FOXD3	27022	Transcription factor
GA binding protein transcription factor, beta subunit 2	GABPB2	2553	Transcription factor
Homeo box 1	HMX1	3166	Transcription factor
v-myc myelocytomatosis viral oncogene homolog	c-MYC	4609	Transcription factor
Retinoblastoma binding protein 8	RBBP8	5932	BRCA1 transcriptional regulation
MADS box transcription enhancer factor 2	RFXANK	8625	Transcription factor
Sine oculis homeobox homolog 1	SIX1	6495	Transcription factor
Sine oculis homeobox homolog 2	SIX2	10736	Transcription factor
Sphingosine kinase type 1-interacting protein	SKIP	80309	Transcriptional co-activator
Transcription factor 4	TCF4	6925	Wnt signaling pathway
Zinc finger protein, Y-linked	ZFY	7544	Transcriptional activator
Zinc finger protein 143	ZNF143	7702	Transcriptional activator

Distal Erα Binding Domains Act as Transcriptional Enhancers

Combining ChIP with tiled microarrays, which contain the complete nonrepetitive sequence of human chromosomes 21 and 22, has identified previously undocumented ERα-chromatin interaction sites in MCF7 cells (53). In this study, a total of 57 estrogen-stimulated ERα binding sites within 32 discrete clusters were found. Almost all of which were not in promoter proximal regions, but rather existed up to 150 kb from putative target genes. These distal ERα binding domains functioned as binding sites for RNA polymerase II and the p160 cofactor, AIB-1, in an estrogen-dependent manner. In two tested cases, the distal enhancer physically interacted with the promoter of target genes following estrogen addition, suggesting that the distal regions function by recruiting the proteins at the enhancers in contact with the promoter (60). This finding shifts the paradigm that functional EREs are not necessarily located in the promoter regions, but also in distal cis -regulatory enhancer regions. Nevertheless, over two-thirds of the ERα binding clusters did not correlate with gene transcription following estrogen stimulation and their functions remain unclear.

Forkhead Proteins Play an Essential Role in ERα Binding and ERα-Mediated Transcription

ERα was shown to bind to significantly less target sites than the predicted EREs on human chromosomes 21 and 22, suggesting that the presence of ERE within DNA is insufficient to determine all putative ERα sites *(53)*. The search for enriched motifs from 57 ERα binding sites found that in addition to EREs, a Forkhead protein named FoxA1, was recruited to almost half of these binding sites *(53)*. Interestingly, FoxA1 was generally associated with chromatin in the absence of estrogen and dissociated from the DNA upon the addition of estrogen. The specific knockdown of FoxA1 in breast cancer cells inhibited ERα association with the chromatin followed by abrogated estrogen-mediated transcription. A ChIP-chip study using promoter arrays has confirmed the essential role of FoxA1 in ERα-mediated transcription *(52)*. It was further demonstrated that targeted silencing of FoxA1 inhibited re-entry into the cell cycle upon estrogen stimulation. Thus, FoxA1 may serve as a licensing factor to propagate a specific domain of estrogen response in breast cancer cells *(52)*. A correlation between FoxA1 and ERα expression in breast cancers further supports the importance of this interaction in breast cancer development *(61)*.

COMPUTATIONAL MODELING REVEALS CIS-REGULATORY MODULES

The combinatorial theory of gene regulation by transcription factors states that transcription factors act cooperatively to mediate target gene activation *(44)*. Accordingly, the identification of the putative *cis*-regulatory modules in ERα-responsive promoters enables the discovery of interacting transcription factors that are involved in crosstalk with the ERα signaling pathway. Our ChIP-chip analysis revealed concerted action of ERα binding and histone modifications in estrogen-responsive promoters in breast cancer cells *(47)*. Maximal ERα binding occurred at 3 h after estrogen treatment and returned to near basal levels at 12- and 24-h time periods. Consistent with this dynamic binding pattern, the enrichment levels of either acetylation or dimethylation of histone 3-lysine 9 were maximal at 3 h after treatment and then subsided. Based on recent studies demonstrating a correlation between acetylation and dimethylation at histone 3-lysine 9 with gene activation and repression, respectively *(15, 62–64)*, we categorized the ERα-responsive promoters into activated and repressed groups according to their histone modification status. The promoter sequences with the predicted TFBSs were retrieved by specific databases *(65, 66)*. Based on the discovery of overrepresented TFBSs by computational modeling, five *cis*-regulatory modules, i.e., ERE + CRX, ERE + MYB, ERE + c-MYC, ERE + SMAD3, and ERE + E47, were identified for activated targets, whereas two modules (ERE + HNF3α and ERE + E47+ETS-1) were identified for repressed targets *(47)*. These transcription factor partners have been previously documented to be involved in breast carcinogenesis *(61, 67–69)*.

ERα and C-Myc Co-regulate Estrogen-Responsive Genes

c-MYC is an important transcription factor known to be involved in estrogen-stimulated proliferation and survival of breast cancer cells *(70, 71)*. In one of the *cis*-regulatory modules, c-MYC binding sites were found to be located near (13 to

214 bp) EREs of 13 ER-responsive promoters *(47)*. Estrogen stimulation enhanced the c-MYC-ERα interaction and facilitated the association of ERα, c-MYC, and TRAAP with these estrogen-responsive promoters. TRRAP was previously identified as a c-MYC-interacting co-activator that mediates histone acetyltransferase recruitment and c-MYC-dependent oncogenesis *(72)*. The deletion of c-MYC binding site in ERα-responsive promoter abolished estrogen responsiveness. Furthermore, targeted silencing of both ERα and c-MYC in breast cancer cells abrogated estrogen-mediated transcription, supporting the notion that ERα and c-MYC co-regulate responsive targets at the transcription level *(47)*. Collectively, these results suggest that ERα and c-MYC physically interact to stabilize the ERα-co-activator complex, thereby permitting other signal transduction pathways to fine-tune ERα signaling.

ERα SIGNALING NETWORK IN BREAST CANCER

While most studies focus on the mechanisms of ERα-mediated gene activation, transcriptional repression by ERα is poorly understood. Recently, a new algorithm for identifying ERα-related TFBSs has been developed for analyzing ChIP-chip dataset *(48)*. The results of this advanced analysis showed that rather than randomly distributed within ERα-responsive promoters, these binding sites form distinctive patterns of both target promoters up- and downregulated by ERα. ERα-activated promoters were specified by the binding pairs DBP + MYC and DBP + MYC/MAX. In contrast, ERα-repressed promoters were characterized with the binding pair of DBP + CETS-1. CETS-1 acts as a transcriptional repressor or activator depending on the promoter context *(73)* and is overexpressed in various primary and metastatic tumors *(74)*. Thus, the *cis*-regulatory element CETS-1 might be required for ERα-mediated gene repression. Hence, the use of integrated ChIP-chip and bioinformatics analyses has led

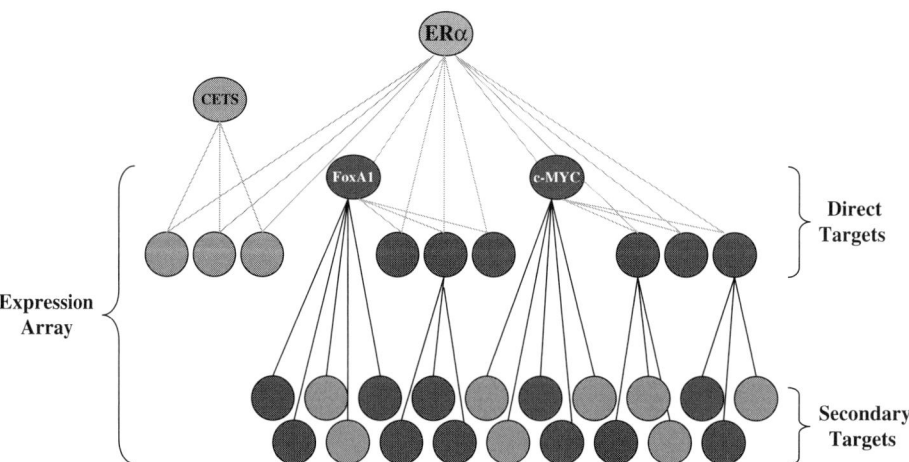

Fig. 3. Graph representation of a partial ERα signaling network. Circle represents either activated (red) or repressed (green) gene in response to ERα. If a transcription factor binds to the regulatory region of a target gene, the two genes are connected by a line. ERα forms feedforward loop with transcription factor partners (FoxA1 or c-MYC) to activate transcription of subsets of target genes (blue lines). Multi-input motif is formed by ERα and CETS to repress expression of target genes (orange lines). While the expression array cannot distinguish between primary and secondary genes, direct targets bound by ERα are identified by ChIP-chip.

to the discovery of transcription partners (FoxA1, c-MYC, and CETS-1) involved in ERα signaling. Based on these findings, a model is constructed to show the complex network of ERα signaling (Fig. 3).

It demonstrates the complexity of ERα signaling and its multiple regulatory interactions with other signal transduction pathways. Crosstalk between ERα and growth factor receptor pathways is considered to be a fundamental factor in endocrine therapy resistance in breast cancer.[11] The full understanding of the ERα signaling network will support the development of combinatorial therapies targeted against ERα and the key regulators involved in the network.

POTENTIAL OF CHIP-CHIP TO IMPROVE UNDERSTANDING OF ENDOCRINE SYSTEMS

The findings on ERα signaling exemplify the power of combining both ChIP-chip and bioinformatics methods to identify regulatory elements in complex signaling network. The integrative approach can be broadly applicable to the elucidation of regulatory networks of other endocrine systems. Most endocrine systems are under genetic control and that the genetic underpinnings of both normal and deranged conditions are polygenic rather than specified by a single gene. The entire repertoire of genes modulated by a particular hormone or regulatory protein on the transcriptional level can be revealed by microarray. However, it will be essential to move from the descriptive expression profiling to understanding the combinatorial interactions that control gene regulatory networks.[46] This functional information may be obtained by combining genomic location maps with maps of gene expression changes after selective removal of the protein of interest. The role of epigenetic regulation on most endocrine systems is largely unexplored. A complete picture of gene regulatory network will require the integration of genome-wide location maps of histone modifications and DNA methylation. The availability of improved high-throughput techniques like ChIP-chip and computational methods will allow us to not only initiate the systematic identification of the building components but also to define their functional interactions in the global genetic and epigenetic networks of any endocrine systems.

CONCLUSIONS

Estrogen signaling has been intensively studied for many years, and the complexity of this pathway is beginning to be unraveled. Integrated ChIP-chip and bioinformatics analyses have revealed the specific target genes of ERα and their regulatory mechanisms. The discovery of transcription factor partners has defined the regulatory elements in ERα signaling network, which might orchestrate the growth of breast cancer cells in response to the mitogenic estrogen. The knowledge of *cis*-regulatory targets and *trans*-acting factors of ERα action will support the development of precise molecular targeting regimen for breast cancer.

ACKNOWLEDGEMENTS

This work was supported in part by National Institutes of Health/National Cancer Institute grants U54 CA113001, R01 CA069065, and R33 CA094441, and funds from The Ohio State University Comprehensive Cancer Center – Arthur G. James Cancer Hospital and Richard J. Solove Research Institute.

REFERENCES

1. Yager, J. D. & Davidson, N. E. (2006). Estrogen carcinogenesis in breast cancer. *N Engl J Med* **354**, 270–82.

2. McKenna, N. J. & O'Malley, B. W. (2002). Combinatorial control of gene expression by nuclear receptors and coregulators. *Cell* **108**, 465–74.

3. Xu, J. & Li, Q. (2003). Review of the in vivo functions of the p160 steroid receptor coactivator family. *Mol Endocrinol* **17**, 1681–92.

4. Revankar, C. M., Cimino, D. F., Sklar, L. A., Arterburn, J. B. & Prossnitz, E. R. (2005). A transmembrane intracellular estrogen receptor mediates rapid cell signaling. *Science* **307**, 1625–30.

5. Simoncini, T., Hafezi-Moghadam, A., Brazil, D. P., Ley, K., Chin, W. W. & Liao, J. K. (2000). Interaction of oestrogen receptor with the regulatory subunit of phosphatidylinositol-3-OH kinase. *Nature* **407**, 538–41.

6. Simoncini, T., Mannella, P., Fornari, L., Caruso, A., Varone, G. & Genazzani, A. R. (2004). Genomic and non-genomic effects of estrogens on endothelial cells. *Steroids* **69**, 537–42.

7. Frasor, J., Danes, J. M., Komm, B., Chang, K. C., Lyttle, C. R. & Katzenellenbogen, B. S. (2003). Profiling of estrogen up- and down-regulated gene expression in human breast cancer cells: insights into gene networks and pathways underlying estrogenic control of proliferation and cell phenotype. *Endocrinology* **144**, 4562–74.

8. Coser, K. R., Chesnes, J., Hur, J., Ray, S., Isselbacher, K. J. & Shioda, T. (2003). Global analysis of ligand sensitivity of estrogen inducible and suppressible genes in MCF7/BUS breast cancer cells by DNA microarray. *Proc Natl Acad Sci U S A* **100**, 13994–9.

9. Lin, C. Y., Strom, A., Vega, V. B., Kong, S. L., Yeo, A. L., Thomsen, J. S., Chan, W. C., Doray, B., Bangarusamy, D. K., Ramasamy, A., Vergara, L. A., Tang, S., Chong, A., Bajic, V. B., Miller, L. D., Gustafsson, J. A. & Liu, E. T. (2004). Discovery of estrogen receptor alpha target genes and response elements in breast tumor cells. *Genome Biol* **5**, R66.

10. DeNardo, D. G., Kim, H. T., Hilsenbeck, S., Cuba, V., Tsimelzon, A. & Brown, P. H. (2005). Global gene expression analysis of estrogen receptor transcription factor cross talk in breast cancer: identification of estrogen-induced/activator protein-1-dependent genes. *Mol Endocrinol* **19**, 362–78.

11. Osborne, C. K., Shou, J., Massarweh, S. & Schiff, R. (2005). Crosstalk between estrogen receptor and growth factor receptor pathways as a cause for endocrine therapy resistance in breast cancer. *Clin Cancer Res* **11**, 865s–70s.

12. Lund, A. H. & van Lohuizen, M. (2004). Epigenetics and cancer. *Genes Dev* **18**, 2315–35.

13. Kuo, M. H. & Allis, C. D. (1999). In vivo cross-linking and immunoprecipitation for studying dynamic protein: DNA associations in a chromatin environment. *Methods* **19**, 425–33.

14. Das, P. M., Ramachandran, K., vanWert, J. & Singal, R. (2004). Chromatin immunoprecipitation assay. *Biotechniques* **37**, 961–9.

15. Kondo, Y., Shen, L. & Issa, J. P. (2003). Critical role of histone methylation in tumor suppressor gene silencing in colorectal cancer. *Mol Cell Biol* **23**, 206–15.

16. Ren, B. & Dynlacht, B. D. (2004). Use of chromatin immunoprecipitation assays in genome-wide location analysis of mammalian transcription factors. *Methods Enzymol* **376**, 304–15.

17. Buck, M. J. & Lieb, J. D. (2004). ChIP-chip: considerations for the design, analysis, and application of genome-wide chromatin immunoprecipitation experiments. *Genomics* **83**, 349–60.

18. Rodriguez, B. A. & Huang, T. H. (2005). Tilling the chromatin landscape: emerging methods for the discovery and profiling of protein-DNA interactions. *Biochem Cell Biol* **83**, 525–34.

19. Ren, B., Robert, F., Wyrick, J. J., Aparicio, O., Jennings, E. G., Simon, I., Zeitlinger, J., Schreiber, J., Hannett, N., Kanin, E., Volkert, T. L., Wilson, C. J., Bell, S. P. & Young, R. A. (2000). Genome-wide location and function of DNA binding proteins. *Science* **290**, 2306–9.

20. Lieb, J. D., Liu, X., Botstein, D. & Brown, P. O. (2001). Promoter-specific binding of Rap1 revealed by genome-wide maps of protein-DNA association. *Nat Genet* **28**, 327–34.

21. Iyer, V. R., Horak, C. E., Scafe, C. S., Botstein, D., Snyder, M. & Brown, P. O. (2001). Genomic binding sites of the yeast cell-cycle transcription factors SBF and MBF. *Nature* **409**, 533–8.

22. Weinmann, A. S., Yan, P. S., Oberley, M. J., Huang, T. H. & Farnham, P. J. (2002). Isolating human transcription factor targets by coupling chromatin immunoprecipitation and CpG island microarray analysis. *Genes Dev* **16**, 235–44.

23. Cross, S. H., Charlton, J. A., Nan, X. & Bird, A. P. (1994). Purification of CpG islands using a methylated DNA binding column. *Nat Genet* **6**, 236–44.

24. Antequera, F. & Bird, A. (1993). Number of CpG islands and genes in human and mouse. *Proc Natl Acad Sci U S A* **90**, 11995–9.

25. Ioshikhes, I. P. & Zhang, M. Q. (2000). Large-scale human promoter mapping using CpG islands. *Nat Genet* **26**, 61–3.

26. Hannenhalli, S. & Levy, S. (2001). Promoter prediction in the human genome. *Bioinformatics* **17 Suppl 1**, S90–6.

27. Wells, J., Yan, P. S., Cechvala, M., Huang, T. & Farnham, P. J. (2003). Identification of novel pRb binding sites using CpG microarrays suggests that E2F recruits pRb to specific genomic sites during S phase. *Oncogene* **22**, 1445–60.

28. Mao, D. Y., Watson, J. D., Yan, P. S., Barsyte-Lovejoy, D., Khosravi, F., Wong, W. W., Farnham, P. J., Huang, T. H. & Penn, L. Z. (2003). Analysis of Myc bound loci identified by CpG island arrays shows that Max is essential for Myc-dependent repression. *Curr Biol* **13**, 882–6.

29. Leu, Y. W., Yan, P. S., Fan, M., Jin, V. X., Liu, J. C., Curran, E. M., Welshons, W. V., Wei, S. H., Davuluri, R. V., Plass, C., Nephew, K. P. & Huang, T. H. (2004). Loss of estrogen receptor signaling triggers epigenetic silencing of downstream targets in breast cancer. *Cancer Res* **64**, 8184–92.

30. Kondo, Y., Shen, L., Yan, P. S., Huang, T. H. & Issa, J. P. (2004). Chromatin immunoprecipitation microarrays for identification of genes silenced by histone H3 lysine 9 methylation. *Proc Natl Acad Sci U S A* **101**, 7398–403.

31. Heisler, L. E., Torti, D., Boutros, P. C., Watson, J., Chan, C., Winegarden, N., Takahashi, M., Yau, P., Huang, T. H., Farnham, P. J., Jurisica, I., Woodgett, J. R., Bremner, R., Penn, L. Z. & Der, S. D. (2005). CpG Island microarray probe sequences derived from a physical library are representative of CpG Islands annotated on the human genome. *Nucleic Acids Res* **33**, 2952–61.

32. Li, Z., Van Calcar, S., Qu, C., Cavenee, W. K., Zhang, M. Q. & Ren, B. (2003). A global transcriptional regulatory role for c-Myc in Burkitt's lymphoma cells. *Proc Natl Acad Sci U S A* **100**, 8164–9.

33. Odom, D. T., Zizlsperger, N., Gordon, D. B., Bell, G. W., Rinaldi, N. J., Murray, H. L., Volkert, T. L., Schreiber, J., Rolfe, P. A., Gifford, D. K., Fraenkel, E., Bell, G. I. & Young, R. A. (2004). Control of pancreas and liver gene expression by HNF transcription factors. *Science* **303**, 1378–81.

34. Schreiber, J., Jenner, R. G., Murray, H. L., Gerber, G. K., Gifford, D. K. & Young, R. A. (2006). Coordinated binding of NF-{kappa}B family members in the response of human cells to lipopolysaccharide. *Proc Natl Acad Sci U S A* **103**, 5899–904.

35. Kapranov, P., Cawley, S. E., Drenkow, J., Bekiranov, S., Strausberg, R. L., Fodor, S. P. & Gingeras, T. R. (2002). Large-scale transcriptional activity in chromosomes 21 and 22. *Science* **296**, 916–9.

36. Cawley, S., Bekiranov, S., Ng, H. H., Kapranov, P., Sekinger, E. A., Kampa, D., Piccolboni, A., Sementchenko, V., Cheng, J., Williams, A. J., Wheeler, R., Wong, B., Drenkow, J., Yamanaka, M., Patel, S., Brubaker, S., Tammana, H., Helt, G., Struhl, K. & Gingeras, T. R. (2004). Unbiased mapping of transcription factor binding sites along human chromosomes 21 and 22 points to widespread regulation of noncoding RNAs. *Cell* **116**, 499–509.

37. Bernstein, B. E., Kamal, M., Lindblad-Toh, K., Bekiranov, S., Bailey, D. K., Huebert, D. J., McMahon, S., Karlsson, E. K., Kulbokas, E. J., 3rd, Gingeras, T. R., Schreiber, S. L. & Lander, E. S. (2005). Genomic maps and comparative analysis of histone modifications in human and mouse. *Cell* **120**, 169–81.

38. Kim, T. H., Barrera, L. O., Zheng, M., Qu, C., Singer, M. A., Richmond, T. A., Wu, Y., Green, R. D. & Ren, B. (2005). A high-resolution map of active promoters in the human genome. *Nature* **436**, 876–80.

39. Boyer, L. A., Lee, T. I., Cole, M. F., Johnstone, S. E., Levine, S. S., Zucker, J. P., Guenther, M. G., Kumar, R. M., Murray, H. L., Jenner, R. G., Gifford, D. K., Melton, D. A., Jaenisch, R. & Young, R. A. (2005). Core transcriptional regulatory circuitry in human embryonic stem cells. *Cell* **122**, 947–56.

40. Lee, T. I., Jenner, R. G., Boyer, L. A., Guenther, M. G., Levine, S. S., Kumar, R. M., Chevalier, B., Johnstone, S. E., Cole, M. F., Isono, K., Koseki, H., Fuchikami, T., Abe, K., Murray, H. L., Zucker, J. P.,

Yuan, B., Bell, G. W., Herbolsheimer, E., Hannett, N. M., Sun, K., Odom, D. T., Otte, A. P., Volkert, T. L., Bartel, D. P., Melton, D. A., Gifford, D. K., Jaenisch, R. & Young, R. A. (2006). Control of developmental regulators by polycomb in human embryonic stem cells. *Cell* **125**, 301–13.

41. Macisaac, K. D., Gordon, D. B., Nekludova, L., Odom, D. T., Schreiber, J., Gifford, D. K., Young, R. A. & Fraenkel, E. (2006). A hypothesis-based approach for identifying the binding specificity of regulatory proteins from chromatin immunoprecipitation data. *Bioinformatics* **22**, 423–9.

42. Liu, X. S., Brutlag, D. L. & Liu, J. S. (2002). An algorithm for finding protein-DNA binding sites with applications to chromatin-immunoprecipitation microarray experiments. *Nat Biotechnol* **20**, 835–9.

43. Li, W., Meyer, C. A. & Liu, X. S. (2005). A hidden Markov model for analyzing ChIP-chip experiments on genome tiling arrays and its application to p53 binding sequences. *Bioinformatics* **21 Suppl 1**, i274–82.

44. Wasserman, W. W. & Sandelin, A. (2004). Applied bioinformatics for the identification of regulatory elements. *Nat Rev Genet* **5**, 276–87.

45. Martone, R., Euskirchen, G., Bertone, P., Hartman, S., Royce, T. E., Luscombe, N. M., Rinn, J. L., Nelson, F. K., Miller, P., Gerstein, M., Weissman, S. & Snyder, M. (2003). Distribution of NF-kappaB-binding sites across human chromosome 22. *Proc Natl Acad Sci U S A* **100**, 12247–52.

46. van Steensel, B. (2005). Mapping of genetic and epigenetic regulatory networks using microarrays. *Nat Genet* **37 Suppl**, S18–24.

47. Cheng, A. S., Jin, V. X., Fan, M., Smith, L. T., Liyanarachchi, S., Yan, P. S., Leu, Y. W., Chan, M. W., Plass, C., Nephew, K. P., Davuluri, R. V. & Huang, T. H. (2006). Combinatorial analysis of transcription factor partners reveals recruitment of c-MYC to estrogen receptor-alpha responsive promoters. *Mol Cell* **21**, 393–404.

48. Li, L., Cheng, A. S., Jin, V. X., Paik, H. H., Fan, M. Y., Li, X. M., Zhang, W., Robarge, J., Balch, C., Davuluri, R. V., Kim, S., Huang, T. H. & Nephew, K. P. (2006). A mixture model based discriminate analysis for identifying ordered transcription factor binding site pairs in gene promoters directly regulated by estrogen receptor-α. *Bioinformatics* **22**, 2210–6.

49. Yue, W., Wang, J. P., Li, Y., Bocchinfuso, W. P., Korach, K. S., Devanesan, P. D., Rogan, E., Cavalieri, E. & Santen, R. J. (2005). Tamoxifen versus aromatase inhibitors for breast cancer prevention. *Clin Cancer Res* **11**, 925s–30s.

50. Lobenhofer, E. K., Bennett, L., Cable, P. L., Li, L., Bushel, P. R. & Afshari, C. A. (2002). Regulation of DNA replication fork genes by 17 beta-estradiol. *Mol Endocrinol* **16**, 1215–29.

51. Cunliffe, H. E., Ringner, M., Bilke, S., Walker, R. L., Cheung, J. M., Chen, Y. & Meltzer, P. S. (2003). The gene expression response of breast cancer to growth regulators: patterns and correlation with tumor expression profiles. *Cancer Res* **63**, 7158–66.

52. Laganiere, J., Deblois, G., Lefebvre, C., Bataille, A. R., Robert, F. & Giguere, V. (2005). From the cover: location analysis of estrogen receptor alpha target promoters reveals that FOXA1 defines a domain of the estrogen response. *Proc Natl Acad Sci U S A* **102**, 11651–6.

53. Carroll, J. S., Liu, X. S., Brodsky, A. S., Li, W., Meyer, C. A., Szary, A. J., Eeckhoute, J., Shao, W., Hestermann, E. V., Geistlinger, T. R., Fox, E. A., Silver, P. A. & Brown, M. (2005). Chromosome-wide mapping of estrogen receptor binding reveals long-range regulation requiring the forkhead protein FoxA1. *Cell* **122**, 33–43.

54. Jin, V. X., Leu, Y. W., Liyanarachchi, S., Sun, H., Fan, M., Nephew, K. P., Huang, T. H. & Davuluri, R. V. (2004). Identifying estrogen receptor alpha target genes using integrated computational genomics and chromatin immunoprecipitation microarray. *Nucleic Acids Res* **32**, 6627–35.

55. Jin, V. X., Sun, H., Pohar, T. T., Liyanarachchi, S., Palaniswamy, S. K., Huang, T. H. & Davuluri, R. V. (2005). ERTargetDB: an integral information resource of transcription regulation of estrogen receptor target genes. *J Mol Endocrinol* **35**, 225–30.

56. O'Lone, R., Frith, M. C., Karlsson, E. K. & Hansen, U. (2004). Genomic targets of nuclear estrogen receptors. *Mol Endocrinol* **18**, 1859–75.

57. Klinge, C. M., Jernigan, S. C., Mattingly, K. A., Risinger, K. E. & Zhang, J. (2004). Estrogen response element-dependent regulation of transcriptional activation of estrogen receptors alpha and beta by coactivators and corepressors. *J Mol Endocrinol* **33**, 387–410.

58. Blais, A., Tsikitis, M., Acosta-Alvear, D., Sharan, R., Kluger, Y. & Dynlacht, B. D. (2005). An initial blueprint for myogenic differentiation. *Genes Dev* **19**, 553–69.

59. Prall, O. W., Rogan, E. M. & Sutherland, R. L. (1998). Estrogen regulation of cell cycle progression in breast cancer cells. *J Steroid Biochem Mol Biol* **65**, 169–74.
60. Carroll, J. S. & Brown, M. (2006). Estrogen receptor target gene: an evolving concept. *Mol Endocrinol.* **20**, 1707–14.
61. Lacroix, M. & Leclercq, G. (2004). About GATA3, HNF3A, and XBP1, three genes co-expressed with the oestrogen receptor-alpha gene (ESR1) in breast cancer. *Mol Cell Endocrinol* **219**, 1–7.
62. Roh, T. Y., Cuddapah, S. & Zhao, K. (2005). Active chromatin domains are defined by acetylation islands revealed by genome-wide mapping. *Genes Dev* **19**, 542–52.
63. Peterson, C. L. & Laniel, M. A. (2004). Histones and histone modifications. *Curr Biol* **14**, R546–51.
64. Peters, A. H., Kubicek, S., Mechtler, K., O'Sullivan, R. J., Derijck, A. A., Perez-Burgos, L., Kohlmaier, A., Opravil, S., Tachibana, M., Shinkai, Y., Martens, J. H. & Jenuwein, T. (2003). Partitioning and plasticity of repressive histone methylation states in mammalian chromatin. *Mol Cell* **12**, 1577–89.
65. Wingender, E., Chen, X., Hehl, R., Karas, H., Liebich, I., Matys, V., Meinhardt, T., Pruss, M., Reuter, I. & Schacherer, F. (2000). TRANSFAC: an integrated system for gene expression regulation. *Nucleic Acids Res* **28**, 316–9.
66. Palaniswamy, S. K., Jin, V. X., Sun, H. & Davuluri, R. V. (2005). OMGProm: a database of orthologous mammalian gene promoters. *Bioinformatics* **21**, 835–6.
67. Rushton, J. J., Davis, L. M., Lei, W., Mo, X., Leutz, A. & Ness, S. A. (2003). Distinct changes in gene expression induced by A-Myb, B-Myb and c-Myb proteins. *Oncogene* **22**, 308–13.
68. Matsuda, T., Yamamoto, T., Muraguchi, A. & Saatcioglu, F. (2001). Cross-talk between transforming growth factor-beta and estrogen receptor signaling through Smad3. *J Biol Chem* **276**, 42908–14.
69. Lincoln, D. W., 2nd & Bove, K. (2005). The transcription factor Ets-1 in breast cancer. *Front Biosci* **10**, 506–11.
70. Rodrik, V., Zheng, Y., Harrow, F., Chen, Y. & Foster, D. A. (2005). Survival signals generated by estrogen and phospholipase D in MCF-7 breast cancer cells are dependent on Myc. *Mol Cell Biol* **25**, 7917–25.
71. Dubik, D., Dembinski, T. C. & Shiu, R. P. (1987). Stimulation of c-myc oncogene expression associated with estrogen-induced proliferation of human breast cancer cells. *Cancer Res* **47**, 6517–21.
72. Park, J., Kunjibettu, S., McMahon, S. B. & Cole, M. D. (2001). The ATM-related domain of TRRAP is required for histone acetyltransferase recruitment and Myc-dependent oncogenesis. *Genes Dev* **15**, 1619–24.
73. Goldberg, Y., Treier, M., Ghysdael, J. & Bohmann, D. (1994). Repression of AP-1-stimulated transcription by c-Ets-1. *J Biol Chem* **269**, 16566–73.
74. Seth, A. & Watson, D. K. (2005). ETS transcription factors and their emerging roles in human cancer. *Eur J Cancer* **41**, 2462–78.

7

Gene Expression Analysis of the Adrenal Cortex in Health and Disease

Anelia Horvath, PhD and *Constantine A. Stratakis, MD, DMedSci*

CONTENTS

INTRODUCTION
METHODS: APPROACHES IN THE STUDY OF ADRENAL
   GENE EXPRESSION
CONCLUSIONS
REFERENCES

Abstract

A cure for adrenocortical cancer remains elusive despite rapid advances in the molecular understanding of many biological processes underlying the function of these steroid-producing cells. With the promise of state-of-the-art molecular technologies and the tools provided by the human genome project, a number of investigators are trying to identify molecular targets of adrenocortical tumorigenesis. One path in this endeavor was the identification by positional cloning of genes that are mutated in rare adrenocortical tumors. The subject of this chapter is an updated summary of the results of experiments in the second path that was followed by us and others: that of using genome-wide expression analysis of adrenocortical cells in normal and various disease states. Transcriptomic analysis is a rapidly evolving technology; one would think that the data would be hard to summarize in a chapter that will be quickly outdated. However, there are a limited number of such studies of the adrenal cortex, and their results are surprisingly reproducible. This chapter summarizes all that has been published so far on this subject and points out the most important genes and molecular pathways that have been identified in both normal and diseased adrenal cortex.

Key Words: Adrenal cortex; microarrays; serial analysis of gene expression (SAGE); oligonucleotides; complementary DNA (cDNA); messenger RNA (mRNA); gene expression; adrenal hyperplasia

INTRODUCTION

Although intensively studied, the genetics of the development of adrenocortical tumors remains poorly characterized. Clonal composition analyses and comparative genomic hybridization (CGH) experiments have suggested that adrenocortical tumorigenesis is a multistep process that follows a sequence of genetic alterations that lead

From: *Contemporary Endocrinology: Genomics in Endocrinology:*
DNA Microarray Analysis in Endocrine Health and Disease
Edited by: S. Handwerger and B. Aronow © Humana Press, Totowa, NJ

to malignancy *(1–4)*. However, the rarity of adrenal cancer compared to the relative frequency of benign lesions suggests that either the accumulation of mutations most often does not lead to cancer, or that there are additional pathways to adrenal carcinogenesis. Just like CGH in the past allowed an instant and clear comparison of the DNA from tumor and normal tissues, the examination of the adrenal transcriptome allows for the detection of expressed genes in normal and abnormal cells and provides an even higher resolution image of the differences between physiology and pathology. Expression analysis does not replace DNA genomic approaches; it supplements other tools developed by the human genome project and should be used in conjunction with such techniques as CGH and traditional cytogenetics. For example, expression arrays, due to their higher resolution, can often lead to the identification of DNA alterations that would otherwise have been missed by contemporary molecular cytogenetics. At the end, an analysis of the transcriptome can only be meaningful when it is compared to genomic and proteomic studies. In this chapter, we have summarized the latest findings in adrenocortical whole-genome expression analysis; these relatively few studies are quite similar in their results and provide a solid basis for the next step, that of the proteomic analyses.

METHODS: APPROACHES IN THE STUDY OF ADRENAL GENE EXPRESSION

Currently, transcriptional profiling of any tissue is performed applying various types of microarrays and the alternate technology of generating libraries of expressed sequence tags (ESTs). An advanced expansion of EST libraries, especially in terms of high-throughput and transcript quantitation, is serial analysis of gene expression (SAGE). SAGE is based on generating, cloning, and sequencing concatenated short sequence tags, each representing a single transcript derived from mRNA from target tissue *(5)*. Analysis of the transcriptional data gained by the above methods is most commonly performed using clustering algorithms that group genes and samples on the basis of expression profiles, and statistical methods scoring the genes on the basis of their relevance to the clinical manifestations.

The method of choice in consideration for global expression profiling depends on several factors including technical, labor, price, time and effort involved, and, most importantly, the type of information that is sought. When comparing microarrays to EST libraries, an appreciable advantage of the latter is its inherent ability to identify transcripts without prior knowledge of the genes' coding sequences; hence it is important for cloning and sequencing novel transcripts and genes. On the other hand, recent technical advances in the development of expression arrays, their abundance and commercial availability, and the relative speed with which analysis can be done, are all factors that make arrays more useful in routine applications. In addition, array content can now be readily customized to cover from gene clusters and pathways of interest to the entire genome: some studies examine series of tissue specific transcripts and/or genes known to be involved in particular pathology; others directly use arrays covering the whole genome.

Another factor that needs to be considered prior to embarking on any high-throughput approach is whether individual or pooled samples will be investigated. Series of pooled samples reduce the price, the time spent, and the number of the experiments down to the most affordable. Investigating individual samples, however, is important for identifying unique expression ratios in a given type of tissue or even individual cells.

Two EST libraries and five different microarray studies on human adrenals have been published so far. In the first library, normal, non-diseased human adrenals were profiled *(6)*. For this study, RNA from two adult individuals was used. Totally 20 626 ESTs were assembled into 9175 gene clusters (3979, 3074, and 4116 clusters in hypothalamus, pituitary, and adrenal glands, respectively). Of these clusters, 30.3% corresponded to known genes, 44.8% to dbESTs, and 24.3% did not correspond to known coding sequence. One of the main outcomes of this study was cloning and sequencing of 200 previously unknown transcripts, 97 of which were adrenal-specific.

The second adrenal global profiling study compared two SAGE libraries, one from a normal adrenal gland and another from a form of bilateral adrenocortical hyperplasia—primary pigmented nodular adrenocortical disease (PPNAD), both from adolescent females *(7)*. 14 846 and 16 698 unique mRNAs from the normal adrenal tissue and PPNAD, respectively, were catalogued and their expression quantified. Compared to the previous study, this one found only 842 (~6% of the total) sequences that did not match any known expressed sequences and, thus corresponded to novel genes and transcripts. Interestingly, however, most unknown transcripts were also among the most abundant or most adrenal-specific transcripts, indicating, like the previous study, that there are quite a number of adrenal-specific genes with unknown function and with limited expression elsewhere. Our laboratory is now attempting to clone the most significant adrenal-specific sequences from this library.

The first of the microarray studies that was published addressed primarily normal adrenal tissue and compared fetal and adult adrenal expression *(8)*. Total RNA from 18 fetal (between 15 and 20 weeks of gestation) and 12 adult adrenal glands were studied in this analysis. The RNA samples were pooled and hybridized to five independent microarrays containing between 7075 and 9182 cDNA elements.

Our laboratory used the same approach to study another form of adrenal hyperplasia—corticotropin (ACTH)-independent, bilateral adrenocortical hyperplasia or massive macronodular adrenocortical disease (MMAD); these data were recently published by Bourdeau et al. *(9)*. This analysis was performed on eight tissues (three of them from patients with food-dependent Cushing syndrome) and compared them with the expression profile of pooled adrenal gland RNA that was obtained from 62 healthy individuals and is commercially available. The arrays utilized for this study contained approximately 10 000 human PCR-amplified cDNA clones and were made at the National Cancer Institute (Bethesda, MD, USA) core facility. As a criterion of differential expression, a cutoff of two along with alteration in at least 75% of the patients was applied. According to these stringent criteria, 82 and 31 genes were found to be consistently up- and downregulated, respectively.

Two other microarray studies compared adrenocortical adenomas and carcinomas *(10, 11)*. The first among them generated transcriptional profiles of eleven carcinomas, four adenomas, three normal adrenal cortices, and one macronodular hyperplasia using Affymetrix HG_U95Av2 oligonucleotide arrays that represented 10 500 unique genes. The second investigated a series of 230 candidate adrenal-specific genes, an "adrenochip," that included 187 cancer-related genes (including genes encoding cell cycle control proteins, growth factors, growth factor receptors, transcription factors, cell adhesion molecules, and proteins involved in cell invasion, angiogenesis, and chemoresistance that were previously shown to be expressed in the adrenal and were suspected to do so), 34 adrenal cortex-specific genes (including genes encoding hormone receptors, components of the cAMP signaling pathway, steroidogenic enzymes, and components of the *IGF*2 system), and nine control genes, most of them classic housekeeping sequences. This gene set was examined in a series of 57 well-characterized human sporadic adrenocortical tumors (33 adenomas and 24 carcinomas). Quite amazingly and despite differences in their design, these two studies revealed highly concordant results.

Finally, a recent study attempted to catalogue the genes expressed in aldosterone-producing adenomas and, more specifically, to determine the degree of alteration of G-protein-coupled receptor's expression, when compared to the normal adrenal cortex *(12)*. For this analysis, RNA from an adult adrenal and a pool of three RNA samples from aldosterone-producing adenomas were hybridized to an Affymetrix human HG-U133+2 oligonucleotide microarray set containing 54 675 probe sets representing approximately 40 500 independent human genes.

A requirement of all high-throughput approaches is confirmation of findings (expression level of a given gene/sequence) by other methods. A select group of genes are tested usually—these genes are picked from the series of sequences that were analyzed either because they were found to have significant changes or due to their particular interest with regard to their expression in adrenocortical tissue or their previously identified relationship to pathology or developmental stage. The confirmation process attempts to support the findings on three different levels: (a) reliability of the high-throughput experiment—for this purpose, the same samples examined by the EST libraries or microarrays are used; (b) trustfulness of the observations in general—to achieve that, larger number of samples are examined, assessment of which by high-throughput approaches is often unaffordable (price- or labor-wise); and (c) verification of the expression changes at the protein level.

All the above-mentioned studies used at least one conventional method, and occasionally two or even three, to confirm the high-throughput findings. Most utilized the recently evolved quantitative real-time reverse transcription PCR (Q-RT-PCR), although a classic Northern blot could also be used. The latter, unfortunately, requires several fold higher amounts of mRNA which is often hard for adrenocortical studies. For verification at the protein level, immunohistochemistry (IHC) and Western blot are the two most commonly chosen techniques. IHC is not quantitative but has the advantage of allowing for the observation of the exact localization of a signal within a cell (cytoplasmic versus nuclear) and the tissue (identifying histologically the tissue that is stained). Modern Western blot methods require smaller amount of protein lysate than older techniques, and have the advantage of offering high-resolution quantitation of expression without the use of radioactivity. Application of these confirmation approaches in adrenal high-throughput expression studies has shown a high degree of reliability for both microarrays and EST libraries. Recent reviews have pointed out the significance and reproducibility of these data *(13)*.

Genes Expressed in Normal Adrenals

The global expression profile of normal human adrenals reflects this steroidogenic tissue's functional characteristics *(6, 7, 14)*. Genes involved in fundamental biological processes, including protein synthesis and processing, energy metabolism, and cell structure, are prevalent. And accordingly, the most adrenal-specific gene cluster included in all studies deals with genes involved in steroidogenesis, such as steroidogenic acute regulator (*STAR*), 11-beta-hydroxylase (*CYP11B1*), 17-alpha- monooxygenase (hydroxylase) (*CYP17A1*), 21-monooxygenase (hydroxylase) (*CYP21A2*), 3 beta- and steroid delta-isomerase 2 (*HSD3B2*), and other enzymes. Developmental genes with known function in the adrenal cortex were also among the most adrenal-specific expressed transcripts, including the nephroblastoma overexpressed gene (*NOV*), chromogranin B (*CHGB*), and delta-like 1 homolog (*DLK1*).

The overall analysis of the existing studies revealed STAR as the most highly expressed, tissue-specific gene for normal human adrenal cortex. It is noteworthy that a number of adrenal ESTs represented rare transcripts of otherwise ubiquitously expressed genes. These included β2-microglobulin (*B2M*), plasminogen (*PLG*), myosin light polypeptide 6 (*MYL6*), and many more. For these genes, more than one transcript has been identified, and the adrenal preferential expression of this transcript did not correspond to the most common and ubiquitously expressed sequence. This finding is suggestive of an extensive presence of alternatively spliced transcripts of known genes in adrenal cells *(15)*.

In our SAGE library, a high proportion of the identified adrenal-specific ESTs did not correspond to known genes and thus represent potentially novel genes or alternative transcripts. To assess the degree of differentiation of the adrenocortical cells, criteria for tissue specificity previously employed for SAGE libraries by Velculescu et al. *(16)* were applied. In brief, the proportion of the transcripts present at 10 and more copies/cell and expressed at very low levels or absent in all other tissues was assessed. This percentage was 1.71% of the transcripts in the normal adrenal tissue library, ranking adrenal expression among the highest in specification for normal human tissue-specific gene expression (in the prostate, for example, it stands at 0.05%) and suggesting a high level of differentiation for adrenocortical cells.

The age-related differentiation and functional development of the adrenals have been studied by comparing the gene expression profiles of human fetal adrenal and adult adrenal gland. Sixty-nine transcripts were found to have a greater than 2.5-fold difference in expression between fetal and adult adrenals *(8)*. Interestingly, the vast majority of these transcripts have not been studied before with regard to adrenal function. The largest differences were observed for transcripts that encode insulin-like growth factor 2 (*IGF2*) (25-fold higher in fetal adrenal) and 3β-hydroxysteroid dehydrogenase (*HSD3B*) (24-fold higher in adult). The enhanced fetal expression of *IGF2* transcripts (second only to the liver) is well documented *(17)*. This finding probably reflects the ability of the adult adrenal to secret aldosterone, cortisol, and dehydroepiandrosterone (DHEA) sulfate, while the mid-gestational fetal adrenal secretes substantial amounts of DHEA sulfate but minimal amounts of cortisol or aldosterone. It is noteworthy that the vast majority of the differentially expressed transcripts have not been studied before with regard to adrenal function.

A group of *IGF2*-related transcripts that were expressed higher in the fetal adrenal were those related to growth and development. The fact that adrenal *IGF2* expression as well as the expression of other growth-related transcripts drops dramatically after birth has led to the hypothesis that *IGF2* could play an important role in the unique functions and/or hyperplasia of the fetal adrenal gland *(18)*. The size of the fetal adrenal (similar to the kidney at 18 weeks) has caused investigators to search for altered growth factor expression that might regulate its hyperplasia. In addition, growth factors have been studied as potential paracrine regulators of steroidogenic capacity of the fetal adrenal. In vitro studies have shown that *IGF2*, acting through the *IGF1* receptor, increases expression of the enzymes in steroid hormone biosynthesis *(19)*.

Another group of transcripts that were found highly expressed in the fetal adrenal was those involved in de novo cholesterol biosynthesis. Human fetal adrenal has been found to possess the greatest ability to produce cholesterol of all fetal tissues *(20)*. In addition, elevation in the expression of low density lipoprotein (LDL) receptor was observed. A number of studies support the idea that the fetal adrenal rapidly metabolizes any available circulating LDL to provide cholesterol for steroidogenesis *(21,22)*. Transcripts related to cellular immunity and signal transduction were preferentially expressed in the adult adrenal. In addition, this study outlined two potentially important adrenal function genes with differential expression between fetal and adult adrenal: (a) *NGFIB*—an orphan nuclear receptor that appears to be an important regulator of transcription of certain steroid-metabolizing enzymes *(23)* and (b) *KIAA0018*—one of the most highly expressed transcripts in the fetal adrenal gland. GenBank searches indicated that the only protein with structural similarity to KIAA0018 is the plant enzyme, DIMINUTO/DWARF1, which is involved in plant steroid biosynthesis and is critical for plant reproduction *(24)*. The role of these and others highly expressed in the normal adrenals transcripts identified by microarrays and other high-throughput techniques warrant further study.

Genes Expressed in Adrenal Hyperplasia

The two forms of ACTH-independent cortisol-producing bilateral adrenal hyperplasia that have been comprehensively profiled so far are PPNAD and MMAD. PPNAD is characterized by small to normal-sized adrenal glands containing multiple small cortical pigmented nodules *(25, 26)*. PPNAD may occur in an isolated form or associated with a multiple neoplasia syndrome, the complex of spotty skin pigmentation, myxomas, and endocrine overactivity, or Carney complex (CNC), in which Cushing's syndrome is the most common endocrine manifestation *(27)*. PPNAD, isolated or associated with CNC, is caused mostly by inactivating mutations of the *PRKAR1A* gene, which codes for regulatory subunit type I of protein kinase A (PKA), the main mediator of cAMP signaling in mammals *(28–30)*. The PPNAD SAGE library was generated from PPNAD from a patient carrying a germline inactivating mutation of the *PRKAR1A* gene and compared with that of control healthy adrenal gland from an age- and gender-matched individual.

MMAD is a rare condition in which cortisol secretion may be mediated by non-ACTH circulating hormones such as gastric inhibiting polypeptide (GIP) [leading to food-dependent Cushing's syndrome(S)], vasopressin, catecholamines, luteinizing hormone, serotonin, angiotensin-II, or leptin. A form of MMAD may develop during the first year of life with McCune–Albright syndrome, whereas the majority present as the isolated type in the fifth decade of life *(31,32)*. Although the most frequent clinical presentation of MMAD is CS, cases with excess gluco-mineralocorticoid, cortisol-estrogen, and -androgen secretion have also been reported *(33–35)*. Usually, computed tomography reveals clear enlargement of both adrenal glands. Histological examination is characterized by non-pigmented nodules composed of two types of cells, those with a clear cytoplasm (lipid-rich) that form cordon nest-like structures, and others with a compact cytoplasm (lipid-poor) that form small nest or island-like structures (36, 37).

Analysis of the global expression profiles in these two adrenal hyperplastic tissues showed significant differences but also some important similarities (7, 9). An interesting finding was the involvement of the Wnt signaling pathway in both types of hyperplasia. The expression levels of several members of the pathway were found significantly elevated; these included catenin (cadherin-associated protein)-like 1 (CTNNAL1), disheveled, dsh homolog 2 (Drosophila; DVL2), casein kinase 1 (CSNK1E), axin 1 (AXIN1), catenin-β1 (CTNNB1), WNT1-inducible signaling pathway protein 2 (WISP2), and glycogen synthase kinase-3β (GSK3B).

Wnt signaling transduction pathway regulates many important cellular and developmental processes, including proliferation, cell-to-cell adhesion, cell fate decisions, and differentiation. The involvement of this pathway in adrenocortical tumorigenic processes was recently supported by finding somatic activating mutations in β-catenin (CTNNB1) in as high as approximately 30% in both adrenal adenomas and carcinomas (38); all these mutations were observed in tumors with abnormal β-catenin accumulation.

Another finding in common for the two hyperplastic tissues was the increased degree of expression of PKA-related genes, specifically PRKAR2B, which is known to be involved in adrenocortical tumors (30). This observation was consistent with upregulation of the PKA signaling pathway in human tissues bearing PRKAR1A-inactivating mutations, as well as in Prkar1a-deficient mice (34–43).

Finally, another similarity between the two hyperplastic conditions was the increased presence of transcripts involved in cell cycle, adhesion, growth- and proliferation-regulating signal transduction pathways, suggesting an elevated potential for growth and proliferation for these lesions.

An important group of genes that were found inversely regulated in PPNAD and MMAD were those involved in steroidogenesis, and particularly, in cortisol production. In PPNAD, as expected from a high cortisol producing tissue, CYP11B1, CYP17A1, CYP21A2, and HSD3B2, were overexpressed. In contrast, the MMAD expression levels of CYP11B1, CYP17A1, and CYP21A2 were found downregulated, a fact that was confirmed by immunohistochemistry studies (44) and was proposed to reflect the relatively low efficiency with which MMAD cells produce cortisol compared to the volume of the tissue and the degree of their hypertrophy in EM studies (25).

Genes Expressed in Adrenal Adenomas and Carcinomas

Studies on individual genes and pathways that may be differentially expressed in adrenal tumors have repeatedly found several genes/gene clusters activated or repressed. One of the most commonly present alterations in the adrenal carcinomas was IGF2 overexpression. It has been identified in as many as 85% of the carcinomas in examination on selected markers for adrenal malignancy (45, 46). In a search for related and other genes involved in adrenal carcinogenesis, two different microarray studies examining a total of 75 adrenocortical tumors have been published. The first among them generated transcriptional profiles of eleven carcinomas, four adenomas, three normal adrenal cortices, and one macronodular hyperplasia using Affymetrix HG_U95Av2 oligonucleotide arrays representing 10 500 unique genes *(10)*. The other one covered a series of 230 adrenal-specific candidate genes ("adrenochip") that included 187 cancer-related genes (including genes encoding cell cycle control proteins, growth factors, growth factor receptors, transcription factors, cell adhesion molecules, and proteins involved in cell invasion, angiogenesis, and chemoresistance), 34 adrenal cortex-specific genes (including genes encoding hormone receptors, components of the cAMP signaling pathway, steroidogenic enzymes, and components of the *IGF*2 system), and nine control genes *(11)*. Using this gene set in a series of 57 well-characterized human sporadic adrenocortical tumors (33 adenomas and 24 carcinomas) not only confirmed the *IGF2* increased transcription levels in carcinomas but also identified the cosegregation of additional overexpressed and related genes, forming the so-called *IGF2* cluster. Interestingly, overexpression of this cluster, combined with underexpression of a second cluster (steroidogenesis cluster) appeared to be extremely powerful discriminator between adrenal adenomas and carcinomas.

The *IGF2* cluster contained eight genes that encode growth factors [*IGF2* and transforming growth factor beta (*TGFB2*)], growth factor receptors [fibroblast growth factor receptor type 1 (*FGFR1*), *FGFR4*, macrophage stimulating 1 receptor (*MST1R*), and *TGFBR1*), as well as *KCNQ1OT1* (also known as *LIT1*, i.e., long QT internal transcript 1] and glyceraldehyde-3-phosphate dehydrogenase (GAPD), a presupposed housekeeping gene. It is noteworthy that fibroblast growth factor receptors *FGFR1* and *FGFR4* are the two of the family of four most strongly expressed in the adrenal cortex tyrosine kinase FGF receptors *(47, 48)*. They bind fibroblast growth factors FGF-1 and FGF-2, which are known as the most powerful mitogens for adult steroidogenic adrenocortical cells *(49)* as well as for the human adrenocortical tumor cell line NCI-H295R *(50)*. They are also known to stimulate the proliferation of endothelial and mesenchymal cells. Thus, the overexpression of the above two *FGFRs* in adrenocortical cancers is likely to reflect their participation in the cellular increased proliferation and vascularization of this tissue during tumorigensis.

The second cluster (steroidogenesis cluster) contained 14 genes, six of them encoding proteins directly involved in the steroid biosynthetic pathway: *STAR*, *CYP11A*, *HSD3B1*, *CYP11B1*, *CYP21A2*, and *CYP17*. It also contained protein phosphatase 1A (*PPM1A*), *S100B* (S100 calcium-binding protein, β-chain), glypican 3 (*GPC3*), cAMP response element modulator (*CREM*), retinoblastoma 1 (*RB1*), and *TGFBR3*. It is noteworthy that the correlation between the level of expression of these different enzymes in each individual tumor and their steroid secretion profile was not trivial. Why the steroidogenesis cluster is underexpressed in adrenal malignant tumors remains to be clarified, but one hypothesis is that it reflects the overall decreased level of tissue-specific functional differentiation that is commonly associated with advanced stages of carcinogenesis.

This study also identified 14 genes, whose expression levels were able to strongly predict the probability of recurrence of the adrenal carcinomas: the group of recurring carcinomas could be separated from the group of non-recurrent tumors on the basis of overexpression of six genes, including *ISGF3G*, *IL2RG*, *GZMA*, *PTPN2*, *ITGB2*, and *ATF1*, and underexpression of eight genes, including *GAPD*, *ACTG1*, *TLN1*, *PRKCSH*, *VIL2*, *ECE1*, *CDKN2A*, and *FOS*. However, this latter group of carcinomas was not large enough to provide significant data; additional validation in a larger sample set and by other techniques is required.

The second microarray study on adrenal adenomas and carcinomas utilized the Affymetrix HG_U95Av2 chip with 10 500 unique human genes. One of the most important findings again was the elevated expression of *IGF2*: significantly increased expression of *IGF2* was observed in 10 of 11 (90.9%) adrenocortical carcinomas (compared to the mean of the normal cortex/adrenal adenomas cohort) for all three probe sets representing *IGF2* gene in the U95A expression chip. Only one carcinoma did not show increased *IGF2* expression and this was one of the two carcinomas identified by all the statistical means as outliers in that study.

A second important finding in agreement with the previous studies was the significant elevation of *FGFR1* in adrenocortical carcinomas as opposed to adenomas. *FGFR1*, which was represented four times on the array, showed 5.6-, 2.1-, and 2.8-fold increased expression in three of the probe sets (with statistical significance $P = 0.00009$, $P = 0.0003$, and $P = 0.003$, respectively). In contrast, no other receptor tyrosine kinases were found to be differentially expressed. These included *EGFR, ERB-B2, HER3, PDGFRA, PDGFRB, KIT, FGFR4, IGF1R, IGFR2, INSR , ESR1, ESR2, PGR*, and others.

The application of hierarchical clustering and principal component analysis on the three diagnostic groups (normal cortex, adrenocortical adenomas plus micronodular hyperplasia, and adrenocortical carcinomas) revealed, in addition to *IGF2* and *FGFR1*, 89 other genes that displayed at least threefold differential expression between adrenocortical carcinomas and the normal cortex, and the adrenocortical adenomas ($P < 0.01$). Included in these genes were, along with those already known to be upregulated in adrenocortical tumors such as *IGF2, FGFR1, TOP2A, Ki-67*, and also some novel and differentially expressed genes such as osteopontin (SPP), serine threonine kinase 15 (*STK15*), angiopoietin 2 (*ANGPT2*), *UBE2C*, and others. A particular gene is worth mentioning: the ectodermal-neural cortex-1 (*ENC1*) gene was overexpressed in carcinomas; interestingly, *ENC1* appears to be upregulated in colorectal carcinoma and was recently identified as a potential target of the β-catenin/T-cell factor complex *(51)*. Upregulation of *ENC1* transcripts in adrenocortical carcinomas is yet another link to the Wnt signaling pathway.

Finally, a very recent microarray study focused on the expression profiles of G protein-coupled receptors in aldosterone-producing adenomas *(12)*. Although the results indicated greater than 25-fold over normal expression for several genes, including *CD47* antigen receptor, the putative purinergic receptor *FKSG79*, and the cholinergic receptor *CHRM1*, major overexpression was observed for the luteinizing hormone receptor (*LHR*) and the aldosterone synthase genes (*CYP11B2*).

Further analysis of the *CYP11B2* and *LHR* expression levels by means of real-time quantitative PCR in 20 normal adrenals and 18 aldosterone-producing adenomas indicated that, as expected, all aldosterone-producing samples expressed levels of the *CYP11B2* gene that were outside the range seen in normal adrenals. In addition, nine of the 18 aldosterone-producing adenomas also exhibited elevated levels of *LHR*. To determine whether LH treatment is able to increase the expression of *CYP11B2*, transient transfections were performed in H295R adrenocortical cells with expression vectors for LH receptor and reporter constructs for the *CYP11B2* promoter. It was shown that LH is indeed able to mediate *CYP11B2* promoter activity in the presence of LH receptor with a maximal 25-fold induction, indicating that LH receptor levels in the adrenal can drive aldosterone production in the presence of LH. Why aberrant expression of the *LHR* would cause hypercortisolism in some cases and hyperaldosteronism in others is still unclear and further studies are necessary to answer this question.

CONCLUSIONS

To date, a total of 91 adrenal tissue samples have been profiled by various genome-wide expression studies, investigating thousands of genes and transcripts: one of the most significant findings, consistent with a large body of evidence, both from high-throughput and individual marker studies, is increased *IGF2* expression as a characteristic transcriptional event in the pathogenesis of adrenocortical carcinomas. The significance of this finding is twofold: not only it identifies *IGF2* as an important marker for adrenocortical cancer, but it also points to its uniqueness since there is really no other such major and consistent signal transduction-related alteration. This suggests that interruption of *IGF2*-induced signal transduction may lead to a significant therapeutic advance in the treatment of adrenocortical cancer. A second, and equally important discovery is the involvement of the Wnt-signaling pathway in both benign and malignant tumors of the adrenal cortex regardless of their molecular or histologic background, from hyperplasia to malignancy. Finally, the fact that a significant number of novel transcripts were also identified among the most abundant and tissue-specific ones, emphasizes the gaps in current knowledge about adrenocortical cell function and differentiation. We conclude that adrenal expression studies provided reproducible and important data in the search of better understanding of the process of this tissue's differentiation, normal function, and tumorigenesis. As this chapter went into press, three more relevant studies, including one on the Y1 mouse adrenal cells, were published, which largely supported the above data *(52–54)*.

REFERENCES

1. Beuschlein, F., Reincke, M., Karl, M., Travis, W. D., Jaursch-Hancke, C., Abdelhamid, S., Chrousos, G. P. & Allolio, B. (1994). Clonal composition of human adrenocortical neoplasms. *Cancer Res* 54, 4927–32.
2. Gicquel, C., Leblond-Francillard, M., Bertagna, X., Louvel, A., Chapuis, Y., Luton, J. P., Girard, F. & Le Bouc, Y. (1994). Clonal analysis of human adrenocortical carcinomas and secreting adenomas. *Clin Endocrinol (Oxf)* 40, 465–77.
3. Kjellman, M., Kallioniemi, O. P., Karhu, R., Hoog, A., Farncbo, L. O., Aucr, G., Larsson, C. & Backdahl, M. (1996). Genetic aberrations in adrenocortical tumors detected using comparative genomic hybridization correlate with tumor size and malignancy. *Cancer Res* 56, 4219–23.
4. Sidhu, S., Marsh, D. J., Theodosopoulos, G., Philips, J., Bambach, C. P., Campbell, P., Magarey, C. J., Russell, C. F., Schulte, K. M., Roher, H. D., Delbridge, L. & Robinson, B. G. (2002). Comparative genomic hybridization analysis of adrenocortical tumors. *J Clin Endocrinol Metab* 87, 3467–74.
5. Velculescu, V. E., Zhang, L., Vogelstein, B. & Kinzler, K. W. (1995). Serial analysis of gene expression. *Science* 270, 484–7.
6. Hu, R. M., Han, Z. G., Song, H. D., Peng, Y. D., Huang, Q. H., Ren, S. X., Gu, Y. J., Huang, C. H., Li, Y. B., Jiang, C. L., Fu, G., Zhang, Q. H., Gu, B. W., Dai, M., Mao, Y. F., Gao, G. F., Rong, R., Ye, M., Zhou, J., Xu, S. H., Gu, J., Shi, J. X., Jin, W. R., Zhang, C. K., Wu, T. M., Huang, G. Y., Chen, Z., Chen, M. D. & Chen, J. L. (2000). Gene expression profiling in the human hypothalamus-pituitary-adrenal axis and full-length cDNA cloning. *Proc Natl Acad Sci U S A* 97, 9543–8.

7. Horvath, A., Mathyakina, L., Vong, Q., Baxendale, V., Pang, A. L., Chan, W. Y. & Stratakis, C. A. (2006). Serial analysis of gene expression in adrenocortical hyperplasia caused by a germline PRKAR1A mutation. *J Clin Endocrinol Metab* 91, 584–96. Epub 2005 November 8.

8. Rainey, W. E., Carr, B. R., Wang, Z. N. & Parker, C. R. Jr. (2001). Gene profiling of human fetal and adult adrenals. *J Endocrinol* 171, 209–15.

9. Bourdeau, I., Antonini, S. R., Lacroix, A., Kirschner, L. S., Matyakhina, L., Lorang, D., Libutti, S. K. & Stratakis, C. A. (2004). Gene array analysis of macronodular adrenal hyperplasia confirms clinical heterogeneity and identifies several candidate genes as molecular mediators. *Oncogene* 23, 1575–85.

10. Giordano, T. J., Thomas, D. G., Kuick, R., Lizyness, M., Misek, D. E., Smith, A. L., Sanders, D., Aljundi, R. T., Gauger, P. G., Thompson, N. W., Taylor, J. M. & Hanash, S. M. (2003). Distinct transcriptional profiles of adrenocortical tumors uncovered by DNA microarray analysis. *Am J Pathol* 162, 521–31.

11. de Fraipont, F., El Atifi, M., Cherradi, N., Le Moigne, G., Defaye, G., Houlgatte, R., Bertherat, J., Bertagna, X., Plouin, P. F., Baudin, E., Berger, F., Gicquel, C., Chabre, O. & Feige, J. J. (2005). Gene expression profiling of human adrenocortical tumors using complementary deoxyribonucleic Acid microarrays identifies several candidate genes as markers of malignancy. *J Clin Endocrinol Metab* 90, 1819–29. Epub 2004 December 21.

12. Saner-Amigh, K., Mayhew, B. A., Mantero, F., Schiavi, F., White, P. C., Rao, C. V. & Rainey, W. E. (2006). Elevated expression of luteinizing hormone receptor in aldosterone-producing adenomas. *J Clin Endocrinol Metab* 91, 1136–42. Epub 2005 December 6.

13. Bornstein, S. R. & Hornsby, P. J. (2005). What can we learn from gene expression profiling for adrenal tumor management? *J Clin Endocrinol Metab* 90, 1900–2.

14. Feige, J. J., Vilgrain, I., Brand, C., Bailly, S. & Souchelnitskiy, S. (1998). Fine tuning of adrenocortical functions by locally produced growth factors. *J Endocrinol* 158, 7–19.

15. Dinel, S., Bolduc, C., Belleau, P., Boivin, A., Yoshioka, M., Calvo, E., Piedboeuf, B., Snyder, E. E., Labrie, F. & St-Amand, J. (2005). Reproducibility, bioinformatic analysis and power of the SAGE method to evaluate changes in transcriptome. *Nucleic Acids Res* 33, e26.

16. Velculescu, V. E., Madden, S. L., Zhang, L., Lash, A. E., Yu, J., Rago, C., Lal, A., Wang, C. J., Beaudry, G. A., Ciriello, K. M., Cook, B. P., Dufault, M. R., Ferguson, A. T., Gao, Y., He, T. C., Hermeking, H., Hiraldo, S. K., Hwang, P. M., Lopez, M. A., Luderer, H. F., Mathews, B., Petroziello, J. M., Polyak, K., Zawel, L., Zhang, W., Zhang, X., Zhou, W., Haluska, F. G., Jen, J., Sukumar, S., Gregory, L. M., Gregory, R. G., Vogelstein, B., & Kinzler, K. W. (1999). Analysis of human transcriptomes. *Nat Genet* 23, 387–8.

17. Han, V. K., Lund, P. K., Lee, D. C. & D'Ercole, A. J. (1988). Expression of somatomedin/insulin-like growth factor messenger ribonucleic acids in the human fetus: identification, characterization, and tissue distribution. *J Clin Endocrinol Metab* 66, 422–9.

18. Mesiano, S. & Jaffe, R. B. (1997). Role of growth factors in the developmental regulation of the human fetal adrenal cortex. *Steroids* 62, 62–72.

19. Mesiano, S., Katz, S. L., Lee, J. Y. & Jaffe, R. B. (1997). Insulin-like growth factors augment steroid production and expression of steroidogenic enzymes in human fetal adrenal cortical cells: implications for adrenal androgen regulation. *J Clin Endocrinol Metab* 82, 1390–6.

20. Carr, B. R. & Simpson, E. R. (1982). Cholesterol synthesis in human fetal tissues. *J Clin Endocrinol Metab* 55, 447–52.

21. Parker, C. R. Jr., Carr, B. R., Winkel, C. A., Casey, M. L., Simpson, E. R. & MacDonald, P. C. (1983). Hypercholesterolemia due to elevated low density lipoprotein-cholesterol in newborns with anencephaly and adrenal atrophy. *J Clin Endocrinol Metab* 57, 37–43.

22. Parker, C. R. Jr., MacDonald, P. C., Carr, B. R. & Morrison, J. C. (1987). The effects of dexamethasone and anencephaly on newborn serum levels of apolipoprotein A-1. *J Clin Endocrinol Metab* 65, 1098–101.

23. Wilson, T. E., Mouw, A. R., Weaver, C. A., Milbrandt, J. & Parker, K. L. (1993). The orphan nuclear receptor NGFI-B regulates expression of the gene encoding steroid 21-hydroxylase. *Mol Cell Biol* 13, 861–8.

24. Klahre, U., Noguchi, T., Fujioka, S., Takatsuto, S., Yokota, T., Nomura, T., Yoshida, S. & Chua, N. H. (1998). The Arabidopsis DIMINUTO/DWARF1 gene encodes a protein involved in steroid synthesis. *Plant Cell* 10, 1677–90.

25. Stratakis, C. A. & Kirschner, L. S. (1998). Clinical and genetic analysis of primary bilateral adrenal diseases (micro- and macronodular disease) leading to Cushing syndrome. *Horm Metab Res* 30, 456–63.

26. Young, W. F. Jr., Carney, J. A., Musa, B. U., Wulffraat, N. M., Lens, J. W. & Drexhage, H. A. (1989). Familial Cushing's syndrome due to primary pigmented nodular adrenocortical disease. Reinvestigation 50 years later. *N Engl J Med* 321, 1659–64.

27. Stratakis, C. A., Kirschner, L. S. & Carney, J. A. (2001). Clinical and molecular features of the Carney complex: diagnostic criteria and recommendations for patient evaluation. *J Clin Endocrinol Metab* 86, 4041–6.

28. Kirschner, L. S., Sandrini, F., Monbo, J., Lin, J. P., Carney, J. A. & Stratakis, C. A. (2000). Genetic heterogeneity and spectrum of mutations of the PRKAR1A gene in patients with the Carney complex. *Hum Mol Genet* 9, 3037–46.

29. Groussin, L., Jullian, E., Perlemoine, K., Louvel, A., Leheup, B., Luton, J. P., Bertagna, X. & Bertherat, J. (2002). Mutations of the PRKAR1A gene in Cushing's syndrome due to sporadic primary pigmented nodular adrenocortical disease. *J Clin Endocrinol Metab* 87, 4324–9.

30. Bossis, I. & Stratakis, C. A. (2004). Minireview: PRKAR1A: normal and abnormal functions. *Endocrinology* 145, 5452–8. Epub 2004 August 26.

31. Kirk, J. M., Brain, C. E., Carson, D. J., Hyde, J. C. & Grant, D. B. (1999). Cushing's syndrome caused by nodular adrenal hyperplasia in children with McCune–Albright syndrome. *J Pediatr* 134, 789–92.

32. Lieberman, S. A., Eccleshall, T. R. & Feldman, D. (1994). ACTH-independent massive bilateral adrenal disease (AIMBAD): a subtype of Cushing's syndrome with major diagnostic and therapeutic implications. *Eur J Endocrinol* 131, 67–73.

33. Hayashi, Y., Takeda, Y., Kaneko, K., Koyama, H., Aiba, M., Ikeda, U. & Shimada, K. (1998). A case of Cushing's syndrome due to ACTH-independent bilateral macronodular hyperplasia associated with excessive secretion of mineralocorticoids. *Endocr J* 45, 485–91.

34. Goodarzi, M. O., Dawson, D. W., Li, X., Lei, Z., Shintaku, P., Rao, C. V. & Van Herle, A. J. (2003). Virilization in bilateral macronodular adrenal hyperplasia controlled by luteinizing hormone. *J Clin Endocrinol Metab* 88, 73–7.

35. Malchoff, C. D., Rosa, J., DeBold, C. R., Kozol, R. A., Ramsby, G. R., Page, D. L., Malchoff, D. M. & Orth, D. N. (1989). Adrenocorticotropin-independent bilateral macronodular adrenal hyperplasia: an unusual cause of Cushing's syndrome. *J Clin Endocrinol Metab* 68, 855–60.

36. Aiba, M., Hirayama, A., Iri, H., Ito, Y., Fujimoto, Y., Mabuchi, G., Murai, M., Tazaki, H., Maruyama, H., Saruta, T. (1991). Adrenocorticotropic hormone-independent bilateral adrenocortical macronodular hyperplasia as a distinct subtype of Cushing's syndrome. Enzyme histochemical and ultrastructural study of four cases with a review of the literature. *Am J Clin Pathol* 96, 334–40.

37. Sasano, H., Suzuki, T. & Nagura, H. (1994). ACTH-independent macronodular adrenocortical hyperplasia: immunohistochemical and in situ hybridization studies of steroidogenic enzymes. *Mod Pathol* 7, 215–9.

38. Tissier, F., Cavard, C., Groussin, L., Perlemoine, K., Fumey, G., Hagnere, A. M., Rene-Corail, F., Jullian, E., Gicquel, C., Bertagna, X., Vacher-Lavenu, M. C., Perret, C. & Bertherat, J. (2005). Mutations of beta-catenin in adrenocortical tumors: activation of the Wnt signaling pathway is a frequent event in both benign and malignant adrenocortical tumors. *Cancer Res* 65, 7622–7.

39. Kirschner, L. S., Carney, J. A., Pack, S. D., Taymans, S. E., Giatzakis, C., Cho, Y. S., Cho-Chung, Y. S. & Stratakis, C. A. (2000). Mutations of the gene encoding the protein kinase A type I-alpha regulatory subunit in patients with the Carney complex. *Nat Genet* 26, 89–92.

40. Stratakis, C. A. (2003). Genetics of adrenocortical tumors: gatekeepers, landscapers and conductors in symphony. *Trends Endocrinol Metab* 14, 404–10.

41. Kirschner, L. S., Kusewitt, D. F., Matyakhina, L., Towns, W. H. II, Carney, J. A., Westphal, H. & Stratakis, C. A. (2005). A mouse model for the Carney complex tumor syndrome develops neoplasia in cyclic AMP-responsive tissues. *Cancer Res* 65, 4506–14.

42. Griffin, K. J., Kirschner, L. S., Matyakhina, L., Stergiopoulos, S. G., Robinson-White, A., Lenherr, S. M., Weinberg, F. D., Claflin, E. S., Batista, D., Bourdeau, I., Voutetakis, A., Sandrini, F., Meoli, E. M., Bauer, A. J., Cho-Chung, Y. S., Bornstein, S. R., Carney, J. A. & Stratakis, C. A. (2004). A transgenic mouse bearing an antisense construct of regulatory subunit type 1A of protein kinase A develops endocrine and other tumours: comparison with Carney complex and other PRKAR1A induced lesions. *J Med Genet* 41, 923–31.

43. Griffin, K. J., Kirschner, L. S., Matyakhina, L., Stergiopoulos, S., Robinson-White, A., Lenherr, S., Weinberg, F. D., Claflin, E., Meoli, E., Cho-Chung, Y. S. & Stratakis, C. A. (2004). Down-regulation of regulatory subunit type 1A of protein kinase A leads to endocrine and other tumors. *Cancer Res* 64, 8811–5.

44. Bourdeau, I. & Stratakis, C. A. (2002). Cyclic AMP-dependent signaling aberrations in macronodular adrenal disease. *Ann N Y Acad Sci* 968, 240–55.

45. Gicquel, C., Bertagna, X., Gaston, V., Coste, J., Louvel, A., Baudin, E., Bertherat, J., Chapuis, Y., Duclos, J. M., Schlumberger, M., Plouin, P. F., Luton, J. P. & Le Bouc, Y. (2001). Molecular markers and long-term recurrences in a large cohort of patients with sporadic adrenocortical tumors. *Cancer Res* 61, 6762–7.

46. Gicquel, C., Raffin-Sanson, M. L., Gaston, V., Bertagna, X., Plouin, P. F., Schlumberger, M., Louvel, A., Luton, J. P. & Le Bouc, Y. (1997). Structural and functional abnormalities at 11p15 are associated with the malignant phenotype in sporadic adrenocortical tumors: study on a series of 82 tumors. *J Clin Endocrinol Metab* 82, 2559–65.

47. Hughes, S. E. (1997). Differential expression of the fibroblast growth factor receptor (FGFR) multigene family in normal human adult tissues. *J Histochem Cytochem* 45, 1005–19.

48. Partanen, J., Makela, T. P., Eerola, E., Korhonen, J., Hirvonen, H., Claesson-Welsh, L. & Alitalo, K. (1991). FGFR-4, a novel acidic fibroblast growth factor receptor with a distinct expression pattern. *EMBO J* 10, 1347–54.

49. Feige, J. J. & Baird, A. (1991). Growth factor regulation of adrenal cortex growth and function. *Prog Growth Factor Res* 3, 103–13.

50. Boulle, N., Gicquel, C., Logie, A., Christol, R., Feige, J. J. & Le Bouc, Y. (2000). Fibroblast growth factor-2 inhibits the maturation of pro-insulin-like growth factor-II (Pro-IGF2) and the expression of insulin-like growth factor binding protein-2 (IGFBP-2) in the human adrenocortical tumor cell line NCI-H295R. *Endocrinology* 141, 3127–36.

51. Fujita, M., Furukawa, Y., Tsunoda, T., Tanaka, T., Ogawa, M. & Nakamura, Y. (2001). Up-regulation of the ectodermal-neural cortex 1 (ENC1) gene, a downstream target of the beta-catenin/T-cell factor complex, in colorectal carcinomas. *Cancer Res* 61, 7722–6.

52. Slater, E. P., Diehl, S. M., Langer, P., Samans, B., Ramaswamy, A., Zielke, A. & Bartsch, D.K. (2006). Analysis by cDNA microarrays of gene expression patterns of human adrenocortical tumors. *Eur J Endocrinol* 154, 587–98.

53. Schimmer, B. P., Cordova, M., Cheng, H., Tsao, A., Goryachev, A. B., Schimmer, A. D. & Morris, Q. (2006). Global profiles of gene expression induced by adrenocorticotropin in Y1 mouse adrenal cells. *Endocrinology* 147, 2357–67.

54. Lampron, A., Bourdeau, I., Hamet, P., Tremblay, J. & Lacroix, A (2006) Whole genome expression profiling of glucose-dependent insulinotropic peptide (GIP)- and adrenocorticotropin-dependent adrenal hyperplasias reveals novel targets for the study of GIP-dependent Cushing's syndrome. *J Clin Endocrinol Metab* 91, 3611–8.

III   Endocrine Producing Tissues

8   DNA Microarray Analysis of Human Uterine Decidualization

Ori Eyal, MD and Stuart Handwerger, MD

CONTENTS

INTRODUCTION

BIOLOGY OF DECIDUALIZATION

IN VITRO MODELS OF DECIDUALIZATION

EARLY STUDIES OF DECIDUALIZATION

MICROARRAY STUDIES OF DECIDUALIZATION

RECENT STUDIES BASED ON MICROARRAY FINDINGS

DECIDUALIZATION IN PATHOLOGICAL CONDITIONS OF PREGNANCY

SUMMARY

REFERENCES

INTRODUCTION

During the late luteal phase of the menstrual cycle, human endometrial stromal cells, under the influence of progesterone and estradiol, differentiate to decidual cells by a process known as decidualization. The decidual cells form the maternal component of the maternal-fetal interface and are critical for successful implantation and the maintenance of pregnancy. Until the availability of microarray technology, studies on the biology of human decidualization were performed by immunohistochemistry, in situ hybridization, and selective mRNA and protein analyses. Because of the limitations imposed by these techniques, relatively little was known about the genetic program and molecular mechanisms that direct the decidualization process. However, investigations of human decidualization over the past five years or so using DNA microarray analyses have provided important new insights into the genetic program of decidualization. These studies provide considerable information about genes that are induced and repressed during human decidualization, the sequence of changes in gene expression during decidualization, the signal pathways that are involved in decidualization and changes in gene expression resulting by different factors, and diseases that are known to affect decidualization. This chapter summarizes new information from our laboratory about gene expression during in vitro decidualization of human uterine stromal cells and the transcription factors and other factors involved in the regulation of decidualization.

From: *Contemporary Endocrinology: Genomics in Endocrinology:*
DNA Microarray Analysis in Endocrine Health and Disease
Edited by S. Handwerger and B. Aronow © Humana Press, Totowa, NJ

147

BIOLOGY OF DECIDUALIZATION

The endometrium is a dynamic tissue that undergoes cyclic changes in response to the ovarian sex steroids, estradiol and progesterone as it prepares for possible implantation. The first half of the menstrual cycle, the proliferative phase, is primarily under the influence of estradiol. During this phase, the endometrial cells undergo proliferation that results in development of blood vessels and a glandular network and the proliferation of stromal cells. The second half of the menstrual cycle, the secretory or luteal phase, is primarily under the influence of progesterone and estradiol. Decidualization begins during the late luteal phase, occurring initially in stromal cells adjacent to vascular structures. The stromal cells transform from an elongated fibroblast-like phenotype to the larger, round phenotype of decidual cells. The decidual cells acquire the characteristics of a secretory cell, such as a euchromatic nucleus, numerous Golgi cisternae, dilated rough endoplasmic reticulum, and dense membrane bound secretory granules. Following decidualization, the decidual cells express numerous new cellular products that provide a nutritive and hormonal environment for the developing embryo. These include many proteins that are not expressed by the cells prior to decidualization, such as prolactin and insulin-like growth factor binding protein-1 (IGFBP-1). However, the sequence and molecular events associated with the transformation of stromal cells to secretory decidual cells are incompletely understood.

IN VITRO MODELS OF DECIDUALIZATION

Several in vitro models have been used to study human decidualization, including primary cultures of endometrial stromal cells *(1, 2)*, primary cultures of human uterine fibroblast cells, and cultures of the N5 *(3)* and St-2 *(4)* immortalized human endometrial stromal cell lines. Treatment of endometrial stromal cells with progesterone in combination with estradiol or relaxin or with high levels of cAMP in the absence of exogenous hormones can induce decidualization. Human uterine fibroblast cells undergo decidualization in response to treatment with medroxyprogesterone acetate (MPA) and estradiol in combination with PGE_2 or cAMP, while MPA and estradiol alone have little or no effect on decidualization. As discussed below, most studies from our laboratory of the genetic program that directs human uterine decidualization have utilized uterine fibroblast cells. Treatment of N5 cells with progesterone and estradiol alone or in combination with prostaglandin E2 stimulates the synthesis and release of prolactin *(3)*. Like uterine decidual cells, St-2 cells decidualize in response to stimulation with MPA plus 8-bromo-cAMP, but treatment with progesterone, alone or in combination with estradiol, is without effect *(4)*. Taken together, studies with these model systems have shown that cAMP is a key factor for induction of decidualization *(5)*. The increase in cAMP levels is dependent upon activation of the protein kinase A pathway *(5, 6)*. Sustained elevation of cytoplasmic cAMP levels during decidualization can be induced by factors secreted by the endometrium (CRF, relaxin, PGE_2), ovary (relaxin), and pituitary (gonadotropins) *(5, 7, 8)*.

EARLY STUDIES OF DECIDUALIZATION

Early studies identified several decidualization-specific genes that are induced endometrial stromal cells as the cells undergo decidualization. These include prolactin *(6, 9–11)*, IGFBP-1 *(12)*, laminin, TIMP3, M-CSF, renin, angiotensinogen, and leukemia inhibitory factor (LIF). However, the biological roles for most of these proteins are poorly understood. Since decidual cells express both prolactin and the prolactin receptor and M-CSF and its receptor, it has been suggested that the relatively large amounts of prolactin and M-CSF released by decidual cells may have autocrine/paracrine effects on the local proliferation and differentiation of cells at the maternal-fetal interface. The roles of IGFBP-1 as an IGF-binding protein and as a trophoblast integrin ligand suggest that the binding protein may have multiple roles in endometrial development and in interactions between decidua and the invading trophoblast *(13)*. Laminin selectively decreases the expression of prolactin and IGFBP-1 during in vitro decidualization of human endometrial stromal cells *(14)* and may act to limit the extent of decidualization. The role of the local renin-angiotensin system in the decidua is unknown, but may have a role in the pathogenesis of pre-eclampsia *(15)*. LIF stimulates proliferation of endometrial stromal fibroblasts in vitro; and there are suggestions that decreased LIF expression may contribute to human infertility *(16)*. The observation that TIMP-3 mRNA is upregulated by progesterone both in vitro and in vivo suggests that TIMP-3 plays a critical role in the control of trophoblastic invasion, probably by limiting ECM degradation.

MICROARRAY STUDIES OF DECIDUALIZATION

Studies from our laboratory *(17)* utilized sequential DNA microarray analyses of human uterine fibroblasts decidualized in vitro to identify genes that are dynamically regulated during decidualization and their expression pattern. Decidual fibroblast cells used in the microarray studies responded to estradiol, progesterone and cAMP exposure with a progressive change in morphology and induction of prolactin and IGFBP-1 gene expression. Regulated genes were selected based on exhibition of a twofold increase or 50% decrease in their expression relative to time 0 pre-decidualization reference samples. The findings suggest that decidualization is characterized by "categorical reprogramming" in which there is simultaneous upregulation and downregulation of a set of genes with related function (Table 1). Of the 6918 genes on the microarray, 281exhibited twofold or greater induction or 50% or more repression. The 281 dynamically regulated genes were separated into nine different K-means cluster groups composed of induced, biphasic, and repressed regulatory behaviors. Of these, 127 genes were induced, and 154 were repressed.

Microarray analysis identified genes that were known to be induced by decidualization as well as genes that were not known previously to be induced by decidualization. The most induced annotated genes detected included IGFBP-1, Lefty2, somatostatin, FOX1A, decorin, spermidine/spermine N1-acetyltransferase, EGF-containing fibulin-like extracellular matrix protein 1, IGFBP-4, TIMP3, solute carrier family 16, member 6, and IGFBP-3. The annotated genes whose mRNAs exhibited the largest repression include cysteine-rich angiogenic inducer 61, collagen type I, α1, transgelin, regulator of G protein signaling 4, fibronectin 1, hexabrachion,

Table 1

Categorical reprogramming scheme for decidual fibroblast differentiation.

Category	Early	Middle	Late
Cell regulation	forkhead box O1A inhibitor of DNA binding 3	twist (Drosophila) homolog mitogen-activated protein kinase kinase kinase 5 **high-mobility group protein isoform I-C**	cannabinoid RECEPTOR 1 (brain) Phosphatidic acid phosphatase type 2b **centromere protein F**
Cell and tissue function	IGFBP-1 IGFBP-3 spermidine/spermine N1-acetyltransferase **cysteine-rich, angiogenic inducer, 61**	IGFBP-4 endometrial bleeding associated factor solute carrier family 16, member 6 **IGFBP-5**	**regulator of G-protein signalling 4** complement component 3 monoamine oxidase A prolactin receptor superoxide dismutase 2, mitochondrial **interferon-induced protein 54** **familial intrahepatic cholestasis 1** **sodium channel, voltage-gated, type IX, alpha polypeptide** **sodium channel, voltage-gated, type IX, alpha polypeptide**

Cell and tissue structure		
EGF-containing fibulin-like extracell matrix protein 1	matrix metalloproteinase 11	CD24 antigen
decorin	matrix metalloproteinase 2	
fibromodulin		
cathepsin K		
tissue inhibitor of metalloproteinase 3		
tropomyosin 1 (alpha)	**hexabrachion (tenascin C, cytotactin)**	**integrin, alpha 4 (antigen CD49D)**
tropomyosin 2 (beta)	**EGF-like repeats and discoidin I-likedomains 3**	**collagen, type I, α1**
transgelin	**fibronectin 1**	**laminin, α4**
myosin, light polypeptide kinase	**keratin 18**	
	neuropilin 1	

A dominant feature of decidual fibroblast differentiation program is the loss of mRNAs for numerous integrins, keratins, and collagens mRNAs simultaneous with the induction of matrix metalloproteinases 2 and 11 and cathepsin K. In addition, there is induction of decorin, fibromodulin, and EGF/Fibulin-like extracellular proteinase. This suggests that major extracellular remodeling accompanies decidualization (17). Repressed genes are shown in bold text.

and IGFBP-5. Table 2 shows a complete list of the 100 most regulated genes, along with the accession numbers and fold change from day 0 level of expression.

The 281 dynamically regulated genes were divided into similar functional categories. These include cell regulation (77 genes), cell and tissue function (74 genes), cell and tissue structure (79 genes), unknown function (6 genes), and expressed sequence tags (45 genes). Table 1 shows the most dynamically regulated induced and repressed genes (fourfold changes) in the categories of cell regulation, cell and tissue function, and cell and tissue structure at different stages of differentiation. In each gene category, several members were strongly induced while several others were strongly downregulated. Members of the integrin, collagen, and laminin gene families varied widely in their regulation during decidualization. The microarray studies also indicate striking changes in extracellular matrix protein genes during decidualization of decidual fibroblasts. There was early induction of the extracellular matrix proteins decorin and fibromodulin and early repression of collagen type X α1, collagen type XI α1, collagen type I α1, collagen type XVI α1, collagen type IV α1, laminin-α4, integrin-α4, vimentin, glycan-1, and fibronectin-1. There was also induction of the proteolytic enzymes matrix metalloprotienases-2 and -11 and TIMP3 and repression of plasminogen activator inhibitor type I. Taken together, these findings indicate that striking reprogramming of genes that determine extracellular matrix and tissue strength and integrity occurs during decidualization and suggest that these changes are a prominent feature of decidualization. Dramatic changes were also observed in many members of the IGFBP family. IGFBP-1, which is known to be a marker of decidualization, was the most induced gene in decidual fibroblasts undergoing differentiation, whereas IGFBP-5 was the most repressed gene. IGFBP-5 binds IGF-I and enhance its effects on target cells *(18)*. IGFBP-5 also exerts IGF-independent effects that may be mediated via a distinct receptor and signal transduction pathway *(19)*. IGFBP-5 also has a nuclear targeting sequence, is translocated into the nucleus of actively dividing cells, and may have effects on cell growth *(20)*. IGFBP-5 is also thought to induce apoptosis *(21)*.

Microarray analyses have also provided new information about the signal pathways that are involved in decidualization. Early in decidualization there is an induction of genes involved in the mitogen-activated protein (MAP) kinase signaling pathway as well as the TGF-β1 family member genes Lefty2 and inhibin-β, and later in decidualization there is an induction of several other genes involved in the phospho-inositide signal transduction pathway. A major increase was noted also in c-myc, other oncogenes, and other factors involved in cellular mitogenic response, including members of the cyclin and the retinoic acid receptor families.

Chen and co-worker *(22)* utilized microarray analysis to study early gestational decidua and chorionic villi. They identified 641 genes that were highly expressed in both decidua and villi, 49 genes with higher expression in decidua, and 75 genes with higher expression in chorionic villi. Many of the genes had not been reported previously during pregnancy; these include myeloid leukemia factor 2, lymphotoxin-β receptor, integrin-linked kinase, disintegrin and metalloproteinase domain 12, GTPase-activating protein ras p21, and some cell surface antigens. IGFBP-1, IGFBP-3, IGFBP-4 were found to be highly expressed in decidua both in vitro and in vivo. In contrast to Brar et al., this study showed that IGFBP-5 is also highly expressed in the decidua. IGFBPs bind IGF-I and IGF-II with high affinity and can either enhance or attenuate IGFs effects on target cells *(23)*. IGFBPs can also exert IGF-independent effects *(24)*. The insulin receptor was also found

Table 2

The 100 most strongly up- or down-regulated genes during decidual fibroblast differentiation (17)

Induced Acc. No.	Gene	Functional Group	Ratio	Repressed Acc. No.	Gene	Functional Group	Ratio.
NM_000596	IGFBP-1	polypeptide hormone	131.38	M92642	collagen, type XVI, α1	cytoskeleton organization	0.3
AF081511	endometrial bleeding associated factor	polypeptide hormone	44.39	X54232	glypican 1	extracellular organization	0.3
AW583834	somatostatin	polypeptide hormone	19.8	AF006043	3-phosphogly-cerate dehydrogenase	metabolism	0.3
AF032885	forkhead box O1A	transcription	16.02	NM_002432	myeloid cell nuclear differentiation antigen	transcription	0.3
NM_001920	decorin	extracellular organization	14.93	X98568	collagen, type X, α1	cytoskeleton organization	0.29
AA856763	ESTs	EST	11.59	AA451928	vimentin	cytoskeleton organization	0.29
Z14136	spermidine/spermine N1-acetyltransferase	metabolism	10.44	H63163	kinesin-like 5	cytoskeleton organization	0.29
NM_004105	EGF-containing fibulin-like extracellular matrix protein 1	cell adhesion	10.06	D14134	RAD51	DNA damage	0.29
M62403	IGFBP-4	polypeptide hormone	10.03	M14083	plasminogen activator inhibitor, type I	extracellular proteolysis	0.29
AI245471	tissue inhibitor of metalloproteinase 3	proteolysis	9.62	W49820	hyaluronan synthase 2	metabolism	0.29
U79745	solute carrier family 16, member 6	transporters	8.39	AI923532	proteasome subunit, beta type, 9	proteolysis	0.29

(Continued)

Table 2
(*Continued*)

Induced Acc. No.	Gene	Functional Group	Ratio	Repressed Acc. No.	Gene	Functional Group	Ratio.
AW051824	ESTs	EST	8.38	W56891	guanine nucleotide exchange factor for Rap1	signal transduction	0.29
NM_006472	upregulated by 1,25-dihydroxyvitamin D-3	EST	8.35	AI521645	integrin, α2 (CD49B)	cell adhesion	0.28
M31159	IGFBP-3	polypeptide hormone	8.23	NM_001854	collagen, type XI, α1	cytoskeleton organization	0.28
AL023282	tissue inhibitor of metalloproteinase 3	proteolysis	8.1	M29696	interleukin 7 receptor	inflammatory mediator	0.28
AF007144	deiodinase, iodothyronine, type II	metabolism	8.09	NM_001809	centromere protein A	cell cycle	0.27
Y00472	superoxide dismutase 2, mitochondrial	stress response	7.95	M26576	collagen, type IV, α1	cytoskeleton organization	0.27
NM_000240	monoamine oxidase A	metabolism	6.56	D79997	KIAA0175 gene product	EST	0.27
X54937	cannabinoid receptor 1 (brain)	signal transduction	6.46	NM_003328	TXK tyrosine kinase	signal transduction	0.27
X91662	twist (Drosophila) homolog	transcription	5.95	NM_006094	deleted in liver cancer 1	EST	0.26
NM_002023	fibromodulin	extracellular organization	5.9	X59892	tryptophanyl-tRNA synthetase	translation	0.26
AI745625	CD24 antigen	cell adhesion	5.86	AI436090	ATPase, Ca++ transporting, plasma membrane 1	transporter	0.26
K02765	complement component 3	infammatory mediator	5.55	Z24727	tropomyosin 1 (alpha)	cytoskeleton organization	0.25

NM_001831	clusterin	apoptosis	5.31	M75165	tropomyosin 2 (beta)	cytoskeleton organization	0.25
X82153	cathepsin K	proteolysis	5.04	AB023155	KIAA0938 protein	EST	0.25
NM_000949	prolactin receptor	polypeptide hormone	4.85	NM_005711	EGF-like repeats and discoidin I-like domains 3	cell adhesion	0.24
NM_005923	mitogen-activated protein kinase kinase 5	signal transduction	4.79	AL047603	actinin, α4	cytoskeleton organization	0.24
NM_012342	putative transmembrane protein	membrane receptor	4.65	NM_005196	centromere protein F	cell cycle	0.21
AW156890	synuclein, α	apoptosis	4.61	AF069601	myosin, light polypeptide kinase	cytoskeleton organization	0.21
NM_005940	matrix metalloproteinase 11	extracellular proteolysis	4.43	AI042261	ESTs	EST	0.21
AB000889	Phosphatidic acid phosphatase type 2b	signal transduction	4.43	X76939	laminin, α4	extracellular organization	0.21
M55593	matrix metalloproteinase 2	extracellular proteolysis	4.3	U28749	high-mobility group protein isoform I-C	transcription	0.21
NM_000887	integrin, αX	cell adhesion	3.99	U74612	forkhead box M1	transcription	0.21
AL021154	inhibitor of DNA binding 3, dominant negative helix-loop-helix protein	transcription	3.96	X16983	integrin, alpha 4 (antigen CD49D)	cell adhesion	0.19
NM_003652	carboxypeptidase Z	metabolism	3.88	AF016050	neuropilin 1	membrane receptor	0.19
M77016	tropomodulin	cytoskeleton organization	3.73	M14660	interferon-induced protein 54	inflammatory mediator	0.18
AA505676	ESTs	EST	3.73	X82835	sodium channel, voltage-gated, type IX, alpha polypeptide	transporters	0.18

(Continued)

Table 3
(*Continued*)

Induced Acc. No.	Gene	Functional Group	Ratio	Repressed Acc. No.	Gene	Functional Group	Ratio.
AW157684	cullin 4B	cell cycle	3.72	X12881	keratin 18	keratin	0.17
NM_006486	fibulin 1	extracellular organization	3.71	X82835	sodium channel, voltage-gated, type IX, alpha polypeptide	transporters	0.17
S77035	IGF-2	polypeptide hormone	3.68	D13633	KIAA0008 gene product	EST	0.15
D87258	protease, serine, 11 (IGF binding)	metabolism	3.62	AF038007	familial intrahepatic cholestasis 1	metabolism	0.15
NM_004844	SH3-domain binding protein 5	signal transduction	3.61	Y12084	cysteine-rich, angiogenic inducer, 61	polypeptide hormone	0.15
NM_002662	phospholipase D1, phophatidylcholine-specific	signal transduction	3.61	AF013711	transgelin	cell adhesion	0.14
X85750	monocyte to macrophage differentiation-associated	membrane receptor	3.6	AW577407	collagen, type I, $\alpha 1$	cytoskeleton organization	0.14
AI089791	cyclin-dependent kinase inhibitor 1C (p57, Kip2)	cell cycle	3.59	AA587912	EST, Highly sim. to phospho-serine aminotransfer	EST	0.14
M13436	inhibin, βA (activin A, activin AB alpha polypeptide)	polypeptide hormone	3.35	X02761	fibronectin 1	extracellular organization	0.07
AF052110	decay accelerating factor for complement (CD55)	proteolysis	3.24	L27560	IGFBP-5	polypeptide hormone	0.05

to be highly expressed in decidual cells, both in vitro and in vivo. These findings suggest that IGFBPs and insulin receptor may modulate trophoblast invasiveness and hormone secretion during early pregnancy. TGF-β, interferon-α, TNF-α, TNF-R, angiotensin receptor, PDGF-R and several other hormone or cytokine-related factors (PDGF-Rβ, IL-1 receptor, INF-α induced protein, and neuropeptide Y receptor-1) were also found to be highly expressed in decidua. The genes that were found to be highly expressed in decidua relate to cell growth and cell cycle, apoptosis, hormones and cytokines, stress response, signal transduction, cell surface antigens/cell adhesion, metabolism, transcription factors, cytoskeleton/extracellular matrix, and housekeeping functions.

Microarray studies of in vitro decidualization of endometrial stromal cells obtained from two different subjects in the late proliferative phase by Popovici et al. *(25)* found marked changes in several interleukins such as IL-5, -6, -8, -10, -13, and -14. Changes were also noted in the expression of several growth factor families, such as induction of the insulin receptor, GM-CSF and its receptor and VEGF, and the VEGFR/KDR receptor. Increases in several chemotactic factors and inflammatory cytokines were also noted. Microarray analysis also identified several neurotransmitter receptors and neuromodulators that undergo changes during endometrial stromal cells decidualization. Some members of the renin/angiotensin system were induced as well as several integrins. Integrins regulate cell-cell adhesion as well as adhesion between cells and components of the extracellular matrix and play an important role in the endometrial phenotype change that occurs during the secretory phase of the menstrual cycle. At the beginning of pregnancy, the change in integrin expression is synchronized with trophoblast attachment and embryo invasion of the decidua. Several diseases, including pre-eclampsia and intrauterine growth retardation, are associated with abnormal integrin expression *(26)*. A major increase was noted also in c-myc, other oncogenes, and other factors involved in cellular mitogenic response, including members of the cyclin and the retinoic acid receptor families. TNF-related apoptosis inducing ligand (TRAIL), a killer of activated lymphocytes, was also induced *(25)*.

IL-11 plays a role in decidualization of endometrial stromal cells in vitro. White et al. *(27)* identified genes regulated by IL-11 in decidualizing human endometrial stromal cells in vitro by utilizing microarray. Microarray analysis revealed 16 upregulated and 11 downregulated genes in decidualizing endometrial stromal cells treated with IL-11 compared with decidualizing endometrial cells not treated with IL-11. There were 11 genes and 1 expressed sequence tag (EST) upregulated and 10 genes downregulated in the presence of exogenous IL-11. Many of these genes were associated with extracellular matrix. The most upregulated gene was IL-1β, which is thought to play a role in implantation. The most down-regulated gene was IGFBP-5.

RECENT STUDIES BASED ON MICROARRAY FINDINGS

Okada et al. *(28)* demonstrated by DNA microarray analysis that fibulin-1, which codes for an extracellular matrix and plasma glycoprotein, is induced by progesterone in human endometrial stromal cells in short-term culture prior to the morphological cellular transformation. Fibulin-1 binds calcium and extracellular proteins including fibronectin, laminin, nidogen, and fibrinogen. A subsequent study *(29)* showed that fibulin-1 mRNA levels in human endometrial tissues were increased significantly during the secretory phase rather than the proliferative phase. Real-time PCR analysis demonstrated that the fibulin-1 mRNA expression level is augmented during the menstrual cycle in human endometrium. Immunohistochemical analysis revealed that fibulin-1 protein expression switched from the glandular epithelium cells during the proliferative phase to the stromal cells during the secretory phase. Fibulin-1 mRNA expression during in vitro decidualization of human endometrial cells was induced by progesterone after three days of culture, as was suggested by the microarray analysis *(28)*.

Based on the microarray finding, we performed a study aimed to investigate the role of the cannabinoid receptor I (CBR-1) in decidualization *(30)*. We found that the CBR-1 agonist R(+)-WIN 55,212-2 mesylate (WIN) causes a dose-dependent inhibition of decidualization of decidual fibroblasts and endometrial stromal cells that was secondary to the inhibition of cAMP. Decidual fibroblasts decidualized in the presence of WIN

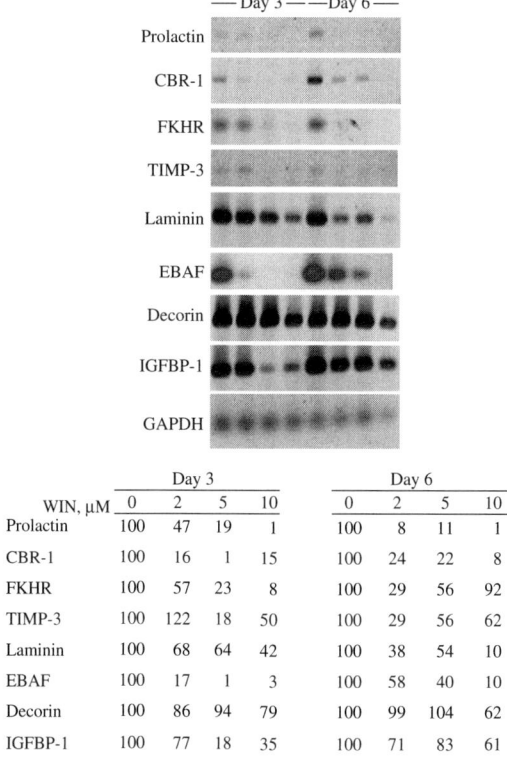

WIN, μM	Day 3				Day 6			
	0	2	5	10	0	2	5	10
Prolactin	100	47	19	1	100	8	11	1
CBR-1	100	16	1	15	100	24	22	8
FKHR	100	57	23	8	100	29	56	92
TIMP-3	100	122	18	50	100	29	56	62
Laminin	100	68	64	42	100	38	54	10
EBAF	100	17	1	3	100	58	40	10
Decorin	100	86	94	79	100	99	104	62
IGFBP-1	100	77	18	35	100	71	83	61

Fig. 1. The CBR-1 agonist WIN inhibits decidualization of decidual fibroblasts. Decidual fibroblasts were exposed for six days to MPA, estradiol and cAMP in the presence and absence of different concentrations of WIN (2, 5, and 10 μM). At the end of days 3 and 6, the relative amounts of the mRNAs for IGFBP-1, prolactin, TIMP3, laminin, decorin, forkhead (FKHR) and GAPDH were determined by RT-PCR. The top figure shows the PCR bands for each mRNA. The bottom table shows the relative changes in the mRNAs for each marker gene following densitometric analyses. The value for each mRNA was normalized to the value for GAPDH mRNA in the same sample. Each of the values for the WIN-treated samples is expressed as a percent of the amount of mRNA in control cells cultured in the absence of WIN *(30)*.

expressed considerably less mRNA levels for genes known to be induced during human decidualization than control cells decidualized in the absence of WIN (Fig. 1). In addition, WIN had a dramatic effect on the morphology of decidual fibroblasts and endometrial stromal cells undergoing decidualization. The CBR-1 agonist caused apoptosis of the fibroblast cells, while the CBR-1 antagonist AM-251 enhanced decidualization.

Numerous transcription factors change in response to decidualization. FOX1A, a member of the forkhead family of transcription factors, was identified by microarray analysis as one of the most induced genes during decidualization (17). To determine whether FOXO1A is critical for decidualization, we examined whether silencing FOX1A gene expression during decidualization would repress the induction of decidualization in response to progesterone, estradiol and PGE₂. We observed that silencing FOX1A gene expression markedly inhibits in vitro decidualization (Fig. 2), strongly suggesting an important role for FOXO1A in the induction of the differentiation process.

DNA microarrays analyses also suggested that the transcription factor ETS1 might play a role in the regulation of human uterine decidualization (17). ETS1 protein expression increased 9- to 20-fold in human endometrial stromal cells undergoing in vitro decidualization in response to progesterone and estradiol (31). Furthermore, computer analysis of the most regulated genes during in vitro decidualization revealed that the promoter regions of 23 of the 25 most induced genes and 8 of the 10 most

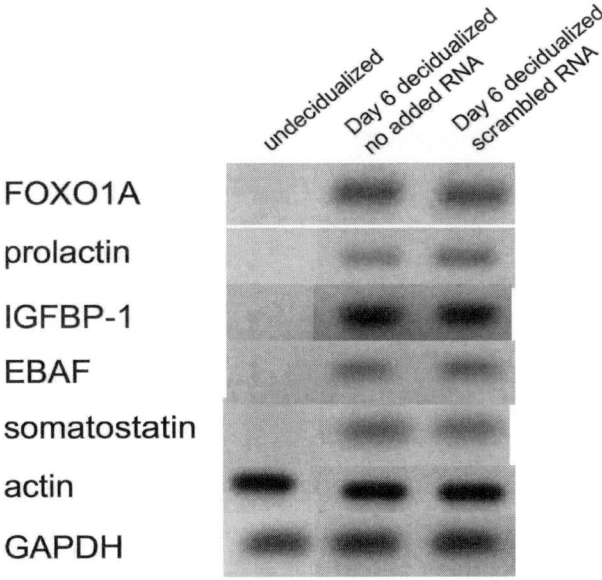

Fig. 2. The effects of a FOXO1A siRNA on selected decidualization-specific marker genes after 6 days of decidualization. Human decidual fibroblast were exposed to FOXO1A siRNA-B and decidualized for 6 days as described in Fig. 1. FOXO1A, IGFBP-1, prolactin, EBAF, and GAPDH mRNA levels were determined at the end of day 6. The results show the means from 4 separate experiments in which triplicate wells were analyzed. Each bar represents the mean of the 4 separate experiments ±SEM. The mRNA levels for each of the marker genes in the scrambled mRNA-treated cells in each of the experiments showed the anticipated increases during the 6 days of in vitro decidualization (data not shown). The effects of the FOXO1A siRNA on each of the mRNAs is expressed as the percent change from the corresponding mRNA level in the control cells treated with scrambled RNA. Each bar represents the mean and standard error of triplicate wells. *, p < 0.05, **, p < 0.01 (44).

repressed genes have one or more ETS binding sites. To examine whether ETS1 plays a critical role in the regulation of human decidualization, we subsequently examined whether silencing of ETS1 gene expression during in vitro decidualization with an ETS1 morpholino antisense oligonucleotide would block the induction of decidualization *(32)*. The ETS1 antisense oligonucleotide markedly blocked the induction of genes that are known to be induced during human decidualization, such as prolactin, IGFBP-1, Lefty2, TIMP-3, decorin, and laminin (Fig. 3).

The peptide hormone relaxin has been shown to induce intracellular cAMP in endometrial stromal cells and hence in vitro decidualization *(33)*. Bartsch et al. *(34)* evaluated the complement of phosphodiesterases in human endometrial stromal cells and their possible role in relaxin-induced in vitro decidualization. By measuring PDE activity and relaxin-stimulated cAMP accumulation in the presence of diverse PDE inhibitors, they identified PDE4 and PDE8 as the principal PDE isoforms involved in human stromal cells.

DECIDUALIZATION IN PATHOLOGICAL CONDITIONS OF PREGNANCY

Golander et al. *(35)* observed that prolactin secretion by decidual explants from patients with pre-eclampsia is significantly less than that from normal controls. Incubation of normal decidual tissue in the presence of serum obtained from pre-eclamptic patients did not induce an inhibitory effect on prolactin production. Since

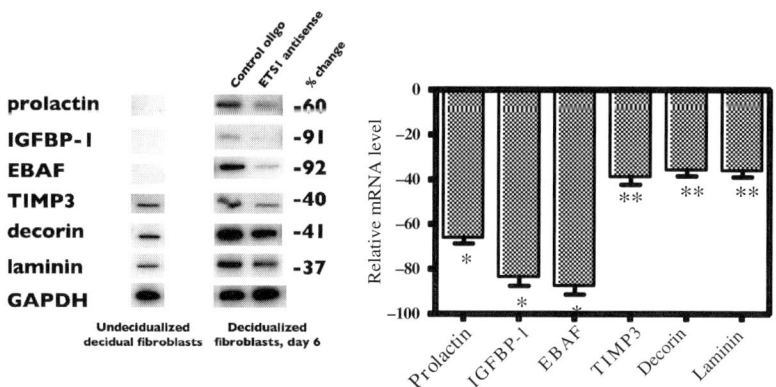

Fig. 3. The effect of ETS1 oligo on markers of decidualization. Decidual fibroblast cells were decidu-alized in vitro for six days following a 3-h incubation with an ETS1 antisense oligonucleotide or a control morpholino oligonucleotide as described in Methods. At the end of the sixth day, mRNA was extracted from the cells and analyzed for the indicated marker mRNAs. The left panel shows the mRNA results from a representative experiment comparing the mRNA bands from undecidualized decidual fibroblast cells and fibroblast cells decidualized for six days in the presence of the ETS1 antisense oligonucleotide or the control oligonucleotide. The numbers indicate the percent decrease in the amount of mRNA for each marker gene in the ETS1 antisense oligo-exposed cells relative to the amount in the control cells that had been exposed to the control oligo. In each instance, the mRNA level for the marker gene was normalized to the GAPDH mRNA level in the same sample. The right panel shows the relative change in mRNA levels in four separate experiments, including the representative experiment depicted in the left panel and three other experiments. The bars indicate the mean decrease in relative mRNA levels for the ETS1 antisense oligonucleotide compared to the control oligonucleotide. The error bars indicate the SEM of the four experiments. *, $p < 0.0001$, **, $p < 0.005$ *(32)*.

prolactin is a marker of decidualization, this study suggests that pre-eclampsia may be characterized by decreased decidualization.

In pre-eclampsia, invasion of the intrauterine decidual blood vessels by placental cytotrophoblast cells is significantly reduced. Gallery et al. *(36)* examined the secretion of matrix metalloproteinase by cultured human decidual endothelial cells from normal and pre-eclamptic pregnancies. They found lower MMP1 expression by decidual endothelial cells from pre-eclamptic women may inhibit endovascular invasion by cytotrophoblast cells. Furthermore, IL-2 expressed in pre-eclamptic decidua might reduce the angiogenic substances arising from trophoblast cells by inducing activation of lymphocytes from decidual lymphocytes, which might be relevant to deranged vasculature of the placenta, a characteristic histology in pre-eclampsia *(37)*. In addition, co-culture experiments showed that decidual endothelial cells downregulated cytotrophoblast migration in placentas from normal pregnancies but not pre-eclampsia placentas. These results suggest dysfunctional interactive regulation of migration and matrix metalloproteinase secretion in pre-eclampsia that could result in abnormal endovascular trophoblast invasion of the maternal vasculature. While Huisman et al. *(39)* confirmed the presence of metalloproteinase-2 and -9 in first trimester placental bed biopsies, no differences in metalloproteinase activity between pre-eclampsia and normal placentas.

Halvorsen et al. *(40)* demonstrated elevated levels of the oxidative stress marker 8-isoprostaglandin F (2 alpha) (8-isoprostane) and lipids in pre-eclamptic decidual tissue. 8-isoprostane is released from tissue phospholipids by phospholipase A2. A subsequent study *(41)* found significantly higher total phospholipase A2 activity in pre-eclamptic decidua compared to control tissue. The investigators speculated that an elevated phospholipase A2 activity in pre-eclamptic decidual tissue could be important in the pathogenesis of acute atherosis, which is frequently observed in pre-eclampsia. Acute atherosis is characterized by an accumulation of lipid-laden macrophages and a mononuclear perivascular infiltrate. However, the mechanisms by which macrophages are activated and recruited to the placental bed in early, uncomplicated pregnancy and in pre-eclampsia are unclear. The presence of monocyte chemoattracant protein-1 (MCP-1), the primary monocyte/macrophage chemoattractant, in cycling endometrium and first trimester decidua suggests that the protein is involved in this infiltration. Lockwood et al. *(42)* recently observed a significant increase in macrophage (CD68-positive cells) in the decidua of pre-eclamptic patients and noted that TNF-α and IL-1β enhance MCP-1, in first trimester decidua. These finding suggests a mechanism by which recruitment of excess macrophages to the decidua impairs endovascular trophoblast invasion.

Gratton et al. *(43)* showed altered IGF-II and IGFBP-1 expression at the fetomaternal interface in pre-eclampsia. IGF-II and IGFBP-1 are abundantly expressed by cells at the maternal-fetal interface and mediate cell-to-cell communication between trophoblasts and decidua.

SUMMARY

The endometrium is a dynamic tissue that undergoes cyclic changes in response to estradiol and progesterone. Endometrial stromal cells differentiate into decidual cells by the process known as decidualization. Decidualization is critical for successful blastocyst implantation. Prior to the availability of microarray technology, studies on human decidualization relied on the investigation of individual genes or gene families. DNA microarray studies have greatly expanded our knowledge about gene expression during human uterine decidualization. These studies have not only identified a large number of genes that are induced during decidualization but also have identified for the first time a large number of repressed genes. Based on these studies, decidualization can be characterized by "categorical reprogramming" in which there is simultaneous upregulation and downregulation of a set of genes with related function. Major gene changes were seen in several functional categories, such as genes that encode for extracellular matrix proteins. This is consistent with our knowledge that extracellular matrix remodeling occurs during decidualization and is essential for embryo implantation. Microarray analysis allows us not only to understand the normal process of decidualization but also to understand the genetic basis of abnormal states, such as pre-eclampsia and intrauterine growth retardation. Studies of pre-eclamptic decidua have shown defects in gene expression compared to control tissue. Recent studies using in vitro model system of decidualization have shown a critical role of FOX1A and ETS1 in the induction of human decidualization, while factors such as cannabinoids inhibit the decidualization process. Further DNA microarray studies should provide additional information about the decidualization process in normal and pathological states.

REFERENCES

1. Frank, G.R., Brar, A.K., Cedars, M.I. and Handwerger, S. (1994) Prostaglandin E2 enhances human endometrial stromal cell differentiation. Endocrinology 134, 258–263.
2. Frank, G., Brar, A., Jikihara, H., Cedars, M. and Handwerger, S. (1995) Interleukin-1 beta and the endometrium: an inhibitor of stromal cell differentiation and possible autoregulator of decidualization in humans. Biol Reprod 52, 184–91.
3. Brar, A.K., Kanda, Y., Kessler, C.A., Cedars, M.I. and Handwerger, S. (1999) N5 endometrial stromal cell line: a model system to study decidual prolactin gene expression. In Vitro Cell Dev Biol Anim 35, 150–4.
4. Brosens, J.J., Takeda, S., Acevedo, C.H., Lewis, M.P., Kirby, P.L., Symes, E.K., Krausz, T., Purohit, A., Gellersen, B. and White, J.O. (1996) Human endometrial fibroblasts immortalized by simian virus 40 large T antigen differentiate in response to a decidualization stimulus. Endocrinology 137, 2225–31.
5. Brar, A.K., Frank, G.R., Kessler, C.A., Cedars, M.I. and Handwerger, S. (1997) Progesterone-dependent decidualization of the human endometrium is mediated by cAMP. Endocrine 6, 301–7.
6. Telgmann, R., Maronde, E., Tasken, K. and Gellersen, B. (1997) Activated protein kinase A is required for differentiation-dependent transcription of the decidual prolactin gene in human endometrial stromal cells. Endocrinology 138, 929–37.
7. Brosens, J.J., Hayashi, N. and White, J.O. (1999) Progesterone receptor regulates decidual prolactin expression in differentiating human endometrial stromal cells. Endocrinology 140, 4809–20.
8. Gellersen, B., Kempf, R. and Telgmann, R. (1997) Human endometrial stromal cells express novel isoforms of the transcriptional modulator CREM and up-regulate ICER in the course of decidualization. Mol Endocrinol 11, 97–113.
9. Tabanelli, S., Tang, B. and Gurpide, E. (1992) In vitro decidualization of human endometrial stromal cells. J Steroid Biochem Mol Biol 42, 337–344.
10. Telgmann, R. and Gellersen, B. (1998) Marker genes of decidualization: activation of the decidual prolactin gene. Hum Reprod Update 4, 472–9.
11. Irwin, J.C., de las Fuentes, L. and Giudice, L.C. (1994) Growth factors and decidualization in vitro. Ann N Y Acad Sci 734, 7–18.

12. Gao, J., Mazella, J., Suwanichkul, A., Powell, D.R. and Tseng, L. (1999) Activation of the insulin-like growth factor binding protein-1 promoter by progesterone receptor in decidualized human endometrial stromal cells. Mol Cell Endocrinol 153, 11–7.

13. Giudice, L.C. and Irwin, J.C. (1999) Roles of the insulinlike growth factor family in nonpregnant human endometrium and at the decidual: trophoblast interface. Semin Reprod Endocrinol 17, 13–21.

14. Brar, A.K., Frank, G.R., Richards, R.G., Meyer, A.J., Kessler, C.A., Cedars, M.I., Klein, D.J. and Handwerger, S. (1995) Laminin decreases PRL and IGFBP-1 expression during in vitro decidualization of human endometrial stromal cells. J Cell Physiol 163, 30–37.

15. Li, C., Ansari, R., Yu, Z. and Shah, D. (2000) Definitive molecular evidence of renin-angiotensin system in human uterine decidual cells. Hypertension 36, 159–64.

16. Vogiagis, D. and Salamonsen, L.A. (1999) Review: the role of leukaemia inhibitory factor in the establishment of pregnancy. J Endocrinol 160, 181–90.

17. Brar, A.K., Handwerger, S., Kessler, C.A. and Aronow, B.J. (2001) Gene induction and categorical reprogramming during in vitro human endometrial fibroblast decidualization. Physiol Genomics 7, 135–48.

18. Jones, J.I., Gockerman, A., Busby, W.H. Jr., Camacho-Hubner, C. and Clemmons, D.R. (1993) Extracellular matrix contains insulin-like growth factor binding protein-5: potentiation of the effects of IGF-I. J Cell Biol 121, 679–87.

19. Andress, D.L. (1998) Insulin-like growth factor-binding protein-5 (IGFBP-5) stimulates phosphorylation of the IGFBP-5 receptor. Am J Physiol 274, E744–50.

20. Schedlich, L.J., Young, T.F., Firth, S.M. and Baxter, R.C. (1998) Insulin-like growth factor-binding protein (IGFBP)-3 and IGFBP-5 share a common nuclear transport pathway in T47D human breast carcinoma cells. J Biol Chem 273, 18347–52.

21. Tonner, E., Barber, M.C., Travers, M.T., Logan, A. and Flint, D.J. (1997) Hormonal control of insulin-like growth factor-binding protein-5 production in the involuting mammary gland of the rat. Endocrinology 138, 5101–7.

22. Chen, H.W., Chen, J.J., Tzeng, C.R., Li, H.N., Chang, S.J., Cheng, Y.F., Chang, C.W., Wang, R.S., Yang, P.C. and Lee, Y.T. (2002) Global analysis of differentially expressed genes in early gestational decidua and chorionic villi using a 9600 human cDNA microarray. Mol Hum Reprod 8, 475–84.

23. Rosenfeld, R.G., Lamson, G., Pham, H., Oh, Y., Conover, C., De Leon, D.D., Donovan, S.M., Ocrant, I. and Giudice, L. (1990) Insulin like growth factor-binding proteins. Recent Prog Horm Res 46, 99–159; discussion 159–63.

24. Jones, J.I. and Clemmons, D.R. (1995) Insulin-like growth factors and their binding proteins: biological actions. Endocr Rev 16, 3–34.

25. Popovici, R.M., Kao, L.C. and Giudice, L.C. (2000) Discovery of new inducible genes in in vitro decidualized human endometrial stromal cells using microarray technology. Endocrinology 141, 3510–3.

26. Merviel, P., Challier, J.C., Carbillon, L., Foidart, J.M. and Uzan, S. (2001) The role of integrins in human embryo implantation. Fetal Diagn Ther 16, 364–71.

27. White, C.A., Dimitriadis, E., Sharkey, A.M. and Salamonsen, L.A. (2005) Interleukin-11 inhibits expression of insulin-like growth factor binding protein-5 mRNA in decidualizing human endometrial stromal cells. Mol Hum Reprod 11, 649–58.

28. Okada, H., Nakajima, T., Yoshimura, T., Yasuda, K. and Kanzaki, H. (2003) Microarray analysis of genes controlled by progesterone in human endometrial stromal cells in vitro. Gynecol Endocrinol 17, 271–80.

29. Nakamoto, T., Okada, H., Nakajima, T., Ikuta, A., Yasuda, K. and Kanzaki, H. (2005) Progesterone induces the fibulin-1 expression in human endometrial stromal cells. Hum Reprod 20, 1447–55.

30. Kessler, C.A., Moghadam, K.K., Schroeder, J.K., Buckley, A.R., Brar, A.K. and Handwerger, S. (2005) Cannabinoid receptor I activation markedly inhibits human decidualization. Mol Cell Endocrinol 229, 65–74.

31. Brar, A.K., Kessler, C.A. and Handwerger, S. (2002) An Ets motif in the proximal decidual prolactin promoter is essential for basal gene expression. J Mol Endocrinol 29, 99–112.

32. Kessler, C.A., Schroeder, J.K., Brar, A.K. and Handwerger, S. (2006) Transcription factor ETS1 is critical for human uterine decidualization. Mol Hum Reprod 12, 71–6.

33. Huang, J.R., Tseng, L., Bischof, P. and J:anne, O.A. (1987) Regulation of prolactin production by progestin, estrogen, and relaxin in human endometrial stromal cells. 121, 2011–2017.

34. Bartsch, O., Bartlick, B. and Ivell, R. (2004) Phosphodiesterase 4 inhibition synergizes with relaxin signaling to promote decidualization of human endometrial stromal cells. J Clin Endocrinol Metab 89, 324–34.

35. Golander, A., Kopel, R., Lazebnik, N., Frenkel, Y. and Spirer, Z. (1985) Decreased prolactin secretion by decidual tissue of pre- eclampsia in vitro. Acta Endocrinol (Copenh) 108, 111–113.

36. Gallery, E.D., Campbell, S., Arkell, J., Nguyen, M. and Jackson, C.J. (1999) Preeclamptic decidual microvascular endothelial cells express lower levels of matrix metalloproteinase-1 than normals. Microvasc Res 57, 340–6.

37. Hamai, Y., Fujii, T., Yamashita, T., Kozuma, S., Okai, T. and Taketani, Y. (1997) Pathogenetic implication of interleukin-2 expressed in preeclamptic decidual tissues: a possible mechanism of deranged vasculature of the placenta associated with preeclampsia. Am J Reprod Immunol 38, 83–8.

38. Campbell, S., Rowe, J., Jackson, C.J. and Gallery, E.D. (2004) Interaction of cocultured decidual endothelial cells and cytotrophoblasts in preeclampsia. Biol Reprod 71, 244–52.

39. Huisman, M.A., Timmer, A., Zeinstra, M., Serlier, E.K., Hanemaaijer, R., Goor, H. and Erwich, J.J. (2004) Matrix-metalloproteinase activity in first trimester placental bed biopsies in further complicated and uncomplicated pregnancies. Placenta 25, 253–8.

40. Halvorsen, B., Staff, A.C., Henriksen, T., Sawamura, T. and Ranheim, T. (2001) 8-iso-prostaglandin F(2alpha) increases expression of LOX-1 in JAR cells. Hypertension 37, 1184–90.

41. Staff, A.C., Ranheim, T. and Halvorsen, B. (2003) Augmented PLA2 activity in pre-eclamptic decidual tissue—a key player in the pathophysiology of 'acute atherosis' in pre-eclampsia? Placenta 24, 965–73.

42. Lockwood, C.J., Matta, P., Krikun, G., Koopman, L.A., Masch, R., Toti, P., Arcuri, F., Huang, S.T., Funai, E.F. and Schatz, F. (2006) Regulation of monocyte chemoattractant protein-1 expression by tumor necrosis factor-alpha and interleukin-1beta in first trimester human decidual cells: implications for preeclampsia. Am J Pathol 168, 445–52.

43. Gratton, R.J., Asano, H. and Han, V.K. (2002) The regional expression of insulin-like growth factor II (IGF-II) and insulin-like growth factor binding protein-1 (IGFBP-1) in the placentae of women with pre-eclampsia. Placenta 23, 303–10.

44. Grinius, L., Kessler, C., Schroeder, J. and Handwerger, S. (2006) Forkhead transcription factor FOXO1A is critical for induction of human decidualization. J Endocrinol 189, 179–87.

9

Large-scale DNA Microarray Data Analysis Reveals Glucocorticoid Receptor-mediated Breast Cancer Cell Survival Pathways

Min Zou, PhD, Wei Wu, MD, PhD, and Suzanne D. Conzen, MD

CONTENTS

INTRODUCTION: THE ROLE OF THE GLUCOCORTICOID RECEPTOR
  ACTIVATION IN BREAST CANCER
MICROARRAY DATA ANALYSIS FOR STUDYING
  HORMONE ACTIVATION
USING MICROARRAY DATA ANALYSIS TO UNCOVER GR-MEDIATED
  BREAST CANCER CELL SURVIVAL PATHWAYS
CONCLUSIONS
REFERENCES

Abstract

Glucocorticoid receptor (GR) activation can inhibit breast epithelial and cancer cells from undergoing programmed cell death in response to diverse apoptotic stimuli. Understanding the mechanisms underlying inappropriate cell survival mechanisms is important for treating breast cancer because if we can reverse these mechanisms, therapies designed to kill tumor cells are likely to be more effective. Recently, genome-wide DNA microarrays have provided a glimpse into the signals and interactions within regulatory pathways of the cell. These arrays enable simultaneous measurement of mRNA abundance of most, if not all, identified genes in a genome under different physiological conditions. Currently, two types of microarray experiments are frequently performed in laboratories. The first is a single time point microarray experiment, and the second is a time course microarray experiment. Single time point microarray experiments are effective in identifying genes regulated by a given treatment, e.g., direct target genes of a hormone treatment. However, because molecular pathways are dynamic processes that take place over time, single time point microarray experiments may not allow us to identify dynamic molecular pathways. This problem can be approached by performing a time course microarray experiment, which measures gene expression changes at various time points following a given treatment. In this chapter, we first describe how to identify target genes of a given treatment using a single time point microarray data analysis. We then present three alternate bioinformatics approaches to uncover molecular mechanisms from time course microarray data. Finally, we present a novel bioinformatics approach for analyzing time course microarray data in order to identify novel GR-mediated breast cancer cell survival pathways.

From: *Contemporary Endocrinology: Genomics in Endocrinology:*
DNA Microarray Analysis in Endocrine Health and Disease
Edited by S. Handwerger and B. Aronow © Humana Press, Totowa, NJ

INTRODUCTION: THE ROLE OF THE GLUCOCORTICOID RECEPTOR ACTIVATION IN BREAST CANCER

Glucocorticoids are steroid hormones synthesized and secreted by the adrenal cortex. Glucocorticoids regulate a variety of physiologic functions and play an important role in maintaining basal and stress-related homeostasis. At the cellular level, the actions of glucocorticoids are mediated by the glucocorticoid receptor (GR), which belongs to the nuclear receptor family of ligand-dependent transcription factors (TFs). Upon hormone binding, the GR undergoes an allosteric change, resulting in its dissociation from hsp90 and other proteins, followed by phosphorylation at five serine residues (positions 113, 141, 203, 211, and 226). In its new conformation, the phosphorylated, ligand-bound GR translocates into the nucleus, where it binds as a homodimer to a GR response element (GRE) located in the promoter region of target genes. The GR then communicates with the basal transcription machinery to either up- or downregulate the expression of target genes, depending on the GRE sequence and promoter context *(1)*. The receptor can also modulate gene expression independently of DNA binding through physical interaction with other TFs, such as activator protein (AP)-1 and nuclear factor (NF)-κB *(2, 3)*. In addition, rapid "nongenomic" effects of activated GR have been described that do not require regulation of gene expression and are extranuclear *(4)*.

Previously, GR activation has been associated with induction of apoptosis in lymphocytes and other cell lines *(5)*. However, we *(6)* and others *(7)* have demonstrated that GR activation can inhibit apoptosis in mammary epithelial and cancer cells. Similar findings have recently been reported in human and rat hepatocytes *(8)*, vascular endothelial cells *(9)*, and osteoclasts *(10)*. Therefore, there appears to be a cell-type-specific role for GR activation in the regulation of cell death *(11)*. Understanding mechanisms that inhibit apoptosis in breast cancer cells is important because if we can reverse these mechanisms, therapies designed to kill tumor cells are likely to be more effective. Therefore, GR-mediated cell survival in breast cancer cells provides an ideal model for identifying novel anti-apoptotic mechanisms in breast cancer cells that may have therapeutic implications.

Because GR is a TF, its activation can induce or repress the expression of specific target genes, which may in turn contribute to GR-mediated cell survival. Therefore, to better understand the direct GR-mediated gene expression changes that might promote cell survival in mammary epithelial cells, we initially used high-density oligonucleotide microarrays to identify GR-activated or repressed genes 30 min after dexamethasone (Dex) treatment in immortalized mammary epithelial cell line MCF10A-Myc. The analysis of "single time point" microarray data has allowed us to identify a group of novel GR-regulated target genes whose expression changes contribute to the mechanism of GR-mediated cell survival. Although somewhat informative, single time point microarray experiments do not reveal much information about dynamic survival pathways that exist downstream of GR activation. Therefore, we also performed "multiple time point" or "time course" microarray experiments in MCF10A-Myc cells at 30 min, 2 h, 4 h, and 24 h following GR activation. The analysis of time course microarray data has allowed us to identify novel mechanisms underlying GR mediated cell survival. In this chapter, we first describe the most widely used bioinformatics methodologies for analyzing both single time point and time course microarray data. We then describe in detail how these methodologies can be used to analyze gene expression data in biological context (e.g., GR-mediated cell survival). Finally, we will discuss possible directions for whereby microarray data analysis can be used to predict dynamic biological pathways.

MICROARRAY DATA ANALYSIS FOR STUDYING HORMONE ACTIVATION

Single Time Point Microarray Data Analysis

Genome-wide DNA microarrays are powerful tools, and provide a glimpse of the signals and interactions within regulatory pathways of the cell. Microarrays allow simultaneous measurement of mRNA abundance of most, if not all, identified genes in a genome under different physiological conditions. Single time point microarray experiments are done by treating cells with a given external stimulant (i.e., hormones) for a certain length of time, and then measuring the expression of genes. In order to examine whether or not genes change their expression levels following a given treatment, a control experiment is also carried out where cells are treated with vehicle (e.g., ethanol). These experiments have to be repeated at least twice in order to ensure statistical significance. For each replicate, the ratio of the expression of a gene in the treated sample to that in the control-treated sample is calculated. If this ratio is larger or equal to a certain cutoff, then the given gene is considered to be upregulated; if smaller than 1/cutoff, then the gene is considered to be downregulated. The statistical likelihood that an up- or downregulation may occur due to chance is then calculated using a Student t-test. The Student t-test generates "p-values," with smaller p-values indicating the probability that gene expression differs in the treated sample from the control due to chance is lower. Genes significantly up- or downregulated in all replicates or with averaged expression values satisfying the up- or downregulation criteria and with significant p-values are selected for further study.

There are three important issues in single time point microarray data analysis that can improve the likelihood of success. First, determination of the criteria for up- or downregulation is a critical component of any microarray data analysis. Inaccurate assignment of up- or downregulation cutoffs may cause real target genes to be missed and false target genes to be identified. Because of physiologically variable systems, there are no fixed cutoffs for determining up- or downregulation. The most widely used cutoffs for determining upregulation are higher than twofold expression change and less than 0.5-fold for downregulation, although the level of expression changes following certain treatments (or in certain cell types) may vary. Expression changes in genes following certain treatments (or in certain cell types) may be very dramatic, while modest changes may result following other treatments or in other cell systems. However, minor changes in gene expression may be physiologically important as well. Therefore, an arbitrary choice of cutoffs may cause some real target genes to be missed. In general, the best way to determine a gene expression cutoff for a specific treatment and cell type is by having an internal standard, such as the expression change of a known physiologically relevant target gene for a given treatment.

The second important issue in single time point microarray data analysis is the method used to select the final set of target genes for further study. Currently, this is done using one of the following two methods. The first method focuses on selecting genes that are up- or downregulated in all replicates and with relatively small p-values (i.e., $p \leq 0.05$ representing that the probability that a gene has differential expression due to chance is less than 5%). The second method selects genes with averaged expression satisfying the up- or downregulated criteria and with significantly small p-values as final target gene set. Both of these methods have been widely used. However, the second method may be more reasonable because the up- or downregulation cutoff used in the first method may artificially delete some genes if the expression of the gene is slightly below the cutoff. For example, if a gene's fold change in one of the replicate experiments is 1.49-fold, while the criterion for upregulation is 1.5-fold, then this gene will be omitted if using the first method. Therefore, the second method may be a more reasonable approach to select the final gene set for further study.

The third issue is the use of a p-value cutoff as an absolute criterion for determining which genes are significantly up- or downregulated. The p-values are important indicators of the probability that an up- or downregulation occurs due to chance. However, a p-value is primarily affected by two factors. First, the standard deviation of the expression values of a given gene in all replicates, and second, the degree of gene expression changes as compared to the control. This means that if there is an error in the measurement of the expression of a gene in any one of the replicates, its standard deviation will be relatively high, and the gene will be excluded due to a high p-value. Similarly, if the expression change of the gene is modest, its p-values may be relatively high, thereby excluding the gene from further consideration.

Since reducing false positives and false negatives is in inherent conflict, one has to maintain a good balance between the two. In order to avoid missing important genes that may have relatively high p-values, several strategies can be adopted. One is to repeat the experiment as many times as possible to reduce a large standard deviation. The other is to use p-values of known target genes of a given treatment, which has modest expression changes as internal standards to determine a reasonable p-value cutoff for a given experimental system.

Time Course Microarray Data Analysis

A time course microarray experiment is composed of data from several time points, each of which measures gene expression changes following a given treatment. Unlike single time point experiments, time course experiments can provide valuable information about dynamic mechanisms (e.g., molecular pathways) leading to a specific phenotype resulting from a given treatment. Usually, for each time point, gene expression levels in the control sample (e.g., vehicle treated) are also measured in order to correct time dependent expression variation. The first step of a time course microarray data analysis is to identify up- or downregulated genes at each time point following treatment. This can be done using the same procedure as for single time point microarray experiments described in Section "Single time point microarray data analysis". The second step is to combine regulated genes from each time point to obtain a complete set of regulated genes following the treatment. In order to discover molecular mechanisms contributing to the phenotype being studied, systematic and relatively unbiased bioinformatics approaches need to be used to analyze time course microarray data. In this chapter, we discuss three bioinformatics approaches used to identify molecular pathways underlying a phenotype from large-scale time course microarray data.

LOCAL CLUSTERING ANALYSIS

Eisen et al. developed a "clustering method" that identifies groups of genes whose expression levels simultaneously rise and fall; in other words, clusters of co-expressed genes *(12)*. This method, however, cannot identify regulatory relationships between genes where a time-delay between the expression of the regulator gene and its target gene(s) exists. Recognizing this limitation, Qian et al. developed another clustering method, called "local clustering analysis" that can identify time-delayed relationships *(13)*. This method uses a degenerate dynamic programming algorithm to discover time-shifted and inverted time-shifted correlations between temporal gene expression profiles, where time-shifted correlation indicates an activation relationship between the corresponding pair of genes, and inverted time-shifted correlation suggests an inhibition relationship.

Suppose there are N time-point measurements in a gene expression profile ($n = 1$, 2, 3...). First, the expression ratio is normalized in "Z-score" fashion, so that for each gene the average expression ratio is 0 and standard deviation is 1. The normalized expression level at time point i for gene x is denoted as x_i, and the expression level for gene y at time point j is denoted as y_j. A simple scoring scheme is assumed where $M(x_i y_i) = x_i y_j$. For simplification, it will be referred as $M_{i,j}$. Then, two $(n+1) \times (n+1)$ sum matrices $\mathbf{E}$ and $\mathbf{D}$ are created, where

$$E_{i,j} = \max(E_{i-1,j-1} + M_{i,j}, 0) \text{ and } D_{i,j} = \max(D_{i-1,j-1} - M_{i,j}, 0).$$

The initial conditions are $E_{0,j}=0$ and $E_{i,0}=0$, where $i,j = 0, 1, 2, ...,n$. The same initial conditions are also applied to the matrix of $\mathbf{D}$. The central idea is to find segments of the expression profiles of two genes that match each other, and which have the maximal accumulated score (i.e., the sum of $M_{i,j}$ in this segment). This can be accomplished by standard dynamic programming as in local sequence alignment of DNA or protein sequences, and results in an alignment of "m" time points, where $m \leq n$.

After scoring matrices $\mathbf{E}$ and $\mathbf{D}$ are filled, an overall maximal value S is found by comparing the maximums for matrices $\mathbf{E}$ and $\mathbf{D}$. This is the match score S for the two expression profiles. If the maximum is "off-diagonal" in its corresponding matrix [Fig. 1(E)], the two expression profiles have a time-shifted correlation. A maximal value from matrix $\mathbf{D}$ indicates that these two profiles have an inverted correlation. At the end of this procedure, one obtains a match score as well as a relationship, designated as "simultaneous," "time-shifted," "inverted," or "inverted time-shifted." For the gene pairs with a very low match score, even though they are also assigned a relationship, they are then classified as "unmatched."

An example illustrating identification of regulatory relationships between genes using this method is presented as below. Fig. 1(A)–(C) shows three time course gene expression profiles corresponding to simultaneous, time-shifted, and inverted relationships, where the number of time points $n = 8$ for each gene expression profile. Fig. 1(D)–(E) shows the corresponding scoring matrices for the gene expression profiles in Fig. 1(A)–(C). Here, matrix $\mathbf{E}$ for the two expression profiles in Fig. 1(C) is not shown, because the maximal accumulated score is not in this matrix, indicating this is an inverted relationship. From Fig. 1(A), the maximal accumulated score for the two expression profiles is $S = 16$, located at the lower-right corner of the matrix (shaded and bolded cell). The fact that the maximal accumulated score is on the diagonal of the matrix indicates that the two expression profiles in Fig. 1(A) are simultaneous. The segments in the two expression profiles in Fig. 1(A) that match each other are represented by a single path starting from this maximal accumulated score and ending at the first encountered "0," represented by the shaded path in Fig. 1(D). Similarly, segments in the two expression profiles in Fig. 1(B) and (C) that match each other are constructed and represented by the shaded paths in Fig. 1(E) and (F), respectively. The fact that the path in Fig. 1(E) is off-diagonal indicates that the two expression profiles in Fig. 1(B) have time-shifted relationship.

Local clustering analysis has been used to analyze yeast cell cycle time course gene expression data, and was able to detect many known gene–gene relationships, which demonstrates its potential in identifying novel gene–gene relationships from time course expression data.

DYNAMIC BAYESIAN NETWORK ANALYSIS

The second method to identify molecular pathways from time course microarray data is the Dynamic Bayesian Network (DBN) analysis *(14–16)*. To our knowledge, Murphy

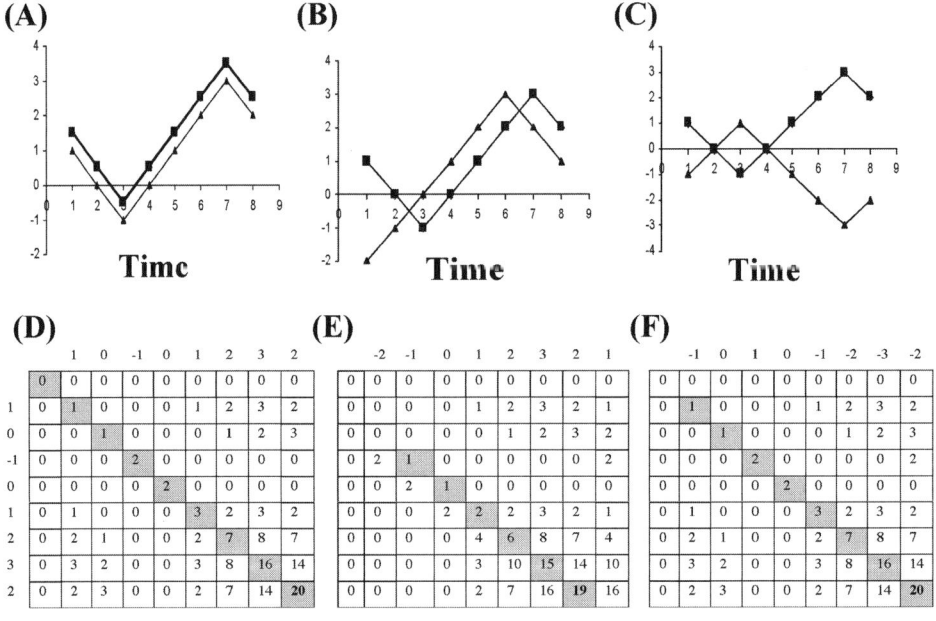

Fig. 1. Three examples showing (A) simultaneous, (B) time-delayed, and (C) inverted relationships in the expression profiles. (D) The corresponding matrix **E** for the expression profile shown in (A). The corresponding profile matrix **D** is not shown because in this case the match score (the maximal score) is form **E** and not **D**. The numbers outside the border of the matrix are the expression ratio shown in (A). The boldly shaded cell contains the overall match score S for these two expression profiles, and the lightly shaded cells indicate the path of the optimal alignment between the expression profiles. The path starts from the match score and ends at the first encountered 0. (E) The corresponding matrix **E** for the expression profile shown in (B). Note the time-shifted relationship and how the length of the overall alignment can be shorter than eight positions. (F) The corresponding matrix **D** for the expression profiles shown in (C). The matrix **E** is not shown because the best match score is not from this matrix in this case. (This figure modified from Figure 1 in *Journal of Molecular Biology* (2001) **314**:1053–1066 with permission).

and Mian *(14)* are to be credited with first employing DBN for modeling time series expression data. In a DBN analysis, regulator–target gene pairs are usually identified based on a statistical analysis of their expression relationships across different time slices. For example, time slices are designated T_1 for the regulator gene and T_2 for the target gene, where T_1 precedes T_2. The time period between the time slices of the regulator and target $(T_2 - T_1)$ is considered to be the transcriptional time lag. Specifically, it is the time it takes for the regulator gene to express its protein product and the transcription of the target gene to be affected (directly or indirectly) by this regulator protein. The statistical analysis of the gene expression relationship between the regulator gene and the target gene focuses on determining how the expression of the target gene changes with that of the regulator gene over time. Conditional probabilities are often used to represent such statistical reliability. For example, suppose we have two genes, gene A and gene D with their transformed time course gene expression data [Fig. 2(A)]. There are many ways to transform time series gene expression data. One popular way is to calculate the averaged gene expression level across all time points, and assign "2" to a gene expression level at a time point if it is larger or equal to the averaged gene expression level, and "1" to a gene expression level if it is less than the average gene expression level. If one measures the time series gene expression levels for genes A and D and estimates the transcriptional time lag between genes A and D as $t = 2$ time units, then gene A's expression occurs two time units earlier than that of gene D. One can next generate a data matrix where the expression of gene A at time 1 (T_1) is aligned with the expression level of gene D at T_3 (due to $t=2$ time units transcriptional time lag between gene A and gene D), and expression level of gene A at T_2 is aligned with expression level of gene D at T_4, and so on. Therefore, you will have a data matrix as depicted in Fig. 2(A).

The next step is to calculate the probabilities of the expression of gene D conditioned on the expression of gene A. This is computed, for example, by first calculating the number of occurrences for expression of gene D = 1 when that of gene A = 1, which

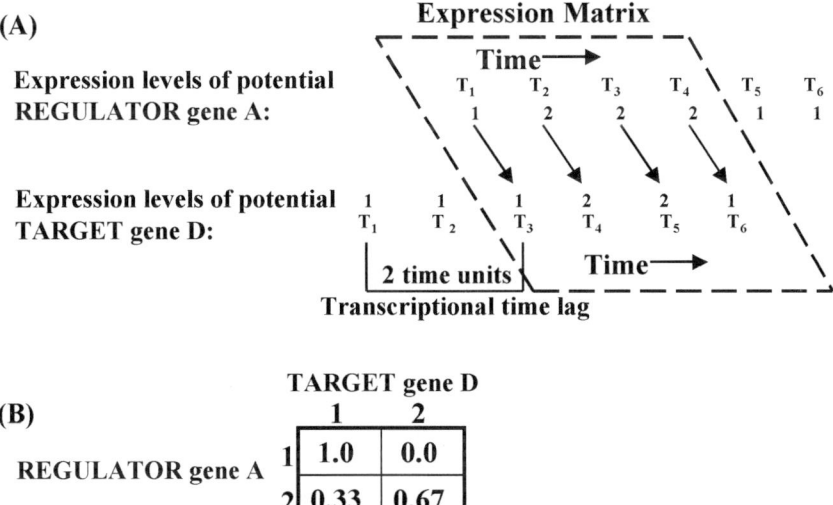

Fig. 2. (A) Transformed expression levels of genes A and D, and the generated data matrix. (B) Conditional probability table of expression of gene D in relation to that of gene A.

is denoted by C (D = 1/A = 1) (C accounts for counts), and C (D = 2/A = 1), C (D = 1/A = 2), and C (D = 2/A = 2). In this hypothetical case, based on the data matrix generated above, C (D = 1/A = 1) = 1, C (D = 2/A = 1) = 0, C (D = 1/A = 2) = 1, and C (D = 2/A = 2) = 2. Therefore, the conditional probabilities for expression of gene D = 1 when that of gene A = 1 is P (D = 1/A = 1) = 1. Similarly, P (D = 2/A = 1) = 0, P (D = 1/A = 2) = 1/3, and P (D = 2/A = 2) = 2/3. This step is illustrated in Fig. 2(B). A marginal likelihood score is then calculated based on the conditional probabilities as calculated earlier. In this hypothetical case, P (D = 1/A = 1) is much larger than P (D = 2/A = 1), and similarly, P (D = 2/A = 2) is much higher than P (D = 1/A = 2). Therefore, the marginal likelihood score for gene pair A and D will be relatively high. If this marginal likelihood score is higher than a given threshold, gene A will be assigned as the regulator gene of gene D.

There is a very important issue in DBN analysis. Estimation of the transcriptional time lag between a potential regulator gene and its target gene is critical for the DBN analysis. This is because an incorrect estimate of the transcriptional time lag directly affects conditional probabilities, and thus causes real gene–gene relationships to be missed and false gene–gene relationships to be identified. However, most existing DBN approaches do not have a sound approach to estimate the transcriptional time lag between potential regulator gene(s) and target gene(s). Therefore, we have developed a new DBN-based approach, which more accurately estimates the transcriptional time lag between a pair of genes and significantly increases the accuracy of predicting gene–gene relationships *(17)*. In this method, we estimate the transcriptional time lag as the time difference between the first expression change of the regulator gene and that of the target gene. Using this more accurate estimation of the transcriptional time lag, the accuracy of the gene–gene relationship prediction is significantly increased.

Both local clustering and DBN analyses are capable of identifying novel gene–gene relationships from time course microarray data with little prior knowledge of the system being studied. This is because both methods discover novel gene–gene relationships solely based on correlation or statistical expression relationships between potential regulator genes and their target genes. However, the disadvantage of these two methods is that both require a relatively large number of time points. Therefore, we have developed another approach that combines time course microarray data analysis with promoter element and phosphorylation site analysis to discover novel downstream target genes of important signaling pathways, which requires fewer time points. This new method is described in detail in the following section.

AN INTEGRATED BIOINFORMATICS APPROACH (TIME COURSE GENE EXPRESSION PROFILING, PROMOTER ELEMENT, AND PHOSPHORYLATION SITE ANALYSIS)

Our laboratory has developed an integrated approach in which (a) time course gene expression profiling and (b) promoter element analysis and/or phosphorylation site analysis are used to identify downstream target genes of signaling pathways activated after a given treatment. This approach is based on the idea that the expression changes of downstream target genes occur subsequent to expression changes of genes encoding proteins, which regulate the downstream genes. This is simply because an upstream gene has to be first translated into protein, and its protein product can only then directly or indirectly affect the expression of the "downstream" target gene.

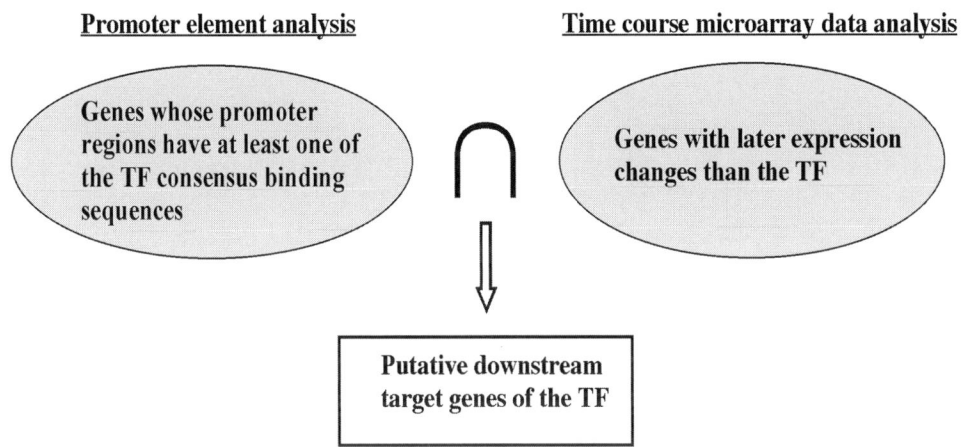

Fig. 3. Schematic representation of identification of putative target genes of a transcription factor (TF) using a combination of promoter element and time course microarray data analyses.

Based on the importance of the "time lag" between the expression changes of the upstream gene and that of the downstream gene as discussed earlier, we first select putative downstream target genes as those whose expression changes occur later than that of the upstream gene. Here, we are more interested in upstream genes that encode either TFs or signaling proteins such as kinases or phosphatases. If the upstream gene encodes a TF, a search for a consensus binding motif of the TF in the predicted proximal promoter regions of the putative target gene is carried out. If the upstream gene encodes a kinase or phosphatase, then a phosphorylation or dephosphorylation site analysis is carried out to identify putative substrates of the given kinase or phosphatase. Substrates (or sometimes secondary substrates) that are TFs are then identified, and a search for the consensus binding motifs of these TFs will be carried out to further select their putative target genes. A schematic representation of this analysis is presented in Fig. 3. This method can identify further downstream target genes of signaling pathways. Another condition for using this method is that the consensus binding motifs of TFs and the phosphorylation/dephosphorylation residues targeted by the kinases and phosphatases, respectively, are known.

USING MICROARRAY DATA ANALYSIS TO UNCOVER GR-MEDIATED BREAST CANCER CELL SURVIVAL PATHWAYS

Identification of GR Direct Target Genes Using Single Time Point Microarray Data Analysis

In this section, we describe an example of a single time point analysis—our identification of direct target genes of the GR in mammary epithelial cells. To better understand these direct GR-mediated gene expression changes that might promote cell survival in MECs, we used high-density oligonucleotide microarrays to identify GR-regulated genes activated or repressed 30 min after Dex or co-treatment of Dex and RU486 (an antagonist of GR) (Fig. 4) *(18)*. Three independent microarray experiments were performed to identify gene expression changes occurring 30 min after Dex treatment.

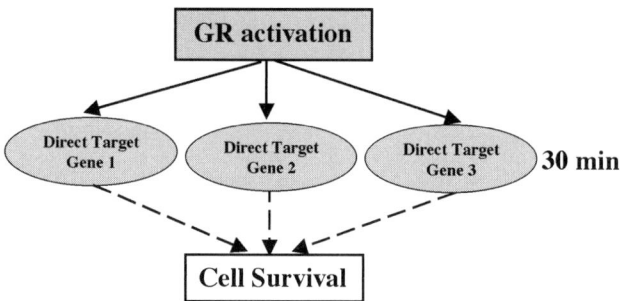

Fig. 4. Schematic representation illustrating activation or inhibition of GR directly regulated target genes may mediate GR-induced cell survival.

Affymetrix 4.0 software calls a gene "present," "marginal," or "absent" depending on the intensity of normalized gene expression defined by a complex algorithm.

In these three experiments, an average of 13.2% of the marginal or present genes met the criteria for 50% downregulation by Dex (1467 ± 144 genes/experiment), whereas an average of 9.7% of these genes met the criteria of higher than 1.5-fold upregulation by Dex (1090 ± 116 genes/experiment). Interestingly, using these fold-change criteria, only 69 downregulated and 95 upregulated genes were found to be commonly regulated in all three experiments (Fig. 5). To examine the consistency of gene regulation by Dex or Dex/RU486 across all three experiments, we performed a paired Student t-test on the expression intensity after hormone treatment (compared with vehicle alone) for each gene. From Fig. 5(C), we found that 45 of the 95 upregulated genes and 30 of the 69 downregulated genes had expression levels that were similar enough to yield a $p \leq 0.05$. These 75 up- and downregulated genes could be further divided into five functional groups: signal transduction, metabolism, putative TF, proapoptotic, or cell cycle/DNA repair genes (Fig. 6). Conversely, 260 genes had significantly similar p-values for expression across all three experiments, but did not meet the fold-change criteria of a 1.5-fold increase or a 50% decrease in all three experiments. This number reflects genes for which small differences in expression, although statistically consistent, were relatively insignificant in magnitude.

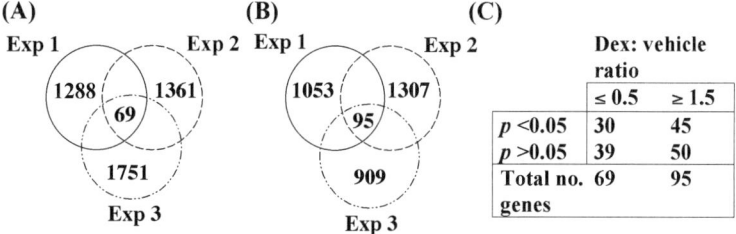

Fig. 5. Venn diagrams showing both the number of genes down-regulated (A) or upregulated (B) by Dex in each of three separate experiments and the number of genes common to all three replicates. (C) *P*-values for down- and upregulated genes. (This figure is modified from Figure 1(a) in *Cancer Research* (2005) **64**:1757–1764, with permission.)

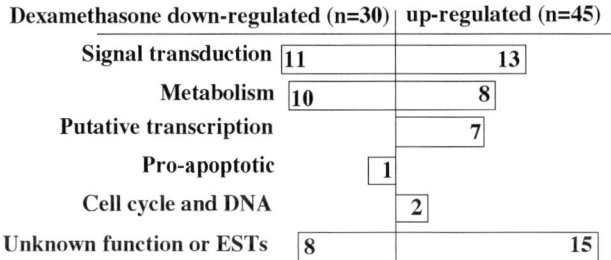

Fig. 6. Functional clustering of Dex-regulated genes that meet both the fold change cutoff (downregulation, ≤0.5; upregulation, ≥1.5) and $p<0.05$. EST, expressed sequence tags. (This figure is modified from Figure 1(b) in *Cancer Research* (2005) **64**:1757–1764, with permission.)

We also identified 34 putative direct GR target genes that were both induced higher than 1.5-fold by Dex and also had their expression inhibited by at least 20% by the addition of the GR antagonist RU486 (Table 1). Of these 34 genes, 19 had consistent enough expression values for Dex upregulation that the intensities across all three experiments achieved $p \leq 0.05$. RU486 caused downregulated gene intensities that were significantly consistent ($p \leq 0.05$) for 11 of these 34 genes. However, 29 of 34 genes had a $p < 0.10$ for upregulation with Dex alone and 24 of 34 had $p < 0.10$ for repression with RU486. Four of these likely GR target genes, (*SGK*), *MKP-1*, inhibitor of nuclear factor κB α (*IκBα*) and growth arrest and DNA damage-inducible protein α (*GADD45α*), were further validated for gene induction following Dex treatment and inhibition following Dex/RU486 treatment by Northern blot analysis.

Unlike previous reports examining GR-induced expression of genes in hepatocytes *(8)*, gastric cancer cells *(19)*, and mouse MECs *(7)*, no prosurvival Bcl-2 family members or inhibitors of apoptosis genes were upregulated in our screen at 30 min, nor proapoptotic genes including death ligands or receptors (e.g., FasL/Fas), caspases (e.g., caspase-3, -8, -9) or proapoptotic Bcl-2 family members were downregulated 30 min after Dex treatment. Instead, gene expression profiling of MCF10A-Myc cells suggests that early GR-regulated genes predominantly include genes that encode signal transduction, metabolism, and TF genes in epithelial cells rather than the cytokine and proapoptotic genes that have been identified in analogous experiments in lymphocytes *(20, 21)*.

Two targets of GR activation identified from the single time point microarray analysis were chosen for further study. First, *SGK-1*, a downstream target of the phosphatidylinositol 3-kinase (PI3K) pathway, is a serine/threonine kinase that was shown previously to mitigate growth factor deprivation-induced apoptosis in neurons *(22)* and MECs *(23)*. Second, upregulation of *MKP-1*, a gene encoding an MAPK phosphatase, results in cell survival in prostate cancer cells *(24)* and mouse fibroblast C3H10T1/2 cells subjected to an apoptotic stimulant *(25)*. Overexpression of MKP-1 is also associated with increased tumorigenicity in breast *(26)*, ovarian *(27)*, and pancreatic cancers *(28)*. We have demonstrated that endogenous SGK-1 and MKP-1 protein levels are increased in breast cancer cell lines treated with Dex. Furthermore, ectopic expression of either SGK-1 or MKP-1 inhibits chemotherapy-induced apoptosis to approximately the same degree as does Dex pretreatment. Using small interfering RNA (siRNA) to decrease SGK-1 and MKP-1 expression after GR activation, we have

Table 1

Genes ($n = 34$) Upregulated by Dexamethasone (Dex) and Repressed by RU486 (RU) in Three of Three Experiments.

GenBank accession no	Gene product	Dex:Vehicle ratio[a]	p	Dex-RU:Vehicle ratio[a]	p[b]
Signal transduction					
S62539	Insulin receptor substrate 1 (IRS-1)	10.4±2.9	0.02	2.5±1.5	0.10
X70340	Transforming growth factor α(TGFα)	9.6±6.0	0.06	3.1±2.1	0.37
U07794	Tyrosine kinase (TXK)	9.5±1.6	0.01	3.8±2.8	0.14
M60047	Heparin-binding growth factor binding protein (FGFBP1)	9.1±4.5	0.05	1.0±0.0	0.05
X68277	CL 100/protein-tyrosine phosphatase (MKP-1)	6.4±1.8	0.04	3.5±1.4	0.08
X75342	SHB adaptor protein (a Src homology 2 protein)	6.3±2.9	0.06	3.7±2.0	0.13
U56387	Proprotein convertase subtilisin/kexin type 5 (PCSK5)	5.4±2.3	0.08	2.0±1.1	0.02
U47011	Fibroblast growth factor 8 (androgen-induced) (AIGF)	4.9±2.4	0.10	0.4±0.3	0.01
M22489	Bone morphogenetic protein 2 precursor (BMP2A)	3.6±1.2	0.07	1.0±0.4	0.15
Y10032	Serum and glucocorticoid regulated kinase (SGK)	3.3±0.7	0.03	1.8±0.3	0.02
M69043	IκB-α (MAD-3/IκB- α)	2.8±0.4	0.02	1.4±0.2	0.09
M60974	Growth arrest and DNA-damage-inducible 45 α(GADD45α)	2.1±0.1	0.01	1.4±0.1	0.01

Accession	Gene				
Metabolism related					
X63359	UDP glycosyltransferase 2 family, polypeptide B10 (UGT2)	16.1±7.9	0.05	2.2±1.2	0.03
U37143	Cytochrome P450, subfamily (CYP2J2)	4.7±0.4	0.03	1.7±0.8	0.20
AJ000503	Dopachrome isomerase, tyrosine-related protein 2 (TYRP2)	4.0±1.8	0.11	2.4±1.4	0.06
L07765	Carboxylesterase 1 (monocyte/macrophage serine esterase 1 (SES1)	3.9±2.2	0.19	2.7±1.6	0.06
X78669	Calcium binding protein (ERC55)	1.9±0.3	0.04	1.1±0.1	0.08
Transcription					
U78722	Zinc finger protein 165 (LD65)	2.3±0.2	0.01	0.4±0.2	0.06
AF025770	Zinc finger protein 189 (ZNF189)	1.9±0.2	0.04	0.7±0.2	0.13
AF003540	Zinc finger protein 195 (ztfp104)	1.8±0.1	0.01	1.1±0.2	0.09
X64318	Nuclear factor, interleukin 3 regulated (NF-IL3A)	1.6±0.1	0.01	1.2±0.1	0.01
Miscellaneous					
AJ01	Tropinin T1, skeletal, slow (ANM)	13.1±6.2	0.09	1.0±0.0	0.11
AJ003125	Cleavage of aminopropeptide of procollagens I, II and V (PCPN1)	8.7±5.7	0.12	2.9±2.3	0.14
AJ011733	Synaptogyrin 4 (syngr4)	3.1±0.3	0.01	1.5±0.3	0.01
Y12851	P2X7 receptor (P2X7)	2.4±0.3	0.02	1.5±0.1	0.02

This table is modified from Table 1 in *Cancer Research* (2005) **64**:1757–1764 and reproduced with permission.

[a] The mean ratio (derived from GeneSpring 4.0) of expression intensity ± SE in three independent experiments.

[b] Dex/RU486 versus Dex.

[c] EST, expressed sequence tag.

shown a corresponding decrease in Dex-mediated protection from chemotherapy in the presence of either SGK-1 siRNA or MKP-1 siRNA. In summary, these observations suggest that GR-mediated transcriptional activation of both SGK-1 and MKP-1 contribute to a GR-mediated signal transduction pathway that ultimately inhibits chemotherapy induced apoptosis.

Identification of Novel GR-mediated Cell Survival Pathways Using Time Course Microarray Data Analysis

In this section, we will describe how we identified further downstream events following activation (or inactivation) of GR directly regulated target genes in MECs using an integrated bioinformatics approach of time course microarray data and promoter element analyses as described in Section "An integrated bioinformatics approach" (Fig. 3). As discussed in Section "Identification of GR direct target genes using single time point microarray data analysis," we have discovered two previously unrecognized GR direct target genes as being associated with cell survival signaling, *Serum and glucocorticoid-inducible kinase (SGK)* and *Map kinase phosphatase-1 (MKP-1)*, using single time point microarray data analysis. In an effort to further dissect the dynamic mechanisms of GR-mediated breast epithelial cell survival, we performed time course microarray experiments in MCF10A-Myc cells at 30 min, 2 h, 4 h, and 24 h following GR activation. In this section, we will focus on how we discovered downstream events following GR-mediated MKP-1 induction from time course microarray data. A similar approach can likely be used to identify dynamic events following other hormone treatment as well. Figure 7 illustrates a schematic representation showing how direct target genes of GR (e.g., genes up- or downregulated 30 min after GR activation) and indirect target genes of GR (e.g., genes up- or downregulated 2 h following GR activation) form molecular pathways that contribute to GR-mediated cell survival.

MKP-1 has previously been shown to dephosphorylate and inactivate MAP kinases such as extracellular signal-regulated kinase 1/2 (ERK1/2) *(29, 30)*, c-Jun *N*-terminal kinase (JNK) *(31)*, and p38 *(32, 33)*. Recently, we have demonstrated that in breast

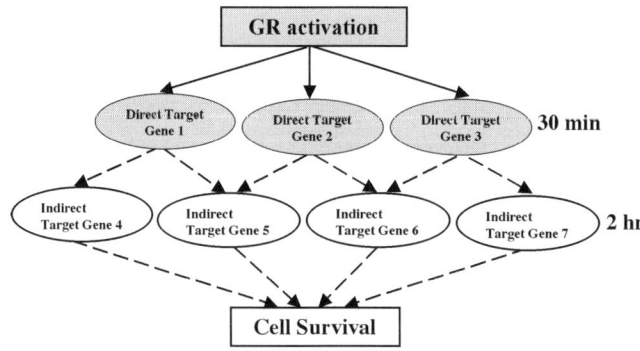

Fig. 7. A schematic representation illustrating how direct target genes of GR (e.g., genes up- or downregulated 30 min after GR activation) and indirect target genes of GR (e.g., genes up- or downregulated 2 h following GR activation) form molecular pathways that contribute to GR-mediated cell survival.

cancer cell lines, GR-mediated MKP-1 induction also results in dephosphorylation and inactivation of ERK1/2 and JNK, while p38 activity is minimally affected *(34)*. Furthermore, inactivation of ERK1/2 and JNK results in inactivation of downstream substrates ELK-1 and c-Jun, both of which are TFs. Inactivation of ELK-1 and c-Jun is sustained for 24 h after Dex treatment. Furthermore, we have shown that ELK-1 and c-Jun inactivation are required for GR-mediated cell survival (Wu and Conzen, unpublished data). Together, these results prompted us to identify further downstream target genes of GR-mediated MKP-1 induction by identifying proapoptotic target genes of ELK-1 and c-Jun, which when downregulated, may contribute to GR-mediated cell survival. In this section, we will describe how we used an integrated bioinformatics approach of time course microarray data and promoter element analyses as described in the Section "An integrated bioinformatics approach" to identify transcriptional target the genes of ELK-1. Similar analyse can also be used to identify transcriptional target genes of c-Jun.

As discussed in the Section "An integrated bioinformatics approach," this analysis is based on the idea that gene expression changes of downstream target genes occur later than that of upstream genes. Our microarray data show a significant induction of *MKP-1* mRNA 30 min following GR activation in MCF10A-Myc cells *(18)*. We hypothesize that because of the time required for translation of *MKP-1* mRNA and subsequent ERK1/2 and ELK-1 dephosphorylation and inactivation, a time lag of at least 2 h may exist between the increase of *MKP-1* mRNA steady-state levels and the subsequent downregulation of ELK-1 target genes. This estimation of time lag is further supported by the phosphorylation studies of p-ELK-1, which shows that p-ELK-1 level starts decreasing at least 2 h after Dex treatment. Therefore, we then identified 907 genes that were downregulated with an averaged fold change (Dex versus vehicle) ≤ 0.7-fold following GR-activation at either 2, 4 or 24 h. However, not every GR-downregulated gene is an ELK-1 target gene; some of these genes are downregulated by GR through other mechanisms. Therefore, we further identified 2860 genes in the human genome that have one or more consensus ELK-1 binding motifs (e.g., CAGGAAG, CCGGAAG, (C/A)GGA(A/T) and GGAANNTTAA where N = A, T, G, or C) between -1000 nt and $+200$ nt. An intersection of the set of GR-downregulated genes and the set of genes in human genome that have at least one ELK-1 binding motifs resulted in 42 putative targets of GR-mediated ELK-1 inactivation (Fig. 8). Among these 42 putative ELK-1 target genes, 17 have known roles in signal transduction. Furthermore, nine of these 17 signal transduction genes have gene expression levels consistent enough in all three replicates to achieve $p \leq 0.05$, while the other eight genes have $p > 0.05$. Due to the fact that gene expression changes following GR activation are relatively modest, p-values of up- or downregulated genes by Dex tend to be relatively high. Therefore, in order to reduce possible type II (false negative) errors, we considered genes with both $p \leq 0.05$ and $p > 0.05$ for further studies as long as the genes encode proteins involved in cell survival. As a result, we focused on three signal transduction genes, Disabled (Drosophila) 2 *(DAB2)* ($p=0.04$), PMA-induced protein 1 *(Noxa)* ($p = 0.05$), and tissue plasminogen activator *(tPA)* ($p = 0.25$) for further investigations.

Tissue-type plasminogen activator *(tPA)* is a serine protease primarily involved in the intravascular dissolution of blood clots. We chose to examine *tPA* as a putative ELK-1 target gene based on published studies showing that *tPA* protein expression can be proapoptotic *(35, 36)*. Low intratumoral *tPA* levels have also been associated

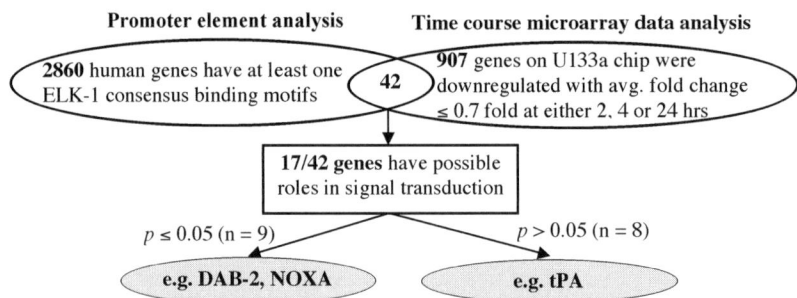

Fig. 8. Identification of putative ELK-1 target genes using a combination of time course microarray data and promoter element analyses. Putative target genes (e.g., *DAB-2*, *Noxa*, and *tPA)* with possible roles in apoptosis were selected for further investigation.

with aggressiveness and poor prognosis in breast cancer patients *(37)*. Furthermore, expression of *tPA* has been shown to be inhibited by Dex in rat mesangial cells *(38)*. In addition, activation of JNK-signaling pathway has been demonstrated to increase expression of *tPA* in endothelial cells *(39)*. In order to validate that the expression of *tPA* is indeed inhibited by GR, we performed quantitative RT-PCR experiments in MCF10A-Myc cells and the results confirmed the downregulation of *tPA* by GR *(34)*. The next step was to test whether *tPA* is a transcriptional target gene of ELK-1. We performed luciferase assay and confirmed that the promoter activity of *tPA* is indeed sensitive to glucocorticoid treatment in an ELK-1 specific manner *(34)*. Further experiments will be carried out to verify specific binding of ELK-1 to the promoter of *tPA* using a ChIP assay. Functional assays will include downregulating *tPA* by siRNA (independently of GR activation) to determine whether or not this is associated with a prosurvival effect in breast cancer cells. In addition, we will ectopically express *tPA* to determine whether or not this can reverse GR-mediated cell survival.

Two additional putative ELK-1 target genes, *Noxa* and *DAB2*, have also been implicated in proapoptotic activities. For example, *Noxa* mediates apoptosis both dependently and independently of p53 activation in a wide range of cancer cell lines, such as head and neck cancer, lung cancer, and breast cancer cells *(40–44)*. *DAB2* is a putative tumor suppressor in breast and ovarian cancers. *DAB2* expression has been shown to be absent or very low in the majority of breast and ovarian cancer cell lines and primary tumors *(45)*. However, the downregulation of *Noxa* and *DAB2* following GR activation in MECs has not been previously demonstrated and *Noxa* and *DAB2* are not known to be ELK-1 target genes. Thus, this approach revealed a hypothetical network that can be tested using traditional molecular biology. The downregulation of both *Noxa* and *DAB2* by GR in MECs has been confirmed by quantitative RT-PCR and we are currently validating whether or not *Noxa* and *DAB2* are indeed ELK-1 transcriptional target genes, and whether or not their downregulation contributed to the prosurvival signaling of the GR. Furthermore, we will test whether downregulation of *Noxa* and *DAB2* contribute to GR-mediated breast epithelial cell survival using similar approaches as for *tPA*.

The integrated bioinformatics approach of time course microarray data and promoter element analysis we have employed earlier appears to be an efficient approach for identifying putative downstream target genes of signaling pathways that involve transcriptional regulation. This is demonstrated by the identification of novel downstream target genes of GR-MKP-1-ERK-ELK-1 signaling pathway. We predict this approach can be applied to discover novel downstream target genes of signaling pathways following many other pharmacological treatments.

CONCLUSIONS

Genome-wide DNA microarrays are powerful tools, and provide a glimpse of the signals and interactions within regulatory pathways of the cell. They enable the simultaneous measurement of mRNA abundance of most, if not all, identified genes in a genome under different physiological conditions. Because signaling pathways are events that take place over time, time course microarray experiments should allow us to identify dynamic molecular mechanisms underlying well-established phenotypes. A wide variety of time course microarray data analysis approaches have been proposed in the past decade, which have succeeded to some extent in discovering novel molecular pathways from time course microarray data. However, the accuracy of pathway prediction using existing approaches is still limited, and this is largely due to our incomplete understanding of complex molecular biological systems. As our basic understanding of gene regulation advances, computational approaches will become increasingly sophisticated and useful for generating testable hypotheses.

REFERENCES

1. Bamberger CM, Schulte HM, Chrousons GP. Molecular determinants of glucocorticoid receptor function and tissue sensitivity to glucorticoids. *Endocr Rev* 1996;**17**:245–261.
2. Jonat C, Rahmsdorf HJ, Park KK, Cato AC, GEbel S, Ponta H, Herrlich P. Antitumor promotion and antiinflammation: down-modulation of AP-1 (Fos/Jun) activity by glucocorticoid hormone. *Cell* 1990;**62**:1189–1204.
3. Scheinman RI, Gualberto A, Jewell CM, Cidiowski JA, Baldwin AS Jr. Characterization of mechanism involved in transrepressin of NF-kaapa B by activated glucocorticoid receptors. *Mol Cell Biol* 1995;**15**:843–953.
4. Lipworth BJ. Therapeutic implications of non-genomic glucocorticoid activity. *Lancet* 2000;**356**: 87–89.
5. Distelhorst CW. Recent insights into the mechanisms of glucocorticosteroid-induced apoptosis. *Cell Death Differ* 2002;**9**:6–19.
6. Moran TJ, Gray S, Mikosz CA, Conzen SD. The glucocorticoid receptor mediates a survival signal in human mammary epithelial cells. *Cancer Res* 2000;**60**:867–872.
7. Schorr K, Furth PA. Induction of bcl-xL expression in mammary epithelial cells is glucocorticoid-dependent but not signal transducer and activator of transcription 5-depdendent. *Cancer Res* 2000;**60**:5950–5953.
8. Bailly-Maitre B, de Sousa G, Boulukos K, Gugenheim J, Rahmani R. Dexamethasone inhibits spontaneous apoptosis in primary cultures of human and rat hepatocytes via Bcl-2 and Bcl-xL induction. *Cell Death Differ* 2001;**8**:279–288.
9. Newton CJ, Xie YX, Burgoyne CH, Adams L, Atkin SL, Abidia A, McCollum PT. Fluvastatin induces apoptosis of vascular endothelial cells: blockade by glucocorticoids. *Cardiovasc Surg* 2003;**11**:52–60.
10. Weinstein RS, Chen JR, Powers CC, Stewart SA, Landes RD, Bellido T, Jilka RL, Parfitt AM, Manolagas SC. Promotion of osteoclast survival and antagonism of bisphosphonate-induced osteoclast apoptosis by glucocorticoids. *J Clin Invest* 2002;**109**:1041–1048.
11. Clark AR. MAP kinase phosphatase 1: a novel mediator of biological effects of glucocorticoids? *J. Endocrinol* 2003;**178**:5–12.
12. Eisen MB, Spellman PT, Brown PO, Botstein D. Cluster analysis and display of genome-wide expression patterns. *Proc Natl Acad Sci* 1998;**95**:14863–14868.

13. Qian J, Dolled-Filhart M, Lin J, Yu H, Gerstein M. Beyond synexpressin relationships: local clustering of time-shifted and inverted gene expressiono profiles identifies new, biologically relevant interactions. *J Mol Biol* 2001;**314**:1053–1066.

14. Murphy K, Mian S. Modeling gene expression data using dynamic Bayesian networks. *Technical Report*, Computer Science Division, University of California, Berkeley, CA, USA, 1999.

15. Kim SY, Imoto S, Miyano S. Inferring gene networks from time series microarray data using dynamic Bayesian networks. *Brief Bioinform* 2003;**4**:228–235.

16. Perrin, BE, Ralaivola L, Mazurie A, Bottani S, Mallet J, D'Alche-Buc F. Gene networks inference using dynamic Bayesian networks. *Bioinformatics* 2003;**19**, Suppl 2:II138–II148.

17. Zou M, Conzen SD. A new dynamic Bayesian network (DBN) approach for identifying gene regulatory networks from time course microarray data. *Bioinformatics* 2005;**21**:71–79.

18. Wu W, Chaudhuri S, Brickley DR, Pang D, Karrison T, Conzen SD. Microarray analysis reveals glucocorticoid-regulated survival genes that are associated with inhibition of apoptosis in breast epithelial cells. *Cancer Res* 2004;**64**:1757–1764.

19. Chang TC, Hung MW, Jiang SY, Chu JT, Chu LL, Tsai LC. Dexamethasone suppresses apoptosis in a human gastric cancer cell line through modulation of bcl-x gene expression. *FEBS Lett* 1997;**415**:11–15.

20. Galon J, Franchimont D, Hiroi N, Frey G, Boettner A, Ehrhart-Bornstein M, O'shea JJ, Chrousos GP, Bornstein SR. Gene profiling reveals unknown enhancing and suppressive actions of glucocorticoids on immune cells. *FASEB J* 2002;**16**:61–71.

21. Franchimont D, Galon J, Vacchio MS, Fan S, Visconti R, Frucht DM, Greenen V, Chrousos GP, Ashwell JD, O'shea JJ. Positive effects of glucocorticoids on T cell function by up-regulation of IL-7 receptor J. *J Immunol* 2002;**168**:2212–2218.

22. Brunet A, Park J, Tran H, Hu LS, Hemmings BA, Greeenberg ME. Protein kinase SGK mediates survival signals by phosphorylating the forkhead transcription factor FKHRL1 (FOXO3a). *Mol Cell Biol* 2001;**21**:952–965.

23. Mkiosz CA, Brickley DR, Sharkey MS, Moran TW, Conzen SD. Glucocorticoid receptor-mediated protection from apoptosis is associated with induction of the serine/threonine survival kinase gene, sgk-1. *J Biol Chem* 2001;**276**:16649–16654.

24. Magi-Galluzzi C, Mishra R, Fiorentino M, Montironi R, Yao H, Capodieci P, Wishnow K, Kaplan I, Stork P, Loda M. Mitogen-activated protein kinase phosphatase 1 is overexpressed in prostate cancers and is inversely related to apoptosis. *Lab Investig* 1997;**76**:37–51.

25. Castillo SS, Teegarden D. Sphingosine-1-phosphate inhibition of apoptosis requires mitogen-activated protein kinase phosphatase-1 in mouse fibroblast C3H10T(1/2) cells. *J Nutr* 2003;**133**:3343–3349.

26. Wang H, Cheng Z, Malbon C. Overexpression of mitogen-activated protein kinase phosphatases MKP1, MKP2 in human breast cancer. *Cancer Lett* 2003;**191**:229–237.

27. Denkert C, Schmitt W, Berger S, Reles A, Pest S, Siegert A, Lichtenegger W, Dietel M, Hauptmann S. Expression of mitogen-activated protein kinase phosphatase-1 (MKP-1) in primary human ovarian carcinoma. *Int J Cancer* 2002;**102**:507–513.

28. Liao Q, Guo J, Kleeff J, Zimmermann A, Buchler M, Korc M, Friess H. Down-regulation of the dual-specificity phosphatase MKP-1 suppresses tumorigenicity of pancreatic cancer cells. *Gastroenterology* 2003;**124**:1830–1845.

29. Steinmetz R, Wagoner HA, Zeng P, Hammond JR, Hannon TS, Meyers JL, Pescovitz OH. Mechanisms regulating the constitutive activation of the extracellular signal-regulated kinase (ERK) signaling pathway in ovarian cancer and the effect of ribonucleic acid interference for ERK1/2 on cancer cell proliferation. *Mol Endocrinol* 2004;**18**:2570–2582.

30. Liu C, Shi Y, Du Y, Ning X, Liu N, Huang D, Liang J, Xue Y, Fan D. Dual-specificity phosphatase DUSP1 protects overactivation of hypoxia-inducible factor 1 through inactivating ERK MAPK. *Exp Cell Res* 2005;**309**:410–418.

31. Liu Y, Gorospe M, Yang C, Holbrook NJ. Role of mitogen-activated protein kinase phosphatase during the cellular response to genotoxic stress. *J Biol Chem* 1995;**270**:8377–8380.

32. Imasato A, Desbois-Mouthon C, Han J, Kai H, Cato AC, Akira S, Li JD. Inhibition of p38 MAPK by glucocorticoids via induction of MAPK phosphatase-1 enhances nontypeable haemophilus influenzae-induced expression of toll-like receptor 2. *J Biol Chem* 2002;**277**:47444–47450.

33. Lasa M, Abraham SM, Boucheron C, Saklatvala J, Clark AR. Dexamethasone causes sustained expression of mitogen-activated protein kinase (MAPK) phosphatase 1 and phosphatase-mediated inhibition of MAPK p38. *Mol Cell Biol* 2002;**22**:7802–7811.
34. Wu W, Pew T, Zou M, Pang D, Conzen SD. Glucocorticoid receptor-induced MAPK phosphatase-1 (MPK-1) expression inhibits paclitaxel-associated MAPK activation and contributes to breast cancer cell survival. *J Biol Chem* 2005;**280**:4117–4124.
35. Tsirka SE, Rogove AD, Strickland S. Neuronal cell death and tPA. *Nature* 1996;**384**:123–124.
36. Flavin MP, Zhao G, Ho LT. Microglial tissue plasminogen activator (tPA) triggers neuronal apoptosis in vitro. *Gila* 2000;**29**:347–354.
37. Corte MD, Verez P, Rodriguez JC, Roibas A, Dominguez ML, Lamelas ML, Vazquez J, Garcia Muniz JL, Allende MT, Gonzalez LO, Fueyo A, Vizoso F. Tissue-type plasminogen activator (tPA) in breast cancer: relationship with clinicopathological parameters and prognostic significance. *Breast Cancer Res Treat* 2005;**90**:33–40.
38. Eberhardt W, Kilz T, Akool el-S, Muller R, Pfeilschifter J. Dissociated glucocorticoids equipotently inhibit cytokine- and cAMP-induced matrix degrading proteases in rat mesangial cells. *Biochem Pharmacol* 2005;**70**:433–445.
39. Gingras D, Nyalendo C, Di Tomasso G, Annabi B, Beliveau R. Activation of tissue plasminogen activator gene transcription by Neovastat, a multifunctional antiangiogenic agent. *Biochem Biophys Res Commun* 2004;**320**:205–212.
40. Flinterman M, Guelen L, Ezzati-Nik S, Killick R, Melino G, Tominaga K, Mymryk JS, Gaken J, Tavassoli M. E1A activates transcription of p73 and Noxa to induce apoptosis. *J Biol Chem* 2005;**280**:5945–5959.
41. Qin JZ, Stennett L, Bacon P, Bodner B, Hendrix MJ, Seftor RE, Seftor EA, Margaryan NV, Pollock PM, Curtis A, Trent JM, Bennett F, Miele L, Nickoloff BJ. P53-independent Noxa induction overcomes apoptotic resistance of malignant melanomas. *Mol Cancer Ther* 2004;**3**:895–902.
42. Yakovlev AG, Di Giovanni S, Wang G, Liu W, Stoica B, Faden AI. BOK and NOXA are essential mediators of p53-dependent apoptosis. *J Biol Chem* 2004;**279**:28367–28374.
43. Seo YW, Shin JN, Ko KH, Cha JH, Park JY, Lee BR, Yun CW, Kim YM, Seol DW, Kim DW, Yin XM, Kim TH. The molecular mechanism of Noxa-induced mitochondrial dysfunction in p53-mediated cell death. *J Biol Chem* 2003;**278**:48292–48299.
44. Jansson AK, Emterling AM, Arbman G, Sun XF. Noxa in colorectal cancer: a study on DNA, mRNA and protein expression. *Oncogene* 2003;**22**:4675–4678.
45. Sheng Z, Sun W, Smith E, Cohen C, Sheng Z, Xu XX. Restoration of positioning control following disabled-2 expression in ovarian and breast tumor cells. *Oncogene* 2000;**19**:4847–4854.

10

Application of Microarrays for Gene Transcript Analysis in Type 2 Diabetes

R. Sreekumar, PhD, C. P. Kolbert, MS,
Y. Asmann, PhD, and K. S. Nair, MD, PhD

CONTENTS

INTRODUCTION
METHODOLOGICAL DEVELOPMENTS
EXPERIMENTAL DESIGN
MICROARRAY DATA ANALYSIS
GENE TRANSCRIPT LEVEL ALTERATIONS IN TYPE 2 DIABETES
REAL-TIME REVERSE TRANSCRIPTION POLYMERASE CHAIN REACTION
    AND ITS APPLICATION IN DIABETES RESEARCH
CONCLUSION
REFERENCES

INTRODUCTION

DNA microarrays, when used as tools for measuring RNA transcript levels, enhance our ability to understand how genes or their transcripts are involved in disease processes and their complications. The measurement of gene transcript levels provides researchers with an estimation of the global gene expression within cells and helps researchers to elucidate pathways of biological or disease processes. Modern microarray techniques have the potential to improve medical treatment by allowing physicians to identify deregulated genes and the molecular and biochemical pathways that they influence. One of the most important contributions of microarray techniques is the development of experimental hypotheses that are based on the analysis of gene transcript profiles. Moreover, the technology provides the opportunity to identify biomarkers of disease and targets for drug treatment. As a result, patient care may be enhanced by the measurement of whole-cell gene transcript levels within an individual person, which may allow physicians to facilitate a personalized treatment approach.

In the last several years, gene transcript measurements have been used to make numerous achievements in the field of endocrine health and disease. For example, in

From: *Contemporary Endocrinology: Genomics in Endocrinology:*
DNA Microarray Analysis in Endocrine Health and Disease
Edited by: S. Handwerger and B. Aronow © Humana Press, Totowa, NJ

1999, the National Institute of Diabetes and Digestive and Kidney Diseases promoted genomic research in Endocrinology by providing funding for the program "Functional Genomics of the Developing Endocrine Pancreas". As a result of this RFA, the Endocrine Pancreas Consortium was formed to create and sequence cDNA libraries of genes related to pancreatic development. One of the products of this work was a glass cDNA microarray, called the PancChip *(1)*. The first release contained 3,400 cDNA clones from IMAGE, dbEST, and Incyte Genomics libraries that are expressed by the mouse pancreas or affected by diabetes. In later studies, the mouse PancChip was updated to include more than 10,000 mouse genes, homologs, and ESTs *(2,3)*, and a human version of the PancChip was assembled, which includes over 13,000 gene transcripts *(3)*. In addition to this work, many other groups, including our own, have made use of the wide variety of microarray platforms to study the expression of genes related to type 2 diabetes *(4,5)*.

Indeed, the ability of microarrays to interrogate virtually all known and putative genes within a cell makes this tool especially useful for the study of complex diseases. The complications associated with type 2 diabetes exhibit an interrelated series of phenotypic and genotypic manifestations. Although much is yet to be learned, greater understanding of the pathophysiology of disease has been afforded by studies that employed microarrays for the measurement of gene transcript levels.

A previous review summarized the technical concepts of microarray design as well as experimental methods to perform gene expression analysis with custom-spotted microarrays *(6)*. The current review will provide an overview of recent advances in the understanding of type 2 diabetes as well as some of the challenges inherent in microarray data analysis. In addition, we will summarize a select group of microarray platforms that are used commonly today.

METHODOLOGICAL DEVELOPMENTS

Several different DNA microarray platforms are available currently to measure gene expression. They can be classified into two broad categories. Spotted microarrays include PCR-amplified complementary DNA (cDNA) clones or oligonucleotides spotted onto solid or flexible substrates (Fig. 1). This platform was developed originally for glass microarrays at Stanford University in 1995 and continues to be used by research labs across the country *(7)*. Historically, cDNA amplification products were used as targets for hybridization. These products usually range in size from 500 bp to 2 kb, thus allowing for great sensitivity during the hybridization process. However, cDNA arrays required that thousands of bacterial clones be grown and maintained. Once the clones are grown, PCR amplification, agarose gel electrophoresis, and DNA purification are necessary to generate fragments suitable for microarray spotting. This process is both costly and time-consuming. As a result, many research groups that prefer to fabricate their own arrays now use oligonucleotide targets that are purchased from commercial vendors. Oligonucleotide targets, which usually range in size from 25 to 80 bp, often provide a higher level of specificity to the hybridization reaction than do cDNA targets, and delivery formats make pre-spotting standardization of the oligo concentrations relatively easy.

The second platform, represented by the GeneChip® array from Affymetrix (Santa Clara, CA, USA; www.affymetrix.com), uses short oligonucleotide probes synthesized

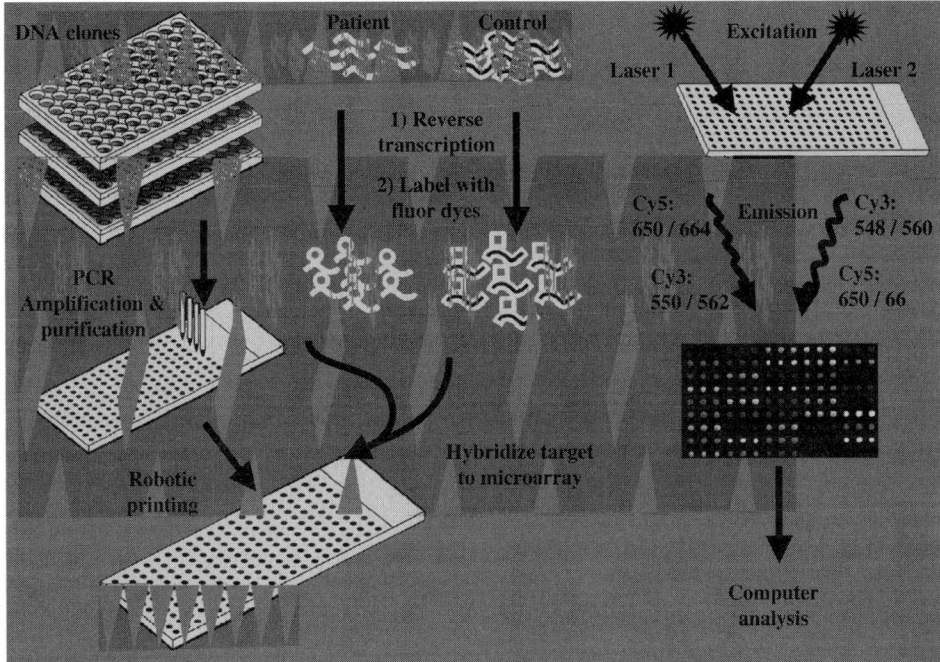

Fig. 1. The steps involved in cDNA microarray analysis. The DNA clones are amplified, purified, and printed onto the glass slides for gene of interest. The total RNA from control and patients samples were separately reverse transcribed, followed by labeling with Cy3 and Cy5 dyes. The labeled RNA is hybridized onto the array and scanned. The image is analyzed for differential expression of transcripts of interest. Adapted from Duggan et al., Jan 1999, Nature Genetics Supp., Vol 21, pp 10–14.

in situ on silicon wafers (8). The Affymetrix array has become one of the most widely used platforms today. Both platforms quantify the levels of gene transcripts within a tissue and allow simultaneous measurement of the expression of tens of thousands of gene transcripts in a single array.

The experimental protocols for performing microarray experiments vary depending on the technology and the detection method. In cDNA microarray analysis, a gene transcript level is quantified by the ratio of the gene-specific intensities between two biological samples (Fig. 2). For Affymetrix GeneChip™ analyses, each biological sample is hybridized to one array. In order to identify gene expression differences within two or more biological samples, intensity data from the arrays is compared. Regardless of the platform, each microarray experiment produces a dataset containing vast numbers of gene expression values and, therefore, requires powerful statistical and analytical tools for data analysis.

Merits and Limitations of Selected Microarray Platforms

CUSTOM-SPOTTED MICROARRAYS

Since the description of custom-spotted arrays in 1995 (7), numerous institutions have established laboratories for the fabrication of microarrays and subsequent labeling and hybridization of experimental samples. Indeed, several courses and seminars are

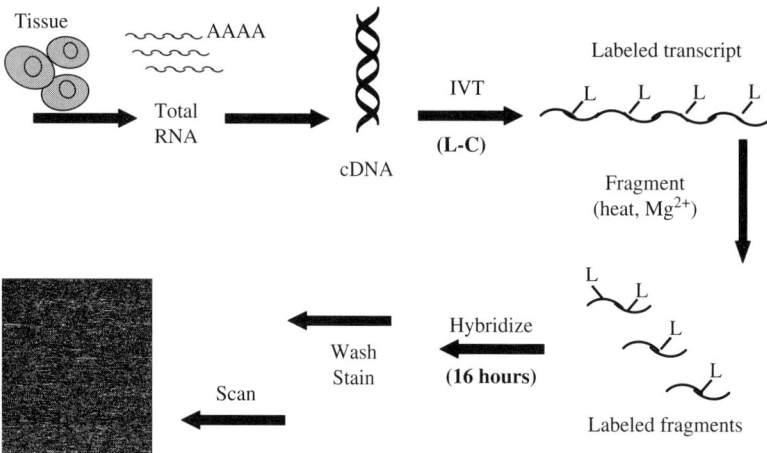

Fig. 2. The steps involved in the Affymetrix gene expression analysis. Total RNA isolated from tissue or serum sample is converted into cDNA followed by in vitro transcription assay, which convert the cDNA to biotin-labeled cRNA. The labeled cRNA is fragmented with heat (45°C) in presence of magnesium chloride. The fragmented cRNA is hybridized onto the Affymetrix GeneChip (e.g., U133 2.0 Plus) and stained with streptavidin–phycoerythrin, washed and scanned to obtain the image which is subsequently analyzed for calculating the differential expression of 54 000 transcripts expression levels.

held each year, which teach prospective microarray researchers how to label and hybridize samples, scan arrays, and analyze the data. Compared to commercial arrays, home-brew arrays may be considerably less expensive, depending on the number of genes included and arrays to be hybridized.

However, considerable expertise is required to design and construct spotted arrays. Labeling and hybridization of custom arrays requires strict attention to detail and superior quality control in order to generate data that can be trusted. The quality of RNA samples should be tested by agarose gel or microcapillary electrophoresis (Agilent 2100 Bioanalyzer, Palo Alto, CA, USA) and proven to be adequate prior to labeling. RNA samples that are deemed to be of poor quality should be removed from the sample set. Labeling efficiency and the dye incorporation rate should be assessed for each sample and numerical cutoff values for acceptance should be assigned. It is necessary to include both positive and negative spotting controls in each run. Housekeeping genes are often included on microarrays as relatively stable elements that can be used as normalization factors. But they can also be used to determine the overall quality of the hybridization. Oligonucleotides derived from nonhomologous species are generally included on spotted microarrays to provide controls for the labeling reactions. Validation studies should include an analysis of the spike controls to show that they are performing as expected. All controls, including the positive controls mentioned above and negative controls, in the form of blank wells and spotting solution without nucleic acid target, should be placed into each block of the array so that the pin-associated array quality can be determined during the process of data analysis. Subsequent to spotting, representative arrays should be scanned to provide an assessment of the overall run quality. Most custom arrays are constructed on glass or nylon substrates. Glass arrays are normally coated with substances such as poly-L-lysine, aminosilane, or aldehyde that allow DNA targets to adhere either by nonspecific binding or by covalent bonds. Fluorescent labeled probes, such as cy3 and cy5 or the Alexa fluors 555 and 647, are often used to detect transcript-specific probe intensity. Several different vendors market laser or CCD-based scanners that can be used for scanning and capturing fluorescent intensities. Radiolabeled or chemiluminescent probes, which require the use of a phosphorimager, are routinely used to detect transcript levels on flexible substrates such as nylon.

RESEARCH GENETICS MAMMALIAN GENEFILTERS

One of the first companies to offer commercial spotted arrays to the research market was Research Genetics (http://www.resgen.com/products/MammGF.php3). Now marketed under the name "Mammalian GeneFilters ®" and distributed by Invitrogen (Carlsbad, CA, USA), the nylon cDNA arrays have been developed for human, rat, and mouse. Recommended RNA input ranges from 1 to 10 µg of total RNA for hybridization of up to four different GeneFilters at a time. Hybridization is detected by radiolabeled nucleotides. Currently, nine different releases for the human GeneFilters array are available, each containing 5184 known genes or ESTs derived from I.M.A.G.E. and Lawrence Livermore National Lab cDNA clones. In addition, specialty arrays are available for genes expressed in bone and skin as well as genes important for pharmacology-related research. In an age of fluorescent and chemiluminescent detection, many labeling protocols for nylon or filter-based arrays still call for radioactive labeled targets. One of the potential advantages of nylon membrane arrays is that they may be stripped and used again. However, this process should be validated with test arrays to ensure complete stripping and adequate hybridization of subsequent targets.

SUPERARRAY GEARRAY BIOSCIENCE

SuperArray (Frederick, MD, USA; http://www.superarray.com) uses a specialized 3D nylon matrix to create membrane arrays that are marketed under the trade name GEArray®. Although many array vendors have focused on providing whole-genome arrays, SuperArray has chosen to provide microarrays for those interested in very specific disease pathways including, but not limited to, diabetes, cardiovascular disease markers, and apoptosis. Indeed, microarrays containing from 100 to 500 cDNA or 60mer oligonucleotide probes have been created for about 75 different pathways. The arrays interrogate genes from three mammalian species, including human, rat, and mouse. Detection of gene expression occurs by chemiluminescent or fluorescent-labeled targets from as little as 100 ng of total RNA. In addition, the company offers a custom array service for investigators who do not have access to specialized laboratory instrumentation.

BD BIOSCIENCES CLONTECH ATLAS

Glass, nylon, and plastic substrate arrays are offered by BD Biosciences Clontech as Atlas™ Microarrays (Mountain View, CA, USA; http://www.clontech.com/clontech/atlas/index.shtml). Glass and plastic arrays contain long oligos (80mers) and are available for human, rat, and mouse applications. Expression-related spot intensity derived from glass arrays can be measured by fluorescence or radio-labeled isotopes (^{33}P). Detection of transcript levels on plastic and nylon arrays is performed with ^{33}P, ^{32}P, or chemiluminescence. Nylon cDNA arrays containing nucleic acid probes ranging from 200 bp to 600 bp are also available for human, rat, and mouse. Glass arrays interrogate 1,081 or 3,800 known genes. Nylon and plastic arrays detect 1,176 and 8,000 genes, respectively.

APPLIED BIOSYSTEMS GENOME SURVEY ARRAYS

Applied Biosystems (Foster City, CA, USA; http://www.appliedbiosystems.com) offers a comprehensive system for microarray analyses, including the 1,700 Chemi-luminescent Microarray Analyzer and a series of spotted microarrays for human, rat, and mouse. The Applied Biosystems Genome Survey arrays consist of 60mer oligonu-cleotides spotted on to a modified glass substrate and then enclosed in a plastic case. The Human Survey Microarray version 2.0 interrogates almost 30,000 different genes from both public and Celera databases. Rat and Mouse Survey microarrays each detect more than 32,000 known genes. Hybridized products are detected with digoxigenin/alkaline phosphatase in the Microarray Analyzer. One of the advantages of this system is that it employs an internal control for assessing spot quality and location. A 24-mer oligonu-cleotide probe, labeled with the fluorescent molecule LIZ, is hybridized along with the gene-specific probe. This molecule is complementary to a fragment that is spotted concurrently on the array, allowing investigators to assess each spot independent of the hybridization of experimental samples. Long known for their real-time PCR appli-cations, Applied Biosystems has also created off-the-shelf PCR-based gene expression assays, thus providing a complete workflow for microarray-based identification and confirmation of candidate genes.

CODELINK BIOARRAY SYSTEMS

In a departure from standard spotted arrays, CodeLink™ Bioarrays (Piscataway, NJ, USA; http://www5.amershambiosciences.com) use a proprietary three-dimensional polyacrylamide matrix containing amine-terminated 30mer oligonucleotide to detect gene expression. According to the vendor, the matrix causes the probe to be held away from the surface of the slide, allowing improved access for hybridization than what is observed for longer probes. Whole genome arrays are available for human (~57 000 transcripts), rat, and mouse. In addition, the company provides 20K arrays for human and mouse, as well as a P450 Human SNP array for SNP genotyping and pharma-cogenetic profiling. As little as 200 ng of total RNA is required in labeling reactions and hybridized products are detected with biotin-labeled dUTP and a streptavidin:cy5 conjugate.

ILLUMINA

Illumina (San Diego, CA, USA; http://www.illumina.com) provides high-throughput platforms for both genotyping and gene expression profiling. For purposes of this review, we will limit our discussion to the gene expression platform. Illumina offers yet another unique gene expression analysis method via its Sentrix® Array Matrix and Sentrix Beadchip arrays. Available for human, mouse, and *Arabidopsis,* these arrays utilize 3-micron-diameter beads coated with thousands of copies of 50mer oligonucleotides and then set into a microwell substrate. The Sentrix® Array Matrix is in a 96-well format that provides high-throughput analyses. Sentrix® Beadchip arrays offer moderate throughput with each chip containing from 6 to 16 unique array sets. Recommended starting input amount ranges from 50 ng to 200 ng of total RNA. Human and mouse arrays interrogate up to 46,000 and 47,000 transcripts, respectively.

AFFYMETRIX GENECHIPARRAYS

One of the most widely used platforms is the GeneChip®, which is produced and distributed by Affymetrix. The Affymetrix platform was one of the early leaders in gene expression technology and continues to hold a large market share. GeneChips® are created by photolithography, consisting of a series of light-directed oligonucleotide syntheses, combined with location-specific masking. Currently, Affymetrix markets ready-made gene expression arrays for 21 different species, including human, rat, mouse, *Arabidopsis*, bovine, canine, *Caenorhabditis elegans,* porcine, and chicken. The Affymetrix GeneChip® uses short oligonucleotide probes (25 bp) and, therefore, are more specific but less sensitive as compared to the cDNA microarrays. A unique feature of the GeneChip® array is to use multiple probes, referred to as a "probe set," to target a single transcript, providing multiple measurements that yield a more robust expression estimation of the transcript *(8)*. Within each probe set are 11 to 20 probe pairs. Each pair represents one probe that is a perfect match to the complementary sequence and another probe that contains a single mismatch. The purpose of this mismatch is to provide a measurement of nonspecific binding that may be subtracted from the intensity derived from the perfect match probe.

EXPERIMENTAL DESIGN

A critical aspect in all microarray studies is the proper design of the experiment. Numerous factors come into play and specific decisions about tissue type, collection, storage, and RNA isolation should be considered carefully. Labeling and hybridization protocols, use of replicates, and analysis methods should be based on the overall experimental design. One of the most important considerations in any design is the quantity and purity of RNA obtained for array hybridization. In some tissue types, the total RNA samples used for microarray hybridizations are extracted from highly homogenous groups of cells, which result in relatively reliable microarray data. However, surgically obtained tissue samples from various cancers often contain both normal and cancer cells and are not homogenously populated, which may also introduce bias.

In this situation, special collection methods, such as Laser Capture Microdissection (LCM), are very useful. The LCM method permits the isolation of specific cells (normal or cancer cells) and can improve sample homogeneity *(9)*. Sample pooling is another method commonly used to improve sensitivity when RNA is limited. However, pooling is seen as "biological averaging." If a given characteristic is common across the multiple samples, then pooling will amplify sensitivity for the gene in question, while decreasing the number of false-positive results for uninteresting genes. But, pooling may confound the interpretation of data in studies that attempt to identify subgroups within a phenotype. In cases where the RNA amount is limited, often the case in human organ studies, pooling is not by choice but by necessity. An alternative approach is to amplify the RNA samples. However, like pooling, there are both advantages and drawbacks to this procedure as well. Although numerous commercial kits are available for this purpose and laboratory protocols are relatively straightforward, amplification may introduce to the data more noise due to nonlinear amplification. In order to eliminate bias, amplified samples in the same microarray experiment should only be compared with other amplified samples, even when these other samples may have enough starting material for standard labeling methods.

MICROARRAY DATA ANALYSIS

Data Acquisition and Quality Assessment

Regardless of the platform, the process of data reduction begins with a high-level analysis meant to extract the raw gene-specific intensity values from the microarray images. In general, preliminary data analyses are conducted in several distinct steps, which include image scanning and acquisition of raw data, background subtraction, normalization, removal of genes obscured by artifact, and outlier removal. Optimal methods for background subtraction, normalization, and calculation of outlier thresholds are the subject of much discussion within the research community and should be chosen carefully, based on the overall experimental design of the study *(10,11)*.

SPOTTED MICROARRAY

For spotted array applications, commercial scanning software, such as Axon GenePix (Molecular Devices, Sunnyvales, CA, USA) and Imagene (Biodiscovery, El Segundo, CA, USA), reduce spot-associated signal to numerical values and provide several algorithms for artifact removal, quality control, and normalization. Freeware applications, such as Scanalyze (http://rana.lbl.gov/), provide image analysis functions that are similar to comparable payware versions, although they may offer fewer analysis options. During image analysis, PMT settings should be balanced to minimize instrument bias of the intensity data for respective lasers, and the overall gain should be set to minimize the number of saturated spots in the image. Array background should be assessed and spots obscured by artifact should be flagged for removal from the analysis. Likewise, negative control intensity values should be monitored from array to array and acceptable thresholds incorporated into the quality assessment scheme. Spike controls also play an important role in determining the quality of microarrays. Normally added to labeling reactions at varying concentrations, spike intensity values can be plotted to ensure that labeling reactions are efficient and to determine the level of array-to-array variability. Once the preliminary assessment is conducted, the data can then be transformed in order to minimize nonbiological variance.

AFFYMETRIX GENECHIP

In some platforms, such as the Affymetrix GeneChip®, the expression values of multiple oligonucleotides designed to interrogate the same gene must be reconciled in order to obtain a single gene expression value. This process is transparent for those using Affymetrix arrays and the proprietary GeneChip® Operating System software (Affymetrix), but not for spotted arrays that contain replicate spots. Upon extraction of data from GeneChip® arrays, data quality is assessed by observing values for background, scale factor, and noise. Multiple probes are available on the arrays for some control genes, which interrogate 3', middle, and 5' locations of the control genes.

The $3'/5'$ ratio of GAPDH, Beta Actin, and the spike controls Lys, Dap, Trp, Phe, and Thre are especially useful for determining how well the transcripts are represented by the data.

DATA REDUCTION AND FILTRATION

Detailed analyses are usually performed on normalized datasets and may include a comparison of sample groups, gene set analysis, functional annotation, and biochemical pathway analysis of candidate genes. Group comparisons usually require the aid of software applications that combine statistical analyses with graphic visualization methods in order to identify differences in gene expression amid thousands of spot-associated intensity values. The goal of such comparisons is to identify genes that are differentially expressed in treatment groups. Sometimes, data are filtered to remove genes that are unchanged. The remaining data points are subject to statistical tests, such as ANOVA or *t*-tests that can determine the validity of the data, and other analysis methods, like hierarchical or K-means clustering, Self-Organizing Maps, and Principle Components Analysis *(10)* Alternatively, Statistical analysis conducted on whole data sets. During cluster analysis, expression patterns may also be colored by intensity to visually enhance data points for genes that are most changed. Numerous applications are currently available for this purpose, including payware versions, such as GeneSpring (Silicon Genetics, Palo Alto, CA, USA) and ArrayAssist (Stratagene, La Jolla, CA, USA), as well as freeware applications, such as Cluster *(12)*. GeneSpring and ArrayAssist provide normalization and visualization algorithms as well as statistical tools for reduction of microarray data. Significance Analysis of Microarrays (SAM) *(13)* employs a supervised learning algorithm to estimate the false discovery rate (FDR) of differentially expressed genes. Depending on the chosen cutoff value for the test statistics, SAM evaluates the expected proportion of false positives in the resulting gene list through a permutation scheme.

Multiple testing errors are common problems when analyzing microarray data. When searching for genes that are differentially expressed between different conditions/phenotypes in microarray studies, a statistical test is often conducted for each gene using a *t*-test, Wilcoxon test, or permutation test. These statistical tests yield a probability (*p-value*) for every gene. However, tens of thousands of such statistical tests are conducted simultaneously, which may result in high false-positive rates. Therefore, individual significant *p*-values no longer correspond to significant findings. Therefore, it is necessary to adjust for multiple testing when assessing the statistical significance of findings in microarray data analysis. One of the algorithms commonly used to make adjustments to the microarray data is the Bonferroni Correction, where adjusted *p*-values = raw *p*-value/number of genes tested simultaneously. For example, if there are 500 genes on the chip, instead of 0.05, 0.0001 (i.e., 0.05/500) is the significant cutoff *p*-value. Because of the statistical concerns described above, the *t*-statistic is no longer an absolute measurement for the identification of the differentially expressed genes in a microarray experiment. Instead, the *p*-values often serve merely as a ranking standard.

FUNCTIONAL ANNOTATION

Once a list of differentially expressed genes is identified, the process of compiling functional annotations and the assignment of coregulated genes to biochemical pathways can proceed. By incorporating biological knowledge, one can hope to detect modest but coordinate expression changes between sets of functionally related genes. There are usually many sets of genes of interest in a given microarray experiment. Examples include genes in biological (e.g., biochemical, metabolic, and signaling) pathways, genes associated with a particular location in the cell, or genes having a particular function or involved in a particular process. For example, the Gene Ontology Consortium (www.geneontology.org) produces a controlled vocabulary that can be applied to gene functions in all organisms. The Kyoto Encyclopedia of Genes and Genomes (KEGG) (www.genome.jp/kegg/) is a suite of databases and associated software integrating current knowledge on molecular interaction networks in biological processes. It contains the information about the universe of genes and proteins, and chemical compounds and reactions. GenMAPP [(14); www.genmapp.org] is a free computer application designed to visualize gene expression and other genomic data on maps representing biological pathways and groups of genes. The idea of "gene set" analysis exploits the idea that alterations in gene expression manifest at the level of biological pathways or co-regulated gene sets, rather than individual genes. Subtle but coordinated changes in expression might be detected more readily (15).

In order to identify differentially expressed gene sets in microarray experiments, we first define the genes included within each set. For example, all genes involved in "ATP biosynthesis" belong to the gene set "ATP biosynthesis." We define a set as being differentially expressed when a relative majority of member genes are up- or downregulated and are distinctly modified with respect to other gene sets represented on the array. There are multiple computational tools for gene set analysis, including GenMAPP and MAPPFinder (14, 16), GOTree Machine (17), the Ingenuity Pathway Analysis (Ingenuity Systems Inc., Redwood City, CA, USA), and Gene Set Enrichment Assay (GSEA) (15).

GENE TRANSCRIPT LEVEL ALTERATIONS IN TYPE 2 DIABETES

Type 2 diabetes is a major cause of blindness, kidney failure, ischemic heart disease, loss of limbs, stroke, and overall mortality. The disease affects approximately 140 million people worldwide (18). Its prevalence in human populations ranges from over 50% among Pima Indians to 2% among Indian tribes in Chile (19). In addition, a substantial increase in the prevalence of diabetes among migrant populations has also been reported (19), suggesting a potential interaction between genetic and environmental factors. Decreased insulin-induced glucose disposal in skeletal muscle (insulin resistance) is a common pathophysiological trait implicated in the development of type 2 diabetes (20, 21). Several lines of evidence clearly point to genetic factors as important determinants of insulin sensitivity (22). It has also been reported that first-degree relatives of people with type 2 diabetes exhibit insulin resistance to glucose disposal without being diabetic, further supporting that insulin resistance has a genetic basis (23–26). A number of candidate genes for insulin resistance have been evaluated, including insulin receptor substrate 1 (27), glycogen synthetase (28), UCP-3 (29), GLUT-4 (30), hexokinase II (31), PI3-kinase (32), MAP kinase (33), serine-threonine kinase (34), rad genes (35), and calpain-10 (36). Although the expression of several of these genes is altered in type 2 diabetes, none has emerged as the leading candidate for causing type 2 diabetes. It is highly likely that type 2 diabetes results from inter-action between many genetic factors and environment. Identification of genes involved will hopefully allow modifications of life style to prevent or delay the onset of diabetes.

Gene Transcript Levels in Skeletal Muscle of Type 2 Diabetes

Genomic analysis techniques offer powerful tools to decipher the pathophysiology of type 2 diabetes at the molecular level, thereby leading to a better fundamental understanding of muscle insulin resistance and providing new therapeutic targets for treatment of this common disease. Several recent investigations have examined human skeletal muscle gene transcript profiles in patients with type 2 diabetes, since muscle is the predominant site of both insulin-induced glucose disposal and insulin resis-tance in type 2 diabetes (4, 15, 37–39). Overall, patients with type 2 diabetes had impaired transcripts of genes involved in oxidative metabolism (4, 15, 39), glucose metabolism (4), and lipid metabolism (39). We recently measured by Affymetrix GeneChip® array the gene transcript alterations in patients with type 2 diabetes two weeks after withdrawing insulin treatment (D_2-), and then again following 10 days of insulin treatment (D_2+) (Tables 1A–D) (4). The diabetic subjects were compared with control subjects with no family history of diabetes. Insulin treatment for 10 days normalized the impaired transcript levels of many genes involved in energy metabolism, including insulin signaling, transcription factors, and mitochon-drial maintenance, but it altered transcription of genes involved in signal trans-duction, structural and contractile components, cell development, and metabolism of protein and fat (4). This study enabled us to determine the gene transcripts that are altered by poor glycemic control and the effect of insulin treatment to achieve euglycemia. Transcripts that were normalized by insulin treatment may represent the genes that are responsible for diabetes complications resulting from poor glycemic control. Gene transcripts that cannot be corrected by insulin treatment may contribute to the pathophysiology of type 2 diabetes. Several gene transcripts were altered by the treatment. Insulin infusion to normalize blood glucose increased systemic insulin concentrations, which may have altered gene transcript levels. These early studies were performed in a relatively small number of subjects and only reported gene transcript alterations that were common among all participants. Later studies examined larger numbers of subjects and used sophisticated bioinformatics tools to interpret the results (15, 37–39). One of the studies investigated the effect of euglycemic hyperinsulinemic clamp on gene transcript levels in muscle (37), while others studied type 2 diabetes (15, 38, 39). These studies did not investigate whether the altered gene transcripts in skeletal muscle of type 2 diabetes are related to alterations in insulin or glucose levels.

Mootha et al. discussed the importance of Gene Set Enrichment Analysis (GSEA), which is a technique that detects small changes in gene expression between functionally similar sets of genes (15). Gene transcript levels in skeletal muscle from the 43 of 54 participating northern European men of varying glucose tolerance was assessed by Affymetrix GeneChip. Preliminary analysis indicated that, after adjusting for multiple comparison errors, there was no statistically significant genetic distinction between type 2 diabetes and nondiabetic subjects. The authors then used GSEA, a novel statistical

Table 1A

Differences in Gene Transcript Levels of Type 2 Diabetic Patients (D_2-) and Insulin-treated
Type 2 Diabetic Patients (D_2+) in Comparison with Control Subjects

Probe set	Fold Δ		Gene name
	A	B	
Structural/contractile			
HG2743–HT2846	↑ 3.3*	(↑ 1.5)[†]	Caldesmon 1
HG1862–HT1897	↑ 2.0[‡]	(↑ 2.9)[§]	Calmodulin type 1
HG2175–HT2245	↓ 4.9*	(↓ 1.4)[†]	Myosin heavy chain polpeptide
HG2260–HT2349	↓ 3.9[‖]	(↓ 1.4)[†]	Duchenne muscular dystrophy protein
M21984	↓ 3.2*	(↓ 1.7)[†]	Troponin-T
M21665	↓ 3.1*	(↓ 1.7)[†]	MHC-I
HG3514–HT3708	↓ 2.5*	(↓ 1.2)[†]	Tropomysin (cytoskeletal)
X04201	↓ 2.4*	(↓ 1.9)*	Tropomyosin (1.3 kb)
X66276	↓ 2.4*	(↓ 2.0)[§]	Skeletal muscle C-protein
L21715	↓ 2.3[‖]	(↓ 2.1)[§]	Troponin-I fast-twitch
J04760	↓ 2.2*	(↓ 2.0)*	Troponin-I slow-twitch
M20642	↓ 2.1[‡]	(↓ 1.4)[†]	Myosin light chain
M20543	↓ 2.0[‖]	(↓ 1.5)[†]	Skeletal muscle <0001> actin
U35637	↓ 2.0[‡]	(↓ 1.5)[†]	Nebulin
X13839	↓ 1.9[‡]	(↓ 1.8)[†]	Smooth muscle <0001>actin
M33772	↓ 1.9[†]	(↓ 2.1)*	Troponin-C fast-twitch
Mitochondrial maintenance/chaperone			
X93511	↓ 2.4*	(↓ 1.3)[†]	Telomeric DNA-binding protein
X83416	↓ 2.4[§]	(↓ 1.4)[†]	PrP
V00594	↓ 2.2[‡]	(↓ 1.6)[†]	MT
X82200	↓ 2.2[‖]	(↓ 1.8)[†]	Staf50
X65965	↓ 2.2[‖]	(↓ 1.2)[†]	SOD2
HG2855–HT2995	↑ 2.0*	(↑ 3.2)*	Heat shock protein, 70 kDa
Growth factors/tissue development/maintenance			
L27560	↑ 2.5*	(↑ 2.9)*	IGFBP-5
M55210	↓ 3.9*	(↓ 1.7)[†]	Laminin B2 chain
L08246	↓ 3.4[‡]	(↓ 2.2)[§]	MCL1
AB000897	↓ 2.8[‖]	(↓ 2.1)[§]	Cadherin FIB3
Insulin signaling/signal transduction/glucose metabolism			
HG2702–HT2798	↓ 5.6[‖]	(↓ 1.4)[†]	Serine/threonine kinase
D23673	↓ 2.8*	(↓ 1.3)[†]	IRS-1
L10717	↑ 2.9[‡]	(↓ 1.3)[†]	T-cell-specific tyrosine kinase
L33881	↓ 3.1[†]	(↓ 1.4)[†]	Protein kinase C-τ
K03515	↓ 2.3[‡]	(↓1.4)[†]	Neuroleukin
J04501	↓ 2.4[‖]	(↓ 1.6)[†]	Muscle glycogen synthase
U27460	↓ 2.3*	(↓ 1.2)[†]	UDP-glucose pyrophosphorylase
M91463	↓ 2.1[§]	(↓ 1.3)[†]	GLUT4
M32598	↓ 2.0*	(↓ 1.2)[†]	Muscle glycogen phosphorylase
Energy metabolism			

D17480	↑ 3.1*	(↑ 1.6)[†]	Mitochondrial enoyl-CoA hydratase
D10523	↑ 3.1[‖]	(↑ 1.2)[†]	2-Oxoglutarate dehydrogenase
AF001787	↓ 2.7[§]	(↓ 1.4)[†]	UCP-3
X13794	↓ 2.7[‡]	(↓ 1.4)[†]	Lactate dehydrogenase B
M83186	↓ 2.5[‖]	(↓ 1.7)[†]	COX VIIa
M19483	↓ 2.4[‖]	(↓ 1.4)[†]	ATP synthase β submit
U65579	↓ 2.3[‡]	(↓ 2.0)[‡]	NADH dehydrogenase-ubiquinone
X69433	↓ 2.2[‖]	(↓ 1.2)[†]	Mitochondrial isocitrate dehydrogenase
U94586	↓ 2.2[‖]	(↑ 1.2)[†]	NADH-ubiquinone oxidoreductase
HG4747–HT5195	↓ 2.1*	(↓ 1.1)[†]	NADH-qbiquinone oxidoeductase MLRQ
M22760	↓ 2.1[‡]	(↓ 1.3)[†]	COX Va
U09813	↓ 2.0[§]	(↑ 1.2)[†]	ATP synthase subunit 9
X83218	↓ 2.0[§]	(↓ 1.1)[†]	ATP synthase
U17886	↓ 2.0[§]	(↓ 1.2)[†]	Succinate dehydrogenase (SDHB)
L32977	↓ 1.9[§]	(↑ 1.4)[†]	Ubiquinol cytochrome C reductase

Transcription factors/protein metabolism

HG1428–HT1428	↓ 2.1[‡]	(↓ 1.4)[†]	Globin-β submit
HG3635–HT3845	↓ 3.8*	(↓ 1.6)[†]	Zinc finger protein, krupple
X69116	↓ 3.7*	(↓ 1.7)[†]	Zinc finger protein
X16064	↓ 2.2*	(↓ 1.5)[†]	Translationally controlled tumor protein
U37690	↓2.0[‖]	(↓1.3)[†]	RNA polymerase II subunit
U95040	↓ 2.1[‡]	(↓ 1.2)[†]	hKAP1/TIF1B
U65928	↓ 1.9*	(↓ 1.1)[†]	Jun activation domain binding protein
HG3214–HT3391	↑ 2.9*	(↓ 1.8)[†]	Metallopanstimulin 1
U73379	↑ 2.3[‡]	(↓ 1.3)[†]	Cyclin-selective ubiquitin carrier
X03689	↑ 2.3[‡]	(↓ 1.4)[†]	Elongation factor TU
M17886	↓ 2.1*	(↓ 1.7)[†]	Acidic ribosomal phosphoprotein P1
U49869	↓ 2.1*	(↓ 1.5)[†]	Ubiquitin
Z21507	↓ 2.0[‡]	(↓ 1.5)[†]	Elongation factor – 1$\underline{\Delta}$
U14968	↓ 1.9[‡]	(↑ 1.5)[†]	Ribosomal protein L27a
U14973	↓ 1.9[‡]	(↓1.4)[†]	Ribosomal protein S29

A, control versus D_2 (patients; B, control versus D_2 + patients. *$p<0.01$; [†]$p>0.05$; [‡]$p<0.05$; [§]$p<0.0001$; [‖]$p<0.001$, with p-values calculated from t-test on the average difference between control subjects and D_2((A) or D_2+(B) patients.))

approach based on a normalized Kolmogorov–Smirnov statistics and determined that the gene group of "oxidative phosphorylation" are down-regulated in type 2 diabetes *(15)*. The expression of this functional group is increased in locations where insulin-mediated glucose disposal occurs and is modulated by PGC-1α.

Using glass cDNA arrays containing 42,557 unique spots that represented 29,308 Unigene clusters, Rome et al. analyzed RNA preparations from six human subjects (two men and four women) *(37)*. A 3-hour euglycemic hyperinsulinemic clamp was conducted to allow the affect of insulin to be studied. The software application SAM *(13)* was then used to identify 762 unique genes that were differentially expressed. Among these transcripts, 478 showed increased expression and 284 displayed a decreased expression. Hypothetical proteins and ESTs made up 353 of the regulated

Table 1B
Additional Genes Altered in Type 2 Diabetes (D_2- and D_2+)

Probe set	Fold Δ		Gene name	Class/function
	A	B		
HG2239–HT2324	↑ 11[*]	(↑1.5)[†]	Potassium channel protein	Ion channel
U90546	↑ 5.8[*]	(↑ 1.7)[†]	Butyrophilin 4	Immune system
X89267	↑ 3.6[*]	(↑ 1.5)[†]	Uroporphyrinogen decarboxylase	Heme biosynthesis
L24564	↑ 3.2[‡]	(↑ 1.8)[†]	Rad	Ras-oncogene associated with diabetes
D43951	↑ 3.1[§]	(↑ 1.3)[†]	KIAA0099	Unknown
D79986	↑ 2.3[§]	(↑ 1.3)[†]	KIAA0164	Unknown
J02611	↑ 2.1[‡]	(↑ 1.8)[†]	Apolipoprotein D	Lipid transport
L11238	↓ 3.2[§]	(↓ 1.6)[†]	Platelet membrane glycoprotein V	Homoeostasis
X59405	↓ 3.0[∥]	(↓ 1.5)[†]	Membrane cofactor protein	Regulatory glycoprotein
X63575	↓ 2.8[‡]	(↓ 1.5)[†]	Calcium ATPase	Calcium homeostasis
M16714	↓ 2.6[§]	(↓ 1.6)[†]	MHC I lymphocyte antigen	Immune system
X00371	↓2.5[*]	(↓ 1.8)[†]	Myoglobin	Oxygen transport
X91103	↓ 2.5[§]	(↓ 1.2)[†]	Hr44	Immune system
X01060	↓ 2.4[‡]	(↓ 1.6)[†]	Transferrin receptor	Iron uptake
U90313	↓ 2.3[§]	(↓ 1.4)[†]	Flutathione-S-transferase	Glutathione metabolism
X75755	↓ 2.3[‡]	(↓ 1.3)[†]	PR264	Unknown
U33286	↓ 2.2[§]	(↓ 1.2)[†]	Chromosome segregation gene CAS	Cell proliferation/apoptosis
X15729	↓ 2.1[§]	(↓ 1.5)[†]	Nuclear p68 protein	Cell proliferation
X06700	↓ 2.1[§]	(↓ 1.7)[†]	Pro-α(III) collagen	Extracellular matrix protein
AB000220	↓1.9[*]	(↓ 1.4)[†]	Semaphorin E	Immune system

A, control versus D_2(patients; B, control versus D_2+ patients. [*]$p<0.01$; [†]$p>0.05$; [‡]$p<0.05$; [§]$p<0.0001$; [∥]$p<0.001$, with p-values calculated from t-test on the average difference between control subjects and D_2((A) or D_2+ (B) patients.

transcripts. Most of the remaining genes were involved in a variety of different functions, including transcription and translation, metabolic functions, and signaling pathways. Additional functions regulated by insulin were cytoskeletal and vesicle traffic, immune response and cytokine actions, and the ubiquitin-proteasome pathway. Nine of the genes identified as differentially expressed were analyzed by RT-PCR in order to confirm the results of the microarray analysis. Expression values for all but one of the nine genes were confirmed.

Increased fat in the diet tends to downregulate genes involved in oxidative phosphorylation in skeletal muscle mitochondria, simulating the changes observed in diabetes

Table 1C
Gene Transcripts Remain Unaltered by 10 Days of Insulin Treatment
in Type 2 Diabetic Patients

Fold Δ	Gene name
Structural/contractile genes	
↑ 2.9*	Calmodulin Type I
↓2.1*	Troponin I fast-twitch
↓ 2.1†	Troponin C fast-twitch
↓ 2.0*	Skeletal muscle C-protein
↓ 2.0†	Troponin I slow-twitch
↓ 1.9†	Tropomyosin
Stress response/energy metabolism	
↑3.2†	Heat shock protein, 70 kDa
↓2.0‡	NADH dehydrogenase-ubiquinone
Growth factor/tissue development	
↑ 2.9†	IGFBP-5
↓ 2.2*	MCL1
↓ 2.1*	Cadherin FIB3

and insulin resistance *(40)*. The authors used custom oligonucleotide arrays to assess gene transcript levels in 10 healthy young men, finding that 297 unique genes were differentially expressed. Six of these genes were involved in oxidative phosphorylation and were decreased an average of twofold. Similarly, Hulver et al. *(41)* determined that expression of stearoyl-CoA desaturase-1 is increased in obese patients causing abnormally high lipid content in skeletal muscle cells, which may lead to insulin resistance; and Patti et al. *(39)* found that a decrease in PGC-1 expression may downregulate NRF-dependent genes potentially leading to insulin resistance and diabetes. Studies performed in rats clearly show that caloric intake (possibly related to insulin and other hormonal changes) impact gene transcript levels in skeletal muscle of rats *(42)*. Some of the changes related to high caloric intake were partly prevented by antioxidants, suggesting that high calorie-related changes are due to oxidative decrease. Furthermore, these studies indicated that caloric restriction preserves many of the energy metabolism-related gene transcript levels. It remains to be determined whether many changes in muscle gene transcript levels of type 2 diabetes patients are related to oxidation damage, since poor diabetes control results in hypermetabolic state *(4)*.

Gene Transcript Levels in Adipose Tissue of Type 2 Diabetes

Genes for Wnt signaling and adipogenesis show decreased expression levels in nonobese insulin-resistant subjects, suggesting that insulin resistance in human skeletal muscle is related to impaired adipogenesis *(43)*. In addition, adipose cells that are enlarged during insulin resistance correlated inversely to transcript levels for Wnt and adipogenic transcription factors.

Genes involved in adipocyte differentiation show decreased expression in mouse models of obesity and diabetes mellitus *(44)*. Using Affymetrix murine arrays to analyze

Table 1D
Gene Transcripts Altered (which was normal in D_2- with 10 Days of Insulin Treatment
in Type 2 Diabetic Patients

Fold Δ	Gene name
Structural/contractile/growth factor	
↑ 2.2*	ACTN3
↑ 2.1*	CO-029
↓ 2.7†	SM22
↓ 2.6*	Sarcolipin
↓ 2.2*	β-Tropomyosin
↓ 1.9‡	Cytoskeletal gamma-actin
↑ 1.9*	IGF-II
Protein metabolism/Signal transduction	
↑ 2.0*	dIF-4C
↑ 1.9*	Ribosomal protein L21
↓ 2.6†	Ubiquitin carboxyl terminal hydrolase
↓ 2.3*	Ribosomal protein L37a
↓ 2.0*	Elongation factor 1 alpha-2
↑ 2.3*	AMP-activated protein kinase
↑ 2.1*	Protein tyrosine phosphatase, alpha
↓ 2.7†	Tyrosine kinase receptor
↓ 2.5†	Adenylyl cyclase
Immune system/energy metabolism/cell adhesion	
↑ 2.1*	Class II histocompatibility antigen DC-α-chain
↑ 2.3†	HK1
↓ 1.9*	NDUFV3
↓ 3.8*	Fibronectin
↓ 2.4†	HSPG2
Tissue development/fatty acid metabolism	
↑ 2.1†	Cadherin FIB2
↓ 2.4*	Fatty acid binding protein
Transposition/extracellular matrix/unknown	
↑ 2.0*	Transposon-like element
↓ 2.3*	Tenascin-X
↑ 3.0†	GOS2
↑ 2.6*	H4 Histone
↓ 2.1*	SCAMP1
↓ 2.0†	Phosphodieaterase 3B

white adipose tissue from epididymal fat pads, the authors detected 214 transcripts that were differentially expressed as compared to lean mice. Functional groups determined to have decreased expression included hormones, secretory genes, transcription factors, signal transduction, mitochondria-related, and lipid metabolism. Groups identified to have increased expression were those of cytoskeletal functions, cell proliferation, adipose-specific genes, membrane proteins, lysosomal factors, and the immune system. Interestingly, when hyperglycemia was induced, transcript levels were decreased for genes related to signal transduction, protein synthesis, and cytoskeletal function. In addition, offspring of rats fed a high-fat diet develop insulin resistance and abnormal pancreatic β-cell function *(45)*.

Conversely, an energy-restrictive diet improves the ability of rat adipose cells to accumulate triglycerides and reverses the progression of diabetes, suggesting that energy restriction may play a role in the prevention or proper maintenance of diabetes *(46)*.

Gene Alterations in Renal Dysfunction and Vascular Complications during Type 2 Diabetes

In an attempt to define the mechanisms of end-stage renal disease in diabetic nephropathy, Morrison et al. studied the effect of high (25 mM) D-glucose concentrations on rat mesangial cells *(47)*. Spotted microarrays were synthesized from 17,664 rat cDNA clones collected from an archive at the University of Iowa. After a 5-day treatment, 459 different rat transcripts were upregulated, compared to 151 transcripts that showed a decrease in activity. Gene groups that formed the largest functional clusters included those involved cell cycle activity, cytoskeletal functions, energy metabolism, and oxidative stress. Additional transcript clusters were detected for protein synthesis and protein sorting. Many transcripts related to cellular degradation were downregulated.

Mouse models for both type 1 and type 2 diabetes have implicated hydroxysteroid dehdrogenase-3β isotype 4 and osteopontin in diabetic nephropathy, suggesting that these markers may be used for diagnosis of glomerular disease *(48)*. In this study, cDNA arrays containing 9,557 random elements were used to classify samples based by pathologic presentation. Hierarchical clustering showed that more than 300 genes were differentially expressed in each model, but only 40 transcripts were common to both type 1 and type 2 diabetes. Patients with type 2 diabetes are more likely to have systemic vascular complications, according to a microarray study performed by Takamura et al. *(49)*. The authors used a glass microarray containing 1,080 cDNA clones from the IMAGE Consortium (Research Genetics, Huntsville, AL, USA) to analyze expression patterns in liver tissue of 12 type 2 diabetes patients and nine normal controls. Liver functions in all patients were normal. For the diabetic liver samples, 105 genes displayed increased expression compared to normal, including genes for angiogenesis, and bone morphogenic proteins. Decreased expression occurred in 134 genes, such as certain stress-related genes. Hierarchical cluster analysis showed that data from all 12 diabetic patients clustered separately from that of the normal controls, suggesting that the genetic profiles of diabetic patients are different from nondiabetic patients. RT-PCR analyses confirmed that BMP4, endothelin, and glutathione S-transferase T1 were upregulated, as shown by microarray analyses.

REAL-TIME REVERSE TRANSCRIPTION POLYMERASE CHAIN REACTION AND ITS APPLICATION IN DIABETES RESEARCH

Though not a new technology, real-time reverse transcriptase PCR is still considered by many researchers to be the gold standard for confirmation of gene expression studies. Characterized by exquisite sensitivity and specificity, RT-PCR methods are very useful in the detection and quantitation of genes of all abundance levels. Due to the cost of the probes, many researchers begin their work with high-throughput analytical methods, such as commercial or custom microarrays. After candidate genes are identified, RT-PCR is then used to confirm those findings.

Methods for RT-PCR detection of gene expression were borne out of early studies that showed that enzymatic amplification of gene products allowed the detection of specific nucleic acid sequences *(50, 51)*. Later, the development of techniques for competitive RT-PCR and the use of fluorescent reporter molecules enhanced the ability of researchers to quantitate mRNA products *(52, 53)*. Several instrument platforms for RT-PCR are currently on the market. Notably, the Applied Biosystems 5′ nuclease assay Fostercity, CA, USA, Roche Lightcycler (Indianapolis, IN, USA), and the BioRad Icycler (Hercules, CA, USA) have made their mark in molecular research laboratories. During real-time reverse transcription PCR, mRNA products are first converted to cDNA, then the cDNA is amplified via a heat-stabile DNA polymerase and the amplification efficiency is assessed during each PCR cycle. Detection of the amplified products may be performed with the use of fluorogenic probes or with DNA intercalating agents, such as Sybr Green (Invitrogen, Carlsbad, CA, USA).

Several studies using RT-PCR for diabetes research have been published in the scientific literature. The relative ease with which experiments can be performed together with the sensitivity of the assay makes PCR-based gene expression analysis an ideal method for confirmation of the results of large-scale microarray projects as well as a stand-alone assay for low to moderate throughput studies. In the overview, we will discuss the of the use of this technique for conducting experiments to further the knowledge regarding the genetic effects of diabetes. Infusion of insulin into healthy human patients caused an increase in the expression of NAD dehydrogenase IV and cytochrome oxidase IV (COX IV) of 160–180%, as well as an increase in skeletal muscle mitochondrial ATP production, as shown by Stump et al. *(54)*. However, insulin infusion into subjects with type 2 diabetes did not increase the production of muscle mitochondrial ATP. These studies used a custom-designed primer and probe set combination, with amplifications products chosen to span an exon boundary or to amplify only the polyA RNA transcript, to perform the 5′ Nuclease Assay (Applied Biosystems) on RNA samples derived from vastus lateralis muscle from nine healthy subjects. In a study that examined insulin regulation of genes expressed in adipose tissue, Koistinen et al. *(55)* found that insulin infusion increased by 80% the expression of sterol regulatory element binding protein-1c (SREBP-1c) in healthy subjects, but did not increase SREBP-1c expression in subjects with type 2 diabetes. Expression of adiponectin, CAP, and 11β-HSD-1 was not changed by exposure to insulin. The effect of cytokine exposure to pancreatic islet cells from humans, rats, and prediabetic NOD mice was studied by Cardozo et al. *(58)*. In human and rat cells, cytokine exposure increased expression of IP-10, MIP-3α, fractalkine, and IL-15. In NOD mice, expression of IL-15, IP-10, and MIP-3α also increased with the age of the mouse.

CONCLUSION

In summary, microarray technology is a powerful approach to understand the disease-induced alterations to mRNA levels within a small amount of tissue or blood cells. These important tools will allow us to further understand the gene transcript changes, indicating certain metabolic pathways that are altered with disease, such as diabetes, and will help us to develop novel hypothesis-driven research on human type 2 diabetes. The microarray studies performed in human type 2 diabetes, which indicated alterations in energy metabolism pathway, stimulated many metabolic studies in type 2 diabetic patients, resulting in advances in our understanding of the pathophysiology of this disease. In addition, one will be able to identify genetic fingerprints for different diseases using a microarray approach. With the advancement in sensitivity and precision of microarray technologies and bioinformatics tools for data analysis, one will be able to capture the genetic changes in a patient at an early stage of the disease using these approaches, suggesting that disease prevention of genetic diseases, and perhaps cures, may occur in the future.

REFERENCES

1. Scearce LM, Brestelli JE, McWeeney SK, Lee CS, Mazzarelli J, Pinney DF, Pizarro A, Stoeckert CJ, Jr., Clifton SW, Permutt MA, Brown J, Melton DA, Kaestner KH: Functional genomics of the endocrine pancreas: the pancreas clone set and PancChip, new resources for diabetes research. *Diabetes* **51**:1997–2004, 2002.
2. Lantz KA, Vatamaniuk MZ, Brestelli JE, Friedman JR, Matschinsky FM, Kaestner KH: *Foxa*2 regulates multiple pathways of insulin secretion. *J Clin Invest* **114**:512–520, 2003.
3. Kaestner KH, Lee CS, Brestelli JE, Arsenlis A, Le PP, Lantz KA, Crabtree J, Pizarro A, Mazzarelli J, Pinney D, Fisher S, Manduchi E, Stoeckert CJ, Jr., Gradwohl G, Clifton SW, Brown JR, Inoue H, Cras-Meneur C, Permutt MA: Transcriptional program of the endocrine pancreas in mice and humans. *Diabetes* **52**:1604–1610, 2003.
4. Sreekumar R, Halvatsiotis P, Schimke JC, Nair KS: Gene expression profile in skeletal muscle of type 2 diabetes and the effect of insulin treatment. *Diabetes* **51**:1913–1920, 2002.
5. Sreekumar R, Rosado B, Rasmussen D, Charlton M: Hepatic gene expression in histologically progressive nonalcoholic steatohepatitis. *Hepatology* **38**:244–251, 2003.
6. Kolbert CP, Taylor WR, Krajnik KL, O'Kane DJ: Gene expression microarrays. In *Novel Anticancer Drug Protocols* Buolamwini JK, Adjei AA, Eds. Totowa, NJ, Humana Press, 2003.
7. Schena M, Shalon D, Davis RW, Brown PO: Quantitative monitoring of gene expression patterns with a complementary DNA microarray. *Science* **270**:467–470, 1995.
8. Lockhart DJ, Dong H, Byrne MC, Follettie MT, Gallo MV, Chee MS, Mittmann M, Wang C, Kobayashi M, Horton H, Brown EL: Expression monitoring by hybridization to high-density oligonucleotide arrays. *Nat Biotechnol* **14**:1675–1680, 1996.
9. Emmert-Buck MR, Bonner RF, Smith PD, Chuaqui RF, Zhuang Z, Goldstein SR, Weiss RA, Liotta LA: Laser capture microdissection. *Science* **274**:998–1001, 1996.
10. Quackenbush J: Computational analysis of microarray data. *Nature Reviews Genetics* **2**:418–427, 2001.
11. Schuchhardt J, Beule, D, Malik, A, Wolski, E, Eickhoff, H, Lehrach, H, Herzel, H: Normalization strategies for cDNA microarrays. *Nucleic Acids Res* **28**:e47 i–v, 2000.
12. Eisen MB, Spellman PT, Brown PO, Botstein D: Cluster analysis and display of genome-wide expression patterns. *Proc Natl Acad Sci U S A* **95**:14863–14868, 1998.
13. Tusher VG, Tibshirani R, Chu G: Significance analysis of microarrays applied to the ionizing radiation response. *Proc Natl Acad Sci U S A* **98**:5116–5121, 2001.
14. Dahlquist KD, Salomonis N, Vranizan K, Lawlor SC, Conklin BR: GenMAPP, a new tool for viewing and analyzing microarray data on biological pathways. *Nat Genet* **31**:19–20, 2002.
15. Mootha VK, Lindgren CM, Eriksson KF, Subramanian A, Sihag S, Lehar J, Puigserver P, Carlsson E, Ridderstrale M, Laurila E, Houstis N, Daly MJ, Patterson N, Mesirov JP, Golub TR, Tamayo P, Spiegelman B, Lander ES, Hirschhorn JN, Altshuler D, Groop LC: PGC-1alpha-responsive genes involved in oxidative phosphorylation are coordinately downregulated in human diabetes. *Nat Genet* **34**:267–273, 2003.
16. Doniger SW, Salomonis N, Dahlquist KD, Vranizan K, Lawlor SC, Conklin BR: MAPPFinder: using gene ontology and GenMAPP to create a global gene-expression profile from microarray data. *Genome Biol* **4**:R7, 2003.

17. Zhang B, Schmoyer D, Kirov S, Snoddy J: GOTree Machine (GOTM): a web-based platform for interpreting sets of interesting genes using gene ontology hierarchies. *BMC Bioinformatics* **5**:16, 2004.
18. King H. Aubert RE, Herman, WH.: Global burden of diabetes, 1995–2025: prevalence, numerical estimates, and projections. *Diabetes Care* **21**:1414–1431, 1998.
19. Bennett PH: Type 2 diabetes among the Prima Indians of Arizona: an epidemic attributable to environmental change? *Nutr Rev* **57**:S51–S54, 1999.
20. Reaven GM: Pathophysiology of insulin resistance in human disease. *Physiol Rev* **75**:473–486, 1995.
21. Ferrannini E: Insulin resistance is central to the burden of diabetes. *Diabetes Metab Rev* **13**:81–86, 1997.
22. Moller DE, Bjorbek, C, Puig, AV: Candidate genes for insulin resistance. *Diabetes Care* **19**:396–400, 1996.
23. Nyholm B, Mengel, A, Skaerbaek, CH, Moller, N, Alberti, KGMM, Schmitz, O: Insulin resistance in relatives of NIDDM patients: the role of physical fitness and muscle metabolism. *Diabetologia* **39**:813–822, 1996.
24. Warram JH, Martin, BC, Krolewski, AS, Soeldner, JS, Kahn, CR: Slow glucose removal rate and hyperinsulinemia precede the development of type II diabetes in the offspring of diabetic parents. *Ann Intern Med* **113**:909–915, 1990.
25. Shulman GI: Cellular mechanisms of insulin resistance. *J Clin Invest* **106**:171–176, 2000.
26. Kadowaki T: Insights into insulin resistance and type 2 diabetes from knockout mouse models. *J Clin Invest* **106**:459–465, 2000.
27. Carvalho E, Jansson PA, Axelsen, M, Eriksson, JW, Huang, X, Groop, L, Rondinone, C, Sjostrom, L, Smith, U: Low cellular IRS 1 gene and protein expression predict insulin resistance and NIDDM. *FASEB J* **13**:2173–2178, 1999.
28. Thorburn AW, Gumbiner, B, Bulacan, F, Wallace, P, Henry, RR: Intracellular glucose oxidation and glycogen synthase activity are reduced in non-insulin dependent (type II) diabetes independent of impaired glucose uptake. *J Clin Invest* **85**:522–529, 1990.
29. Krook A, Digby, J, O'Rahilly, S, Zierath, JR, Henriksson, HW: Uncoupling protein 3 is reduced in skeletal muscle of NIDDM patients. *Diabetes* **47**:1528–1531, 1998.
30. Choi WH, O'Rahilly, S, Buse, JB, Rees, A, Morgan, R, Flier, JS, Moller, DE: Molecular scanning of the insulin-responsive glucose transporter (GLUT-4) gene in patients with non insulin dependent diabetes mellitus. *Diabetes* **40**:1712–1718, 1991.
31. Pendergrass M, Koval J, Vogt C, Yki-Jarvinen H, Iozzo P, Pipek R, Ardehali H, Printz R, Granner D, DeFronzo RA, Mandarino LJ: Insulin-induced hexokinase II expression is reduced in obesity and NIDDM. *Diabetes* **47**:387–394, 1998.
32. Cusi K, Maezono K, Osman A, Pendergrass M, Patti ME, Pratipanawatr T, DeFronzo RA, Kahn CR, Mandarino LJ: Insulin resistance differentially affects the PI 3-kinase- and MAP kinase-mediated signaling in human muscle. *J Clin Invest* **105**:311–320, 2000.
33. Krook A, Roth, RA, Jiang, XJ, Zierath, JR, Henriksson, HW: Insulin-stimulated Akt kinase activity is reduced in skeletal muscle from NIDDM subjects. *Diabetes* **47**:1281–1285, 1998.
34. Reynet C, Kahn, CR: Rad: a member of the Ras family overexpressed in muscle of type II diabetic humans. *Science* **262**:1441–1444, 1993.
35. Horikawa Y, Oda N, Cox, NJ, Li, X, Orho-Melander, M, Hara, M, Hinokio, Y, Lindner, TH, Mashima, H, Schwarz, PEH, del Bosque-Plata, L, Horikawa, Y, Oda, Y, Yoshiuchi, I, Colilla, S, Polonsky, KS, Wei, S, Concannon, P, Iwasaki, N, Schulze, J, Baier, LJ, Bogardus, C, Groop, L, Boerwinkle, E, Hanis, CL, Bell, GI: Genetic variation in the gene encoding calpain-10 is associated with type 2 diabetes mellitus. *Nat Genet* **26**:163–175, 2000.
36. Permutt MA, Mizrachi, EB, Inoue, H: Calpain-10: the first positional cloning of a gene for type 2 diabetes? *J Clin Invest* **106**:819–821, 2000.
37. Rome S, Clement K, Rabasa-Lhoret R, Loizon E, Poitou C, Barsh GS, Riou JP, Laville M, Vidal H: Microarray profiling of human skeletal muscle reveals that insulin regulates approximately 800 genes during a hyperinsulinemic clamp. *J Biol Chem* **278**:18063–18068, 2003.
38. Yang X, Pratley RE, Tokraks S, Bogardus C, Permana PA: Microarray profiling of skeletal muscle tissues from equally obese, non-diabetic insulin-sensitive and insulin-resistant Pima Indians. *Diabetologia* **45**:1584–1593, 2002.

39. Patti ME, Butte AJ, Crunkhorn S, Cusi K, Berria R, Kashyap S, Miyazaki Y, Kohane I, Costello M, Saccone R, Landaker EJ, Goldfine AB, Mun E, DeFronzo R, Finlayson J, Kahn CR, Mandarino LJ: Coordinated reduction of genes of oxidative metabolism in humans with insulin resistance and diabetes: Potential role of PGC1 and NRF1. *Proc Natl Acad Sci U S A* **100**:8466–8471, 2003.

40. Sparks LM, Xie, H, Koza, RA, Mynatt, R, Hulver, MW, Bray, GA, Smith, SR: A high-fat diet coordinately downregulates genes required for mitochondrial oxidative phosphorylation in skeletal muscle. *Diabetes* **54**:1926–1933, 2005.

41. Hulver M, Berggren, JR, Carper, MJ, Miyazaki, M, Ntambi, M, Hoffman, EP, Thyfault, JP, Stevens, R, Dohm, GL, Houmard, JA, Muoio, DM: Elevated stearoyl-CoA desaturase-1 expression in skeletal muscle contributes to abnormal fatty acid partitioning in obese humans. *Cell Metab* **2**:251–261, 2005.

42. Sreekumar R, Unnikrishnan J, Fu A, Nygren J, Short KR, Schimke J, Barazzoni R, Nair KS: Effects of caloric restriction on mitochondrial function and gene transcripts in rat muscle. *Am J Physiol Endocrinol Metab* **283**:E38–43., 2002.

43. Yang X, Jansson, P, Nagaev, I, Jack, MM, Carvalho, E, Sunnerhagen, KS, Cam, MC, Cushman, SW, Smith, U: Evidence of impaired adipogenesis in insulin resistance. *Biochem Biophys Res Commun* **317**:1045–1051, 2004.

44. Nadler ST, Stoehr, JP, Schueler, KL, Tanimoto, G, Yandell, BS, Attie, AD: The expression of adipogenic genes is decreased in obesity and diabetes mellitus. *Proc Natl Acad Sci U S A* **97**:11371–11376, 2000.

45. Taylor PD, McConnell, J, Khan, YI, Holemans, K, Lawrence, KM, Asare-Anane, H, Persaud, SJ, Jones, PM, Petrie, L, Hanson, MA, Poston, L: Impaired glucose homeostasis and mitochondrial abnormalities in offspring of rats fed a fat-rich diet in pregnancy. *Am JPhysiol* **288**:134–139, 2006.

46. Colombo M, Kruhoeffer, M, Gregersen, S, Agger, A, Jeppesen, P, Oerntoft, T, Hermansen, K: Energy restriction prevents the development of type 2 diabetes in Zuker diabetic fatty rats; coordinated patterns of gene expression for energy metabolism in insulin-sensitive tissues and pancreatic islets determined by oligonucleotide microarray analysis. *Metab ClinExp* **55**:43–52, 2006.

47. Morrison J, Knoll K, Hessner MJ, Liang M: Effect of high glucose on gene expression in mesangial cells: upregulation of the thiol pathway is an adaptational response. *Physiol Genomics* **17**:271–282, 2004.

48. Susztak K, Bottinger, E, Novetsky, A, Liang, D, Zhu, Y, Ciccone, E, Wu, D, Dunn, S, McCue, P, Sharma, K: Molecular profiling of diabetic mouse kidney reveals novel genes linked to glomerular disease. *Diabetes* **53**:784–794, 2004.

49. Takamura T, Sakurai M, Ota T, Ando H, Honda M, Kaneko S: Genes for systemic vascular complications are differentially expressed in the livers of type 2 diabetic patients. *Diabetologia* **47**:638–647, 2004.

50. Saiki RK, Scharf S, Faloona F, Mullis KB, Horn GT, Erlich HA, Arnheim N: Enzymic amplification of b-globin genomic sequences and restriction site analysis for diagnosis of sickle cell anemia. *Science* **230**:1350–1354, 1985.

51. Mullis KB, Faloona FA: Specific synthesis of DNA in vitro via a polymerase-catalyzed chain reaction. *Methods Enzymol***155**:335–350, 1987.

52. Gilliland G, Perrin S, Blanchard K, Bunn HF: Analysis of cytokine mRNA and DNA: Detection and quantitation by competitive polymerase chain reaction. *Proc Natl Acad Sci U S A* **87**:2725–2729, 1990.

53. Wang T, Brown, MJ: mRNA quantification by real time Taqman polymerase chain reaction: Validation and comparison with RNase protection. *Anal Biochem* **269**:198–201, 1999.

54. Stump CS, Short KR, Bigelow ML, Schimke JM, Nair KS: 2003. Effect of insulin on human skeletal muscle mitochondrial ATP production, protein, synthesis. and mRNA transcripts. *Proc Nat Acad Sci* **100**(13):7996–8001.

55. Koistinen HA, Fongren M, Wallberg-Henricksson H, Zierath JR: 2004. Insulin action on expressinn of novel adipose genes in healthy and type 2 diabetic subjects. *Obesity Res.* **12**(1):25–31 .

56. Cardozo AK, Proost P, Gysemans C, Chen M.-C, Mathieu C, Eizirik DL: 2002. ILβ and IFN-α induce the expression of diverse chemokines and IL-15 in human and rat pancreatic islet cells, and in islets from pre-diabetic NOD mice. *Diabetologia* **46**:255–266.

11

DNA Microarray Analysis of Effects of TSH, Iodide, Cytokines, and Therapeutic Agents on Gene Expression in Cultured Human Thyroid Follicles

Kanji Sato, MD, PhD, Kazuko Yamazaki, BS, and Emiko Yamada, MD

CONTENTS

INTRODUCTION
MATERIALS AND METHODS
RESULTS
DISCUSSION
CONCLUSION
REFERENCES

INTRODUCTION

When a large amount of iodide is administered to hyperthyroid patients with Graves' disease, it acutely inhibits thyroid hormonogenesis and the release of thyroid hormones, accompanied by a decrease in blood flow in the hypervascular thyroid gland. In general, TSH ubiquitously stimulates thyroid function via the protein kinase A and C pathways, whereas excess iodide generally exerts an inhibitory effect via multiple mechanisms *(1)*. Furthermore, a number of factors such as proinflammatory cytokines and therapeutic agents modulate thyroid function, angiogenesis, and the autoimmune system *(2)*. To elucidate the mechanism by which these thyroid-modulating factors regulate thyroid function, we studied their effects on gene expression in cultured human thyroid follicles, which maintain thyroid hormonogenesis in response to physiological concentrations of TSH *(3)* and exhibit acute Wolff–Chaikoff effects in the presence of high concentrations of iodide *(4)*. Initially, we used cDNA microarray (MICROMAX, NEN Life Science Products, Boston, MA, USA), which can analyze 2400 genes in a single run *(5)*. Very recently we have been using oligo-DNA microarray, by which 41000 gene spots, i.e., the entire human genome *(6)*, can be analyzed in a single run *(7, 8, 25)*.

From: *Contemporary Endocrinology: Genomics in Endocrinology:*
DNA Microarray Analysis in Endocrine Health and Disease
Edited by S. Handwerger and B. Aronow © Humana Press, Totowa, NJ

MATERIALS AND METHODS

Suspension Culture of Human Thyroid Follicles

Human thyroid follicles were cultured as reported previously *(3–5, 9–12)*. In brief, thyroid tissue (15–30 g) was obtained from patients with Graves' disease during subtotal thyroidectomy and minced into small pieces with scissors *(13)*. The prepared tissues were then digested with 0.3 mg/mL collagenase and 5 mg/mL dispase. After mild centrifugation, thyroid follicles were separated from erythrocytes and mononuclear cells, although a few immunocompetent cells that had infiltrated into the thyroid follicles could not be removed *(4)*. Then, 1 mL of a thyroid follicle suspension (about 200–300 follicles/mL) was added to each of the wells in a 24-multiwell dish, the bottoms of which had been coated with agarose (Fig. 1). After several days of preculture, ^{125}I was added to the supplemented culture medium [RPMI-1640/F-12 (1:1) medium supplemented with bovine serum albumin (2 mg/mL), bovine insulin (5 μg/mL), transferrin (5 μg/mL), hydrocortisone (10^{-8}M), and NaI (10^{-8}M)] containing various concentrations of human or bovine TSH (bTSH). After an additional 3 days of culture, ^{125}I incorporated into thyroid follicles and organic ^{125}I metabolites (mainly de novo-synthesized and secreted ^{125}I-T3 and ^{125}I-T4) was quantified (Fig. 2). To

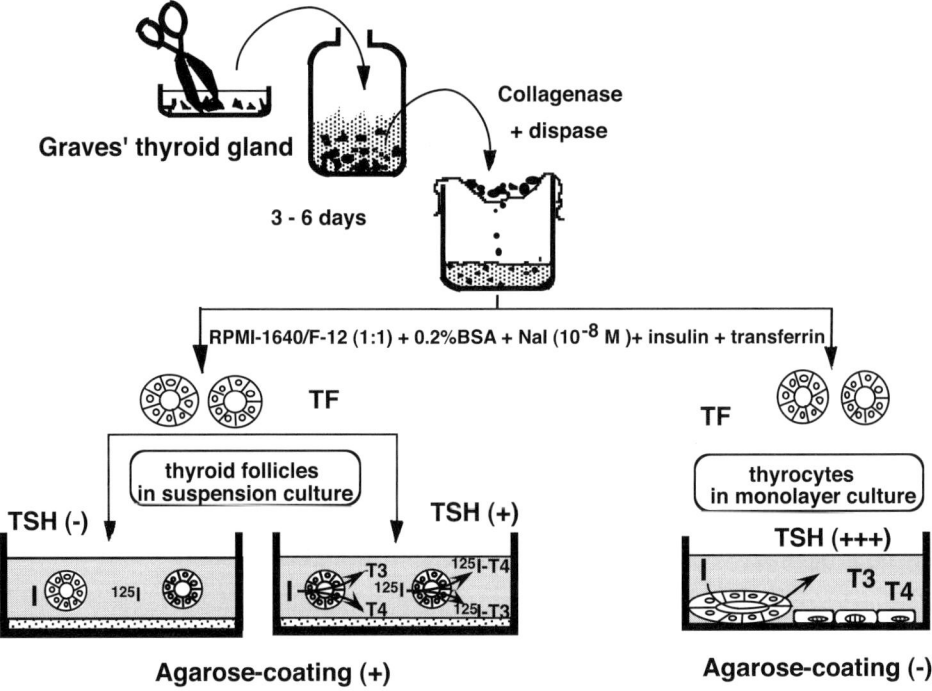

Fig. 1. Thyroid follicles in suspension culture and monolayer culture. Thyroid follicles were cultured in the supplemented medium in 24-multiwell tissue culture dishes that were uncoated or coated with agarose. Thyroid follicles cultured in coated dishes were unable to attach to the bottom and were cultured in suspension, maintaining thyroid function for a couple of weeks. When ^{125}I was added, thyroid follicles incorporated ^{125}I and secreted de novo-synthesized thyroid hormones (^{125}I-T3 + ^{125}I-T4) into the culture medium *(3, 9–12)*.

study Wolff–Chaikoff effects, thyroid follicles were precultured in the supplemented medium in 24-multiwell dishes without agarose coating. The thyroid follicles were then cultured in medium containing TSH (100 μU/mL) and various concentrations of iodide (0–10^{-5}M), and T3 released into the culture medium was determined by radioimmunoassay (Fig. 3).

For microarray analysis, thyroid follicles were cultured in 10-cm dishes (with agarose coating) in RPMI-1640/F-12 (1:1) medium containing 0.5% fetal calf serum, bTSH (30 μU/mL), and iodide (10^{-8} M). After several days of preculture, thyroid follicles were cultured in fresh medium containing bTSH (30 μU/mL) and various concentrations of proinflammatory cytokines (IL-1, TNF-α, or IL-6) and iodide (10^{-8}–10^{-5} M). After an appropriate culture period, total RNAs were prepared, and the test material-induced genes were analyzed by DNA microarray (Whole Human Genome Oligo-Microarray Kit, 17 000 or 41 000 gene spots, Agilent Technologies, Palo Alto, CA, USA) exactly according to the manufacturer's recommended procedure. The results were confirmed by real-time PCR and Northern blot hybridization.

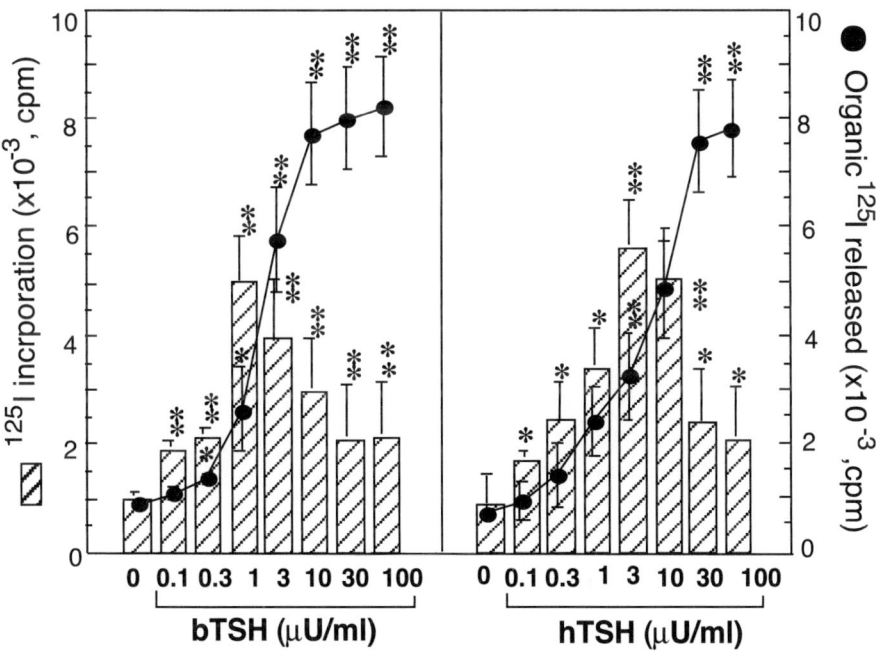

Fig. 2. Effect of TSH on thyroid function in human thyroid follicles in suspension culture. TSH stimulated ^{125}I incorporation into thyroid follicles (columns) and release of de novo-synthesized ^{125}I-T3 and ^{125}I-T4 into the supplemented culture medium (circles) in a concentration-dependent manner. Note that the thyrotropic effects were most remarkable at physiological concentrations of hTSH (normal range; 0.4–4.0 μU/mL) and that bTSH was more potent than hTSH. Data are means ±SD of triplicate cultures. (Reproduced from *(3)*, with permission of the publisher.) *$P < 0.05$, **$P < 0.01$.

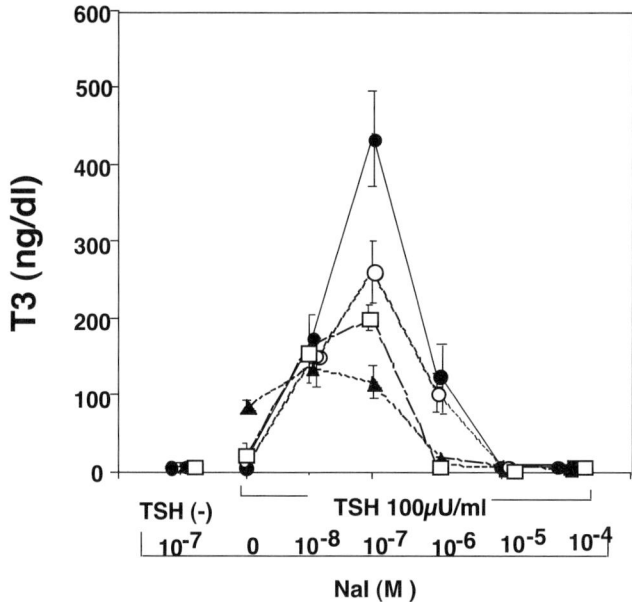

Fig. 3. The Wolff–Chaikoff effect in human thyrocytes. Thyroid follicles were cultured in 24-multiwell dishes (without agarose coating) in the supplemented culture medium containing various concentrations of NaI with or without bTSH (100 μU/mL) for up to 16 days. After 2 days of preculture, the medium was changed every 3 or 4 days for 14 days. Data are means ±SD of triplicate cultures. ▲–▲, days 3–5; □–□, days 6–8; ●–●, days 9–13; o–o, days 14–16. (Reproduced from *(4)*, with permission of the publisher.)

RESULTS

Effect of TSH on Gene Expression in Thyroid Follicles

As expected, the most abundantly expressed genes were thyroglobulin (TG) and thyroid peroxidase (TPO) *(7, 25)*, and TSH increased their levels of expression more than 7- and 15-fold, respectively. TSH also increased the expression levels of sodium iodide symporter (NIS), type I and II deiodinases *(14, 15)*, VEGF-A *(12, 16)*, and enzymes participating in carbohydrate and lipid metabolism, such as ATP synthase, aldolase, creatine kinase, and 3-hydroxy-3-methylglutaryl-coenzyme A reductase *(5)*. Furthermore, prion mRNA was also stimulated by TSH *(17)*.

Effects of Iodide on Gene Expression in Thyroid Follicles

Iodide at high concentration (10^{-5} M) decreased the level of expression of VEGF-A to nearly 1/2 after 48 h of culture *(7)*. Iodide did not significantly (less than 1/2 or more than twofold) modulate the expression levels of other proangiogenic factors such as FGF-1, FGF-2, or angiopoietin-1. Interestingly, iodide at high concentration increased the expression level of urokinase-type plasminogen activator (u-PA). Real-time PCR confirmed that iodide decreased and increased the expression levels of VEGF-A and u-PA, respectively, in a time- and dose-dependent manner. Adenylate kinase 1, which was activated more than fourfold by TSH, was suppressed to 50% by iodide.

Effects of Proinflammatory Cytokines on Gene Expression in Thyroid Follicles

Proinflammatory cytokines, such as IL-1, TNF-α, INF-α, INF-β, INF-γ, and IL-6, inhibit thyroid function *(9–11)*. When thyroid follicles were cultured with or without IL-1 for 24 h, approximately 720 genes and 640 genes among 17 000 gene spots were increased more than twofold or decreased to less than 1/2, respectively (Fig. 4). As expected, the expression levels of IL-6, IL-8, IL-11, and IL-19 increased more than 10-fold, and that of NF-kB nearly threefold, whereas the expression levels of genes related to thyroid hormonogenesis, such as TSH receptor (TSHR), NIS, type 1 and 2 deiodinases, pendrin (SLC26A4), TG, and TPO were suppressed to less than 1/2.

Effects of Methimazole and Amiodarone on Gene Expression in Human Thyroid Follicles

When human thyroid follicles were cultured for 48 h with or without methimazole at 3 µmol/L, which is the level attainable in serum of patients taking a maintenance dose,

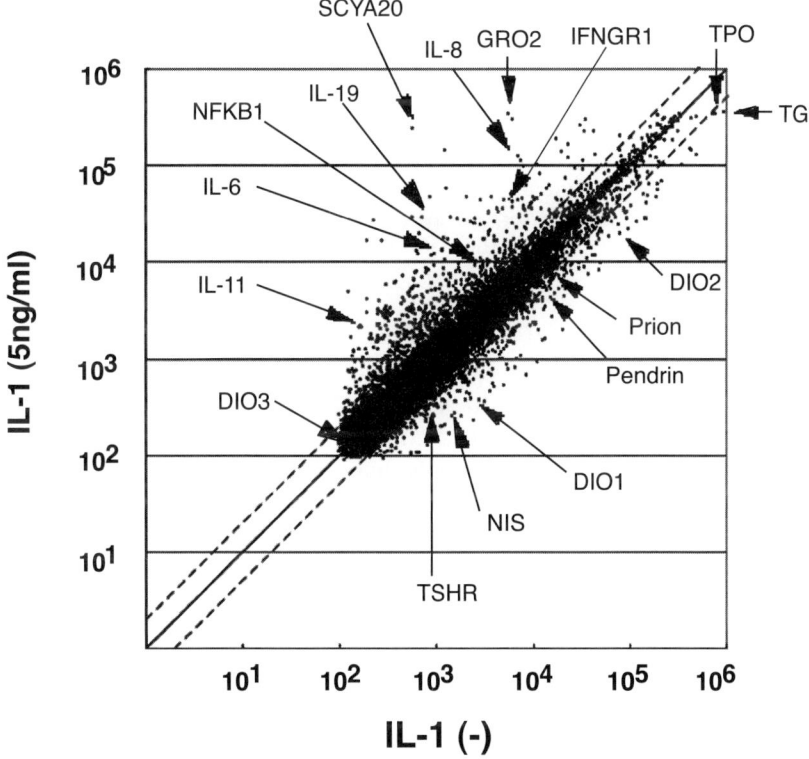

Fig. 4. Scatter plot of the effect of IL-1 on gene expression in human thyroid follicles. Human thyroid follicles were cultured in medium supplemented with bTSH (30 µU/mL) in the presence or absence of IL-1alpha (5 ng/mL). After 24 h of culture, gene expression was analyzed by microarray. TPO, thyroperoxidase; TG, thyroglobulin; DIO1, type I deiodinase; DIO2, type 2 deiodinase; DIO3, type 3 deiodinase; NIS, sodium iodide symporter; SCYA20, cytokine A20; GRO2, macrophage inflammatory protein 2; NFKB1, nuclear factor of kappa light polypeptide gene enhancer; IFNGR1, interferon gamma receptor 1.

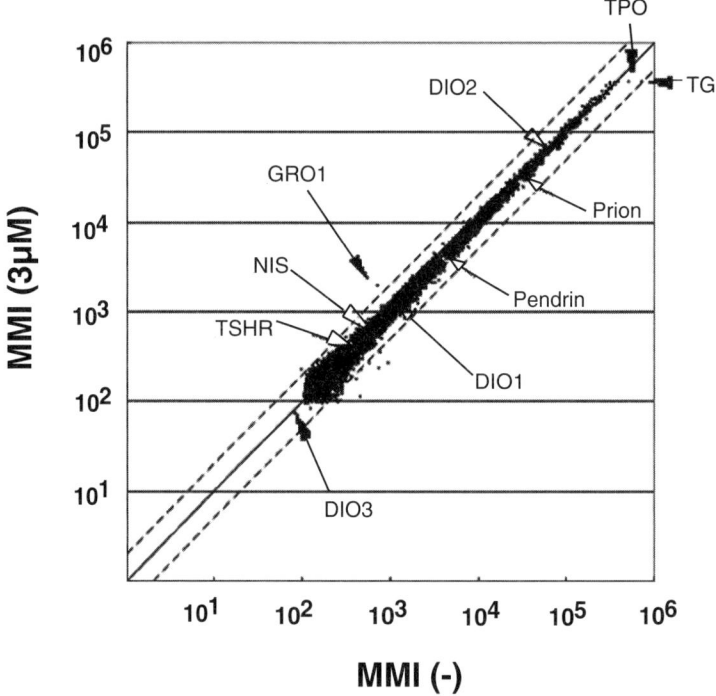

Fig. 5. Scatter plot of the effect of methimazole on gene expression in human thyroid follicles. Human thyroid follicles were cultured in medium supplemented with bTSH (50 μU/mL) in the presence or absence of methimazole (3 μM). After 48 h of culture, gene expression was analyzed by microarray. Symbols are the same as those in Fig. 1.

the expression of only one gene (GRO-1, a CXC chemokine that binds to the interleukin 8 receptor to mobilize intracellular calcium) was increased more than twofold (Fig. 5). No significant change was observed in the expression of 17 000 other genes, suggesting that the antithyroid agent does not damage thyrocytes.

In contrast, when human thyroid follicles were cultured with the iodine-containing antiarrhythmic agent amiodarone at 1–2 μmol/L, which is the level attainable in patients taking a maintenance dose *(18)*, the expression level of NIS was decreased to less than half *(8)*. Furthermore, the expression of 19 and 64 genes was up- or downregulated, respectively, suggesting that the antiarrhythmic agent is far more cytotoxic to thyrocytes than methimazole.

DISCUSSION

When performing microarray studies on endocrine glands, it is very important to use a culture system composed of purified cells that respond to tropic hormones with high sensitivity and behave as they do in vivo. We have established a system for suspension culture of thyroid follicles, which can secrete de novo-synthesized thyroid hormones in response to physiological concentrations of TSH (Fig. 2), and exhibit the Wolff–Chaikoff effect in response to a high concentration of iodide (Fig. 3) *(3–5)*. Therefore, the data obtained in our present in vitro study are physiologically relevant, and would reflect precisely the in vivo phenomena occurring in the thyroid gland of a patient. It should be stressed that thyroid cell lines such as FRTL-5, which grow continuously in a monolayer, are not normal thyrocytes, do not form thyroid follicles and hardly synthesize thyroid hormones, and sometimes exhibit an aberrant or even opposite response to TSH *(19)*. Therefore, in order to investigate the physiological effects of iodide, thyroid follicles containing abundant thyroglobulin should be used.

As we reported previously *(12)*, TSH or TSH receptor-stimulating Graves' IgG stimulates the expression of VEGF-A mRNA and its receptors on endothelial cells, causing the thyroid gland to become hypervascular, whereas iodide decreases the expression of VEGF-A, accompanied by an increase in the expression of u-PA, which produces angiostatin, a potent angiogenesis inhibitor, from plasminogen *(20)*. Therefore, it is highly likely that iodide decreases thyroidal blood flow by decreasing and increasing the expression levels of pro- and anti-angiogenesis factors, respectively *(7)*. These in vitro findings at least partly account for the effect of iodide administration prior to total or subtotal thyroidectomy in patients with Graves' disease, which has been used routinely and empirically by endocrine surgeons for many years to decrease bleeding *(21)*.

Using the present in vitro system, we have been studying the mechanism by which iodide inhibits the release of de novo-synthesized as well as stored thyroid hormone. If it were possible to prolong or prevent the wearing off of the acute Wolff–Chaikoff effect, it might be feasible to develop a new strategy for treatment of patients with severe hyperthyroidism, such as thyroid storm. Interestingly, under the present culture conditions, iodide stimulates the levels of expression of several inflammatory or immunoregulatory cytokines, despite minor contamination with only a few immuno-competent cells *(4)*. Our present microarray approach will clarify the mechanism by which iodide aggravates autoimmune thyroiditis in patients with Hashimoto's disease.

Since amiodarone, a potent antiarrhythmic agent that contains 37% iodine, is used in Japan, the incidence of amiodarone-induced thyrotoxicosis type II (AIT-II) is steadily increasing *(22)*. Usually, patients who are taking a maintenance dose are euthyroid, due to escape from the acute Wolff–Chaikoff effect. The present microarray analysis revealed that the expression level of NIS was decreased to less than 1/2, as demonstrated by Eng et al. in rats fed a large amount of iodine in drinking water *(23)*. In addition, a number of genes (phospholipases, metallothioneins) were up- or downregulated, suggesting that an imbalance between lipid peroxidation and its protective system may be involved in the development of AIT-II. In comparison with amiodarone, methimazole hardly affected the expression of genes in thyroid follicles, suggesting that it is far less cytotoxic to thyrocytes, although an idiosyncratic adverse reaction (i.e., agranulocytosis) may develop in a very small proportion of Graves' patients.

Very recently, Harii et al.*(24)* reported that toll-like receptor 3 is expressed in FRTL-5 thyroid-like cells, and that when the cells were treated with double-stranded RNA [polyinosinic-polycytidylic acid, Poly(I:C)], they showed an increase of IFN-β mRNA. We confirmed these findings and found that treatment of thyroid follicles with Poly(I:C) strongly inhibited thyroid function *(25)*. Such studies may provide clues to the pathophysiological mechanism by which viral infection may be involved in the development of autoimmune thyroiditis.

CONCLUSION

Microarray is an extremely powerful method for elucidating the mechanism of hormone actions in endocrine organs. To obtain physiologically relevant data, however, experiments should be performed using a culture system composed of purified endocrine tissues or cells, which respond to stimulators or inhibitors with high sensitivity. Using human thyroid follicles in suspension culture, we have demonstrated that TSH and iodide respectively increase or decrease the levels of angiogenesis factors, particularly VEGF-A. Our microarray studies will clarify the pathophysiological mechanism of iodide- or virus-induced thyroid dysfunction, since the microarray can analyze all of the genes expressed in human thyroid follicles in a single run.

REFERENCES

1. Wolff J (1989) Excess iodide inhibits the thyroid by multiple mechanisms. *Adv Exp Med Bio* **261**:211–244.
2. Mitchell JC, Parangi S (2005) Angiogenesis in benign and malignant thyroid disease. *Thyroid* **15**:494–510.
3. Yamazaki K, Sato K, Shizume K, Kanaji Y, Ito Y, Obara T, Nakagawa T, Koizumi T, Nishimura R (1995) Potent thyrotropic activity of human chronic gonadotropin variants in terms of ^{125}I incorporation and de novo-synthesized thyroid hormone release in human thyroid follicles. *J Clin Endocrinol Metab* **80**:473–479.
4. Nakajima K, Yamazaki K, Yamada E, Kanaji Y, Kosaka S, Sato K, Takano K (2001) Amiodarone stimulates interleukin-6 production in cultured human thyrocytes, exerting cytotoxic effects on thyroid follicles in suspension culture. *Thyroid* **11**:101–109.
5. Yamazaki Y, Yamada E, Kanaji Y, Yanagisawa T, Kato Y, Takano K, Obara T, Sato K (2003) Genes regulated by TSH and iodide in cultured human thyroid follicles: analysis by cDNA microarray. *Thyroid* **13**:149–158.
6. International Human Genome Sequencing Consortium (2001) Initial sequencing and analysis of the human genome. *Nature* **409**:861–921.
7. Yamada E, Yamazaki K, Takano K, Obara T, Sato K (2006) Iodide inhibits vascular endothelial growth factor (VEGF) – A expression in cultured human thyroid follicles: A microarray search for effects of TSH and iodide on angiogenesis factors. *Thyroid* **16**:545–554.
8. Yamazaki K, Yamada E, Yamada T, Takano K, Obara T, Sato K (2005) Effect of amiodarone on gene expression participating in thyroid hormogenesis in cultured human thyroid follicles. *Folia Endocrinol Jpn* **81** (Suppl.):18, 313.
9. Sato K, Satoh T, Shizume K, Ozawa M, Han D, Imamura H, Tsushima T, Kanaji Y Jr, Ito Y, Obara T, Fujimoto Y, Kanaji Y (1990) Inhibition of ^{125}I organification and thyroid hormone release by interleukin-1, tumor necrosis factor-α, interferon-γ in human thyrocytes in suspension culture. *J Clin Endocrinol Metab* **70**:1735–1743.
10. Yamazaki K, Yamakawa Y, Shizume K, Kanaji Y, Kanaji Y Jr, Obara T, Sato K (1993) Reversible inhibition of interferons alpha and beta of ^{125}I incorporation and thyroid hormone release by human thyroid follicles in vitro. *J Clin Endocrinol Metab* **77**:1439–1441.
11. Yamazaki K, Yamada E, Kanaji Y, Shizume K, Wang D, Maruo N, Obara T, Sato K (1996) Interleukin-6 (IL-6) inhibits thyroid function in the presence of soluble IL-6 receptor in cultured human thyroid follicles. *Endocrinology* **137**:4857–4863.
12. Sato K, Yamazaki K, Shizume K, Kanaji Y, Obara T, Ohsumi K, Yamaguchi S, Shibuya M (1995) Stimulation by thyroid stimulating hormone and Graves' immunoglobulin G of vascular endothelial growth factor mRNA expression in human thyroid follicles in vitro and flt mRNA expression in the rat thyroid in vivo. *J Clin Invest* **96**:1295–1302.
13. Sato K. (2001) Vascular endothelial growth factors and thyroid diseases: review. *Endocr J* **48**:635–646.

14. Dohan O, De la Vieja A, Paroder V, Riedel C, Artani M, Reed M, Ginter CS, Carrasco N (2003) The sodium/iodide symporter (NIS): characterization, regulation, and medical significance. *Endocr Rev* **24**:48–77.

15. Bianco AC, Salvatore D, Gereben B, Berry MJ, Larsen PR (2002) Biochemistry, cellular and molecular biology, and physiological roles of the iodothyronine selenodeiodinases. *Endocr Rev* **23**:38–89.

16. Turner HE, Harris AL, Melmed S, Wass JA. (2003) Angiogenesis in endocrine tumors. *Endocr Rev* **24**:600–632.

17. Yamazaki K, Yamada E, Kanaji Y, Yanagisawa T, Kato Y, Sato K, Takano K, Sakasegawa Y, Kaneko K (2003) Stimulation of cellular prion protein (PrPc) expression by TSH in human thyroid follicles. *Biochem Biophys Res Comm* **305**:1034–1039.

18. Kashima A, Funahashi M, Fukumoto K, Komamura K, Kamakura S, Kitakaze M, Ueno K (2005) Pharmacokinetic characteristics of amiodarone in long-term oral therapy in Japanese population. *Biol Pharm Bull* **28**:1934–1938.

19. Miyagi E, Katoh R, Li X, Lu S, Suzuki K, Maeda S, Shibuya M, Kawaoi A (2001) Thyroid stimulating hormone down-regulates vascular endothelial growth factor expression in FRTL-5 cells. *Thyroid* **11**:539–543.

20. Ramsden JD, Yarram S, Mathews E, Watkinson JC, Eggo MC (2002) Thyroid follicular cells secrete plasminogen activators and can form angiostatin from plasminogen. *J Endocrinol* **173**:475–481.

21. Lal G, Ituarte P, Kebebew E, Siperstein A, Duh QY, Clark OH (2005) Should total thyroidectomy become the preferred procedure for surgical management of Graves' disease? *Thyroid* **15**:569–574.

22. Sato K, Miyakawa M, Eto M, Inaba T, Matsuda N, Shiga T, Ohnishi S, Kasanuki H (1999) Clinical characteristics of amiodarone-induced thyrotoxicosis and hypothyroidism in Japan. *Endocr J* **46**:443–451.

23. Eng PHK, Cardona GR, Fang SL, Previti M, Alex S, Carrasco N, Chin WW, Braverman LE (1999) Escape from the acute Wolff–Chaikoff effect is associated with a decrease in thyroid sodium/iodide symporter messenger ribonucleic acid and protein. *Endocrinology* **140**:3404–3410

24. Harii N, Lewis CJ, Vasko V, McCall K, Benavides-Peralta U, Sun X, Ringel MD, Saji M, Giuliani C, Napolitano G, Goetz DJ, Kohn LD (2005) Thyrocytes express a functional toll-like receptor 3: overexpression can be induced by viral infection and reversed by phenylmethimazole and is associated with Hashimoto's autoimmune thyroiditis. *Mol Endocrinol* **19**:1231–1250

25. Yamazaki K, Suzuki K, Yamada E, Yamada T, Takeshita F, Matsumoto M, Mitsuhashi T, Obara T, Takano K, Sato K (2007) Suppression of iodide uptake and thyroid hormone synthesis with stimulation of the type I interferon system by double-stranded RNA (dsRNA) in cultured human thyroid follicles. *Endocrinology.* **148**:3226–3235.

IV   DISEASES OF HORMONAL SYSTEMS

12 Genomics and Polycystic Ovary Syndrome (PCOS): The Use of Microarray Analysis to Identify New Candidate Genes

Jennifer R. Wood, PhD
and Jerome F. Strauss III, MD, PhD

CONTENTS

INTRODUCTION
MICROARRAY ANALYSIS OF NORMAL AND PCOS THECA CELLS AND
  WHOLE OVARIES
GENES POTENTIALLY CONTRIBUTING TO THE OVARIAN PHENOTYPES
  OF PCOS
GENOMICS AND THE IDENTIFICATION OF NEW PCOS CANDIDATE
  GENES
ACKNOWLEDGMENTS
REFERENCES

Abstract

Transcriptome profiling offers a potentially valuable approach to the identification of candidate genes and pathways that contribute to complex diseases. Here we describe the application of microarray analysis of human theca cells and ovaries to the understanding of the pathophysiology of polycystic ovary syndrome (PCOS), a complex reproductive endocrine and metabolic disorder, and the investigation of selected candidate genes derived from these analyses with respect to association and linkage to PCOS.

Key Words: Polycystic ovary syndrome, theca cells, steroidogenesis, follicle, ovary

INTRODUCTION

The clinical phenotype of polycystic ovary syndrome (PCOS) was first elucidated by Stein and Leventhal in 1935 who described an association between polycystic ovaries, amenorrhea, hirsutism, and obesity *(1)*. It is estimated that 5–10% of reproductive aged women exhibit the symptoms of PCOS, which include those related to increased

From: *Contemporary Endocrinology: Genomics in Endocrinology:*
DNA Microarray Analysis in Endocrine Health and Disease
Edited by S. Handwerger and B. Aronow © Humana Press, Totowa, NJ

circulating androgen levels, anovulatory infertility, and frequently, insulin resistance and hyperinsulinemia (2–5).

Although increased adrenal steroidogenesis contributes to increased androgen levels in a subset of PCOS women (25–60%) (6, 7), the ovary is the primary source of increased circulating androgens (2,3,8). In the normal ovary, androgens are synthesized from cholesterol in the theca cells. Studies using freshly isolated theca cells or theca cells maintained in long-term culture demonstrate that PCOS theca cells produce more progesterone (P_4), 17α-hydroxyprogesterone, dehydroepiandrosterone (DHEA), androstenedione (Δ4A), and testosterone than normal theca cells (9,10). Increased theca cell steroidogenesis in PCOS has been correlated to increased expression and/or activity of certain steroidogenic enzymes. Specifically, the mRNA levels of P450 side chain cleavage (CYP11A1) and P450 17α-hydroxylase, 17/20 lyase (CYP17) are elevated in PCOS as compared to normal theca cells, while steroidogenic acute regulatory protein (StAR), type V 17β-hydroxysteroid dehydrogenase (HSD), 3α-HSD, and 20α-HSD mRNA levels are not different between normal and PCOS theca cells (10,11) (Table 1).

In addition to increased ovarian steroidogenesis, arrested follicle development, which results in the clinical phenotype of anovulatory infertility, is a central feature of PCOS (2, 3, 12). In the normal ovary, follicular growth is driven by processes of recruitment and selection which results in an increase in oocyte size, the proliferation of the granulosa cells, the development of the theca cell layer, the formation of a fluid-filled antrum, and the selection of a dominant follicle for ovulation (13, 14). The recruitment of a follicle from the resting primordial pool to an actively growing follicle is initially regulated by paracrine growth factors, while selection of a dominant follicle is primarily dependent on stimulation by follicle stimulating hormone (FSH) and luetinizing hormone (LH) (14–18). In PCOS, the maturation of the follicle arrests at the small antral stage (5–10 mm) with a lack of selection of a dominant follicle (19). Measurements of gonadotropin levels in PCOS women demonstrate that the levels of luetinizing hormone (LH) are increased (20–22). Thus, the increased secretion of LH and the increased LH/FSH ratio may impact the selection of a dominant follicle and ultimately impair folliculogenesis in PCOS. Likewise, altered levels of several growth factors including growth differentiation factor-9, leukemia inhibitory factor, anti-Müllerian hormone, and/or inhibin may contribute to the phenotype of arrested folliculogenesis in PCOS (23–29). However, the specific molecular mechanism for arrested follicle development in PCOS remains unclear. Stromal hypertrophy is another

Table 1
Stable Steroidogenic Profile of PCOS Theca Cells

Biochemical phenotype	Molecular phenotype
Progesterone synthesis—increased	StAR mRNA and StAR promoter activity—no difference CYP11A1 mRNA—increased 3βHSD activity—increased
17OH-progesterone and testosterone synthesis—increased	CYP17 mRNA, CYP17 promoter activity, and CYP17 enzymatic activity—increased 17βHSD mRNA—no difference

characteristic feature of the PCOS ovary, which may result from high local androgen levels. It is not yet known whether the dense stroma contributes to the pathophysiology of follicular growth arrest of whether it is an unrelated phenomenon.

Women with PCOS are also prone to defects in insulin dynamics as a result of insulin resistance *(3, 8, 30, 31)*. In addition, circulating insulin levels are increased as a result of increased pancreatic insulin secretion and decreased hepatic insulin clearance *(8, 30)*. One of the consequences of hyperinsulinemia is reduced hepatic sex hormone binding globulin production, which increases the free testosterone levels, exacerbating the reproductive disturbance. While there is no change in insulin receptor (IR) abundance or binding affinity, studies by Dunaif et al. demonstrate that there is a defect in IR-dependent signal transduction *(32–34)*. Approximately 50% of PCOS women exhibit basal autophosphorylation of the IR which subsequently inhibits transmission of the insulin signal *(32)*. Inhibition of downstream signaling events including impaired insulin receptor substrate-1 phosphorylation and inhibition of phosphatidylinositol 3-kinase (PI3K) activation represent alternative mechanisms for decreased insulin signaling in PCOS muscle and fibroblasts *(30, 33, 34)*. These insulin-signaling defects in PCOS women are confounded by an increased incidence of obesity. Likewise, insulin resistance, associated abnormalities in glucose metabolism, and obesity lead to an increased risk for type 2 diabetes, altered lipid profiles, and cardiovascular disease *(3, 4, 35)*. Collectively, these studies demonstrate that PCOS is a complex endocrine disease that impacts both the reproductive and metabolic health of affected women.

Epidemiological studies of PCOS in the 1970s and 1980s indicate that the syndrome aggregates in families and suggest that there is a genetic basis to PCOS [*(8)* and references therein]. In 1998, Legro et al. carried out a comprehensive familial study which demonstrated that sisters of PCOS women have a threefold greater risk of being hyperandrogenemic compared with control women *(36)*. Likewise, Kahsar-Miller et al. demonstrated an approximately fivefold increase in the incidence of PCOS among first-degree female relatives of affected patients compared to the general population *(37)*. Given the heterogeneity of the syndrome, it is likely that PCOS is an oligogenic trait, although mutation of a single gene cannot be ruled out. While numerous genes associated with steroid synthesis and insulin signaling have been examined for genetic variation (Table 2), the genetic etiology of PCOS remains unclear *(38–40)*.

In the pregenomic era, biochemical studies and candidate gene approaches were utilized to gain a better understanding of the pathophysiology and genetics of PCOS and focused primarily on known genes which directly affect androgen synthesis and insulin sensitivity. In the postgenomic era, investigators have been and will be able to take advantage of unbiased approaches including gene expression profiling and genome wide scans to identify novel genes which may contribute to the phenotype and/or genotype of PCOS. In particular, microarray analysis has proven to be a valuable tool to (a) define the underlying molecular mechanisms of increased androgen synthesis and altered insulin signaling in PCOS and (b) identify new candidate genes which may contribute to the genetic etiology of the syndrome.

Table 2
Genes Evaluated for Association with PCOS

Gene name (symbol)	Results	Reference(s)
Luetinizing hormone β subunit (*LHB*)	Variation but no causal link	*(38, 39)*
Luetinizing hormone receptor (*LHCGR*)	No association	*(38)*
Follistatin (*FST*)	No clear association or linkage	*(38, 39)*
Dopamine receptor (*DRD3*)	Polymorphism but not association	*(39)*
Gonadotropin releasing hormone receptor (*GNRHR*)	No association	*(39)*
P450 side chain cleavage (*CYP11A1*)	Weak linkage; remains a plausible candidate	*(38, 39)*
P450 17α hydroxylase, 17–20 lyase (*CYP17*)	No linkage or association	*(38, 39))*
P450 aromatase (*CYP19*)	No association	*(38, 39)*
17β-hydroxysteroid dehydrogenase type 3 (*HSD17B3*)	Polymorphism but no association	*(39)*
Androgen receptor (*AR*)	No association	*(39)*
Sex hormone binding globulin (*SHBG*)	VNTR polymorphism; remains a plausible candidate	*(39)*
Steroidogenic factor-1 (*SF-1*)	No association	*(39)*
Steroidogenic acute regulatory protein (*StAR*)	No association	*(39)*
Dosage-sensitive sex reversal-adrenal hypoplasia (*DAX-1*)	No association	*(39)*
Insulin (*INS*)	VNTR polymorphism; conflicting results regarding association	*(38–40)*
Insulin receptor (*INSR*)	No association but strong association for locus (D19S884) near *INSR*	*(38–40)*
Insulin receptor substrate-1 (*IRS1*)	Variants but no clear association	*(39, 40)*
Insulin receptor substrate-2 (*IRS2*)	Variants but no clear association	*(39, 40)*
Insulin-like growth factor 1 (*IGF1*)	No association	*(40)*
Insulin-like growth factor 2 (*IGF2*)	Associated variant in Spanish population	*(40)*
Calpain 10 (*CAPN10*)	Conflicting results; may exhibit ethnicity-specific association	*(39, 40)*
Peroxisome proliferators-activated receptor-γ (*PPARG*)	Polymorphism but no association	*(40)*
Sorbin and SH3 domain-containing 1 (*SORBS1*)	Polymorphism but no association	*(40)*
Paraoxonase-1 (*PON1*)	Associated polymorphism in Spanish population	*(40)*
Leptin (*LEP*)	No association	*(39, 40)*
Leptin receptor (*LEPR*)	No association	*(39, 40)*
Glycogen synthase-2 (*GYS2*)	No association	*(40)*
Resistin (*RETN*)	No association	*(40)*
Adiponectin (*APM1*)	No association	*(40)*

MICROARRAY ANALYSIS OF NORMAL AND PCOS THECA CELLS AND WHOLE OVARIES

Candidate gene approaches have been successfully used to delineate correlations between increased androgen biosynthesis and steroidogenic enzyme gene expression and identify proteins that exhibit impaired function with regards to insulin signaling in PCOS *(10, 11, 23–25, 27, 28, 32–34, 41–44)*. However, this experimental approach is hampered by a limited knowledge of the genes controlling androgen synthesis and insulin-mediated signal transduction. Furthermore, androgen synthesis and impaired insulin signaling are phenotypic endpoints of PCOS, making it difficult to anticipate which upstream gene networks and signaling pathways participate in the manifestation of these PCOS phenotypes. The completion of the human genome project and the advent of microarray chip technology have provided investigators with an unbiased approach to define the gene networks and signaling pathways that contribute to the pathophysiology of PCOS and to identify new candidate genes which may be involved in the genetic etiology of this syndrome.

Gene Expression Profile of PCOS Ovaries and Isolated Theca Cells

Since excess ovarian androgen production is the hallmark characteristic of PCOS and is likely an intrinsic ovarian defect, ovarian tissue including whole ovary and isolated theca cells have been analyzed using microarray genechips *(45–49)*. The gene expression profile of individual normal and PCOS theca cell cultures were defined using the Affymetrix GeneChip platform *(46)*. Hierarchical cluster and principal component analysis of the theca cell microarray data demonstrate that the normal and PCOS theca cells have distinct transcriptomes (Fig. 1). The initial statistical analysis of the theca cell microarray data identified approximately 350 genes with a significant increase or decrease in mRNA abundance in PCOS theca cells *(46)*. Similar results were obtained by Jansen et al. when RNA from normal and PCOS whole ovary were hybridized on Affymetrix chips *(47)*. Specifically, normal and PCOS ovaries exhibit distinct gene expression profiles which can be distinguished using principal component analysis. Furthermore, this array analysis identified 230 genes which have altered mRNA abundance in the PCOS compared to the normal ovary. Likewise, Diao et al. and Oksjoki et al. identified 290 and 44 genes, respectively, which are differentially expressed in the PCOS as compared to normal ovary using cDNA array platforms *(48, 49)*. When the genes which exhibit altered mRNA abundance in the four independent studies were compared, there were very few genes which were identified as differentially expressed in more than one study. These incongruent results are likely a reflection of the use of different tissue types (i.e., theca cells versus whole ovary), small sample sizes, and differences in microarray platforms and experimental design and demonstrate that care needs to be taken when interpreting these microarray results. However, these complementary studies do demonstrate that PCOS theca cells and whole ovarian tissues exhibit a unique molecular fingerprint and suggest that there is a genetic alteration or stable epigenetic imprint intrinsic to the ovary, which is important for manifestation of the ovarian PCOS phenotypes.

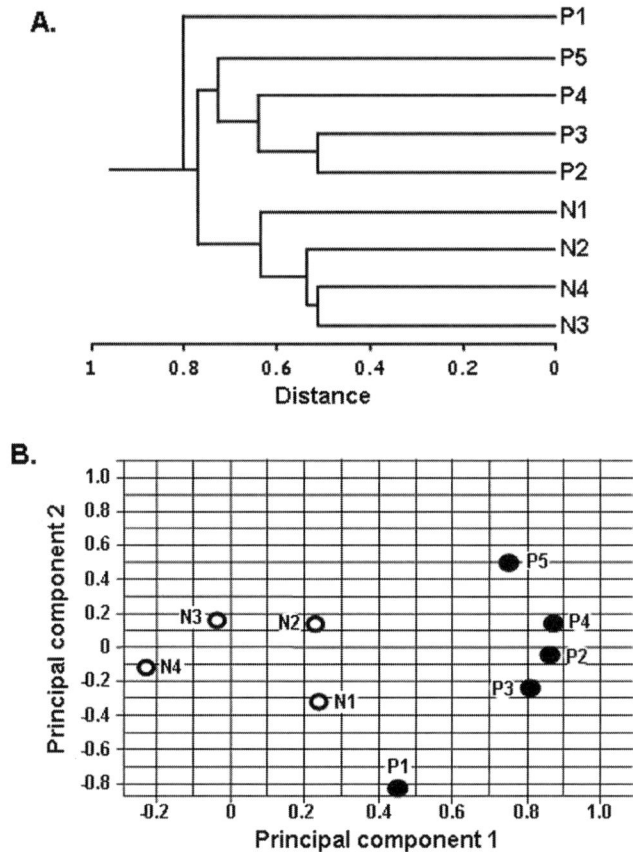

Fig. 1. Gene expression profiling shows that PCOS and normal theca cells have distinct transcriptomes. (A) The gene expression profiles for the four independent normal theca cell samples (N1, N2, N3, and N4) and the five independent PCOS theca cell samples (P1, P2, P3, P4, and P5) were compared using hierarchical clustering. This analysis demonstrates tight association between samples within each experimental group. (B) The gene expression data for samples N1, N2, N3, N4, P1, P2, P3, P4, and P5 was subjected to principal component analysis and graphed accordingly. This analysis also demonstrates tight association between the samples within each experimental group.

Validation of Microarray Analyses

The ability of microarray analyses to reliably predict genes with altered mRNA abundance in PCOS is highly dependent on several factors including experimental design, array controls, normalization parameters, and statistical analysis (50–53). Thus, to determine the validity of the theca cell analysis described above, several control experiments have been carried out. The independent methodology of quantitative, RT-PCR (QPCR) which is commonly used to validate microarray experiments was employed. When the PCOS-to-normal ratio of mRNA abundance for individual genes was determined by QPCR or microarray analysis, significant correlations between the two methodologies were seen (46). In addition to the QPCR validation, several other control experiments including multiple statistical analyses and multiple theca cell culture replicates were carried out. Since there is no consensus approach to

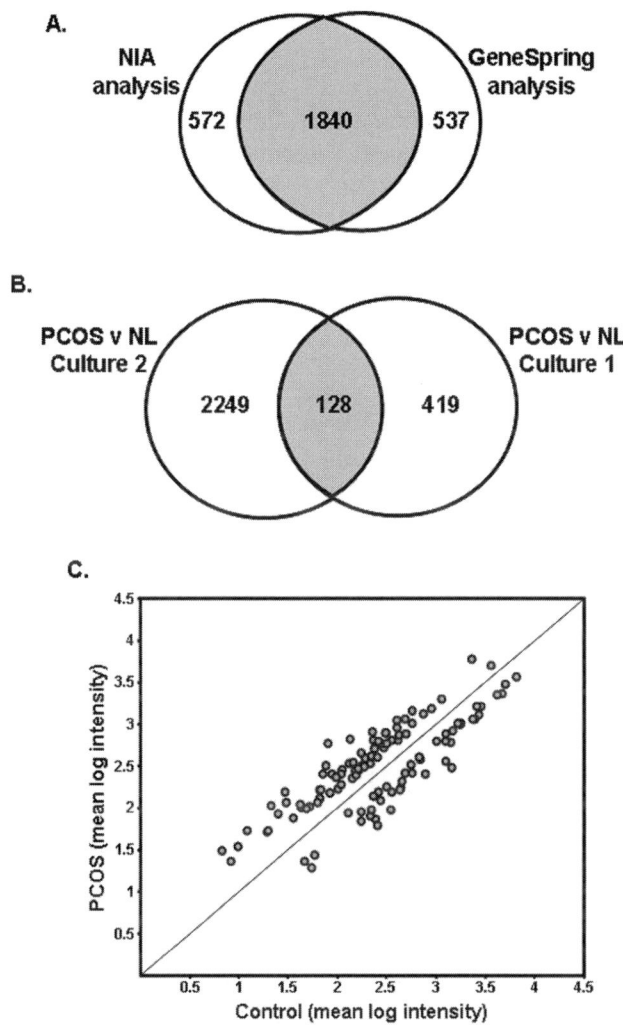

Fig. 2. Multiple analysis tools define 114 transcripts with increased or decreased mRNA abundance in PCOS compared to normal theca cells. (A) The microarray data obtained from the four normal and five PCOS theca cell samples was filtered for statistical significance using either NIA Array Analysis (left circle) or GeneSpring (right circle). Each analysis identified approximately 2400 transcripts with altered mRNA abundance in PCOS compared to normal theca cells. Of these 2400 transcripts, 1840 were identified by both statistical analyses (intersection of two circles). (B) The gene expression profiles of the PCOS theca cells were compared to the gene expression profiles of sibling cultures of normal theca cells which were independently cultured and hybridized (NL-Culture 1 versus NL-Culture 2). There were 547 transcripts differentially expressed in PCOS compared to NL-Culture 1 theca cells (right circle) and 2377 transcripts with altered mRNA abundance in PCOS compared to NL-Culture 2 theca cells (left circle). One hundred and twenty-eight transcripts showed altered mRNA abundance in PCOS theca cells regardless of the specific culture of normal theca cells (intersection of two circles). (C) When the theca cell microarray data were filtered for significance based on both the method of statistical analysis and multiple culture comparisons, there were 114 transcripts which consistently exhibit increased or decreased mRNA abundance in PCOS compared to normal theca cells. For each transcript the mean log fluorescence intensity in the normal theca cells (x-axis) was plotted against the mean fluorescence log intensity in the PCOS theca cells (y-axis). The line denotes a 1:1 ratio of fluorescence intensity.

statistical analysis of microarray data *(50)*, two different statistical approaches were used to identify transcripts with altered mRNA abundance in PCOS theca cells. First, parametric, one-way ANOVA which was corrected using the cross-gene error model was carried out using the GeneSpring software package (Agilent Technologies, Santa Clara, CA, USA). The data were also subjected to one-way ANOVA with variance averaging using software from the National Institutes of Aging Array Analysis *(54)*. Each statistical analysis identified approximately 2400 transcripts with a significant difference ($P < 0.05$) in mRNA abundance between normal and PCOS theca cells and 1840 transcripts which were identified as differentially expressed by both analyses (Fig. 2A). Since the theca cells were maintained in long-term cultures, differences due to culture conditions were also determined. To this end, replicate gene expression profiles of the normal theca cell samples which were obtained from sibling cultures (i.e., two different cell stocks derived from theca collected from the same follicles but cultured at different times) were independently compared to the gene expression profile of the PCOS theca cells. Only 128 transcripts exhibited altered mRNA abundance in both comparisons (Fig. 2B). When the theca cell microarray data were filtered for significance based on both the method of statistical analysis and multiple culture comparisons, there were 114 transcripts which consistently exhibit increased or decreased mRNA abundance in PCOS compared to normal theca cells (Fig. 2C, Table 3). QPCR experiments confirm that 10 of these 114 transcripts have a statistically significant difference in mRNA abundance in PCOS compared to normal theca cells *(45, 46)* and indicate that gene expression profiling of cultured theca cells accurately predicts genes differentially expressed in PCOS theca cells.

Table 3
Genes with Significantly Increased or Decreased mRNA Abundance in PCOS Theca Cells

Accession number	Gene Symbol	Affymetrix number	PCOS :NL ratio	P(NIA)	P(GS)	FDR	Chr
NM_001613	ACTA2	200974_at	0.556	0.00316	0.0324	0.08409	10q23.31
NM_001615	ACTG2	202274_at	0.457	0.00175	0.0194	0.06024	2p13.1
NM_017825	ADPRHL2	223097_at	0.561	0.0034	0.032	0.08817	1p34.3
NM_000693	ALDH1A3	203180_at	1.556	0.0376	0.0173	0.29876	15q26.3
NM_006305	ANP32A	201043_s_at	3.746	0.01282	0.036	0.18642	15q23
NM_001642	APLP2	208702_x_at	0.598	0.00935	0.0368	0.1552	11q24.3
NM_004925	AQP3	39248_at	2.521	0.00237	0.00115	0.05676	9p13.3
NM_015161	ARL6IP	211935_at	1.961	0.00328	0.0328	0.0335	6p12.3
NM_133436	ASNS	205047_s_at	0.200	0.00037	0.00948	0.02389	7q21.3
NM_004776	B4GALT5	221485_at	1.740	0.00482	0.00358	0.10847	20q13.13
NM_018476	BEX1	218332_at	1.539	0.03104	0.0161	0.27509	Xq22.1
NM_004335	BST2	201641_at	4.742	0.0282	0.00669	0.26321	19p13.11
NM_020644	C11orf15	222507_at	0.539	0.00139	0.0267	0.05161	11p15.4
NM_153211	C18orf17	238480_at	1.771	0.00498	0.00733	0.11083	18q11.2
NM_052966	C1orf24	217967_at	0.238	0.00005	0.00619	0.00638	1q25.3
NM_152371	C1orf93	231835_at	0.568	0.00519	0.0385	0.11359	1p36.32
AB047784	C21orf25	212875_s_at	1.683	0.01103	0.00407	0.16992	21q22.3
NM_014165	C6orf66	219006_at	1.813	0.01354	0.00047	0.19071	6q16.1
NM_032310	C9orf89	223398_at	0.583	0.00527	0.0466	0.11436	9q22.31

NM_001226	CASP6	209790_s_at	2.375	0.00009	0.00012	0.009	4q25
NM_001797	CDH11	236179_at	2.399	0.00264	0.00841	0.07578	16q21
		207172_s_at	1.699	0.03196	0.0381	0.27782	
NM_024111	CHAC1	219270_at	0.370	0.0006	0.0115	0.03206	15q15.1
NM_018413	CHST11	226372_at	1.818	0.01697	0.0136	0.21307	12q23.3
NM_001827	CKS2	204170_s_at	3.502	0	0.000008	0.00001	9q22.2
NM_001866	COX7B	202110_at	2.302	0.00002	0.00236	0.00399	Xq21.1
NM_148923	CYB5	207843_x_at	2.049	0.00061	0.00087	0.0322	18q22.3
		215726_s_at	1.528	0.03153	0.0349	0.27616	
NM_133506	DCN	211813_x_at	2.480	0	0.00036	0.00098	12q21.33
NM_016041	DERL2	218333_at	1.694	0.01172	0.0092	0.17735	17p13.2
NM_004675	DIRAS3	215506_s_at	2.333	0.04528	0.0455	0.32857	1p31.2
NM_015621	DKFZP434C171	212886_at	3.316	0.01603	0.00469	0.20661	5q33.1
NM_003315	DNAJC7	202416_at	1.833	0.00283	0.00122	0.07841	17q21.1
NM_004247	EFTUD2	222398_s_at	0.607	0.00639	0.0365	0.12627	17q21.31
NM_001431	EPB41L2	201719_s_at	2.534	0.00001	0.00006	0.00262	6q23.1
NM_004450	ERH	200043_at	2.050	0.00029	0.00014	0.01969	14q24.1
CR592254	EST	225155_at	2.494	0.03829	0.0239	0.30168	6q14.3
	EST	235964_x_at	2.467	0.01304	0.00462	0.18783	20q11.23
		235529_x_at	2.704	0.01196	0.00408	0.17914	
	EST	229885_at	1.728	0.01474	0.0198	0.19874	11q13.5
	EST	240165_at	1.933	0.02213	0.0456	0.24098	12q21.33
NM_004114	FGF13	205110_s_at	3.367	0.00052	0.00257	0.0293	Xq26.3
NM_001456	FLNA	200859_x_at	0.478	0.00001	0.00824	0.00256	Xq28
		214752_x_at	0.554	0.00113	0.0131	0.04532	
		213746_s_at	0.455	0.00002	0.00584	0.00306	
NM_013281	FLRT3	219250_s_at	2.062	0.01757	0.00875	0.21691	20p12.1
NM_005254	GABPB2	205510_s_at	2.539	0.00183	0.00446	0.06144	15q21.2
NM_005257	GATA6	210002_at	2.492	0.00008	0.00409	0.00847	18q11.2
NM_005458	GPR51	209990_s_at	0.484	0.01199	0.0338	0.17914	9q22.33
NM_001001555	GRB10	210999_s_at	0.503	0.00168	0.0175	0.05854	7p12.2
NM_032999	GTF2I	210891_s_at	1.726	0.00546	0.00095	0.11707	7q11.23
		201065_s_at	1.521	0.02999	0.00461	0.2706	
NM_000858	GUK1	200075_s_at	0.573	0.00068	0.0277	0.03415	1q42.13
NM_017445	H2BFS	208579_x_at	1.905	0.01232	0.00823	0.18143	21q22.3
NM_031372	HNRPDL	209067_s_at	1.551	0.02442	0.0228	0.24935	4q21.22
NM_003725	HSD17B6	37512_at	2.629	0.00083	0.00425	0.06434	12q13.3
NM_000204	IF	203854_at	2.799	0.01364	0.00667	0.19084	4q25
NM_000877	IL1R1	202948_at	1.754	0.03368	0.0165	0.28273	2q11.2
NM_005537	ING1	208415_x_at	1.618	0.03032	0.0133	0.2721	13q34
NM_000426	LAMA2	216840_s_at	2.367	0.01393	0.015	0.19331	6q22.33
NM_012134	LMOD1	203766_s_at	0.297	0.01908	0.0293	0.22375	1q32.1
NM_138781	LOC113386	242140_at	0.563	0.00582	0..0396	0.12073	19q13.43
BC036544	LOC284801	225767_at	0.321	0.00534	0.0203	0.11539	20p11.1
NM_001002919	LOC285016	238018_at	2.241	0.00792	0.0153	0.14124	2p25.3
NM_020169	LXN	218729_at	3.488	0.00157	0.000793	0.05635	3q25.32
NM_032272	MAF1	222998_at	0.473	0.0006	0.0138	0.03201	8q24.3
NM_052897	MBD6	227833_s_at	0.4512	0.00008	0.0147	0.00886	12q13.3
NM_005466	MED6	207079_s_at	1.918	0.00912	0.0188	0.1533	14q24.2
NM_053045	MGC14327	225052_at	0.608	0.01218	0.0477	0.18051	9q34.3

(Continued)

Table 3
(Continued)

Accession number	Gene Symbol	Affymetrix number	PCOS :NL ratio	P(NIA)	P(GS)	FDR	Chr
NM_207350	MGC72104	242331_x_at	4.457	0.00353	0.0494	0.08989	20q11.1
NM_007283	MGLL	211026_s_at	0.429	0.00005	0.00448	0.00601	3q21.3
NM_145255	MRPL10	224671_at	0.550	0.01872	0.0476	0.22188	17q21.32
NM_016491	MRPL37	222993_at	0.559	0.00317	0.0306	0.08435	1p32.3
NM_005955	MTF1	205323_s_at	0.654	0.04998	0.0359	0.34424	1p34.3
NM_033375	MYO1C	32811_at	0.529	0.00104	0.017	0.23663	17p13.3
NM_023018	NADK	208918_s_at	0.582	0.00702	0.0256	0.1327	1p36.33
NM_006993	NPM3	205129_at	1.801	0.02726	0.0114	0.26004	10q24.32
NM_000919	PAM	202336_s_at	1.695	0.00882	0.00874	0.15084	15q21.1
		214620_x_at	1.792	0.02037	0.0119	0.23224	
NM_002720	PPP4C	208932_at	0.447	0.00034	0.0147	0.0223	16p11.2
NM_002794	PSMB2	201404_x_at	0.434	0.00027	0.0102	0.0189	1p34.3
BC065228	PSPH	205048_s_at	0.267	0.00711	0.0166	0.13295	7p11.2
NM_031934	RAB34	224710_at	0.515	0.00506	0.0304	0.1119	17q11.2
NM_002882	RANBP1	202483_s_at	1.531	0.02616	0.0395	0.25492	22q11.21
NM_007023	RAPGEF4	205651_x_at	4.369	0.00085	0.00127	0.03858	2q31.1
NM_206963	RARRES1	221872_at	4.141	0.00003	0.00005	0.00408	3q25.32
		206392_s_at	5.054	0.04175	0.0165	0.31535	
NM_002947	RPA3	209507_at	1.541	0.03589	0.00207	0.2919	7p21.3
NM_153225	RPESP	214725_at	2.703	0.01669	0.0359	0.21062	8q13.3
NM_005870	SAP18	208740_at	2.245	0.00122	0.00671	0.04771	13q12.11
NM_016275	SELT	217811_at	2.077	0.00038	0.00242	0.02407	3q25.1
NM_000062	SERPING1	200986_at	2.705	0.00462	0.00493	0.10673	11q12.1
NM_003038	SLC1A4	212810_s_at	0.418	0.00017	0.0104	0.01374	2p14
NM_005628	SLC1A5	208916_at	0.389	0.00128	0.0149	0.04944	19q13.32
NM_017836	SLC41A3	224931_at	0.593	0.00769	0.0324	0.13973	3q21.2
NM_005775	SORBS3	207788_s_at	0.353	0	0.00612	0.00086	8p21.3
NM_002959	SORT1	212807_s_at	0.566	0.00782	0.0341	0.1401	1p13.3
NM_006939	SOS2	211665_s_at	0.461	0.01972	0.0453	0.22775	14q21.3
NM_003128	SPTBN1	226765_at	1.643	0.01732	0.0278	0.21503	2p16.2
NM_006937	SUMO2	208738_x_at	1.777	0.00326	0.00097	0.08579	17q25.1
NM_014220	TM4SF1	209386_at	2.481	0.00023	0.00325	0.01682	3q25.1
		209387_s_at	2.253	0.02634	0.0355	0.25586	
		215034_s_at	2.331	0.00118	0.00172	0.04645	
NM_023003	TM6SF1	219892_at	0.380	0.00004	0.00977	0.00579	15q25.2
NM_014255	TMEM4	202857_at	1.564	0.02124	0.0253	0.2369	12q13.3
NM_014452	TNFRSF21	218856_at	2.494	0.00244	0.0141	0.07293	6p12.3
NM_021158	TRIB3	218145_at	0.279	0.00001	0.00534	0.00228	20p13
NM_004616	TSPAN8	203824_at	7.248	0	0.000004	0	12q15
NM_022717	U1SNRNPBP	205300_s_at	4.804	0.00019	0.00689	0.01505	12q24.31
NM_058167	UBE2J2	225209_s_at	0.496	0.00054	0.0183	0.03018	1p36.33
NM_198536	UNQ501	225003_at	1.681	0.0081	0.0074	0.14257	19p13.2
NM_016089	ZNF589	210062_s_at	0.347	0.00114	0.0177	0.0454	3p21.31
NM_003461	ZYX	215706_x_at	0.418	0.00002	0.00874	0.00326	7q34

National Institute of Aging Array Analysis (NIA); GeneSpring (GS); False Discovery Rate (FDR); Chromosome locus (Chr)

GENES POTENTIALLY CONTRIBUTING TO THE OVARIAN PHENOTYPES OF PCOS

GATA6

Transcriptional regulation of *StAR, CYP11A1, CYP17*, and *3βHSD* by LH, insulin, and other growth factors is essential for temporal regulation of steroid synthesis in the ovary *(42, 55–59)*. Recent reports indicate that GATA6, which is expressed in the gonads and adrenal cortex, also regulates the expression of steroidogenic genes *(60,61)*. Specifically, GATA6 stimulates the promoter activities of the *StAR, CYP11A1, CYP17*, and the steroid sulfotransferase 2A1 genes (Table 4) *(46,62–64)*. Interestingly, microarray analysis shows that GATA6 mRNA abundance is increased in PCOS compared to normal theca cells *(46)*. QPCR analysis demonstrates that both total mRNA and nascent transcript levels of GATA6 are increased in PCOS theca cells (Fig. 3) *(65)*, and suggests that *GATA6* gene transcription is increased in the PCOS as compared to normal theca cells. Western blot analysis also confirms that GATA6 protein levels are increased in the PCOS compared to normal theca cells (Table 4) *(46)*. Taken together, the microarray data and functional assays indicate that increased GATA6 expression can stimulate the expression of steroidogenic enzymes involved in androgen synthesis and contribute to the hyperandrogenic phenotype of PCOS.

Genes Involved in Retinoic Acid Metabolism

While the identification of single genes with altered expression in PCOS theca cells can point to possible molecular mechanisms which contribute to the ovarian phenotypes of PCOS, the true power of microarray analysis is the identification of new signaling pathways which upon altered expression may contribute to the manifestation of increased ovarian androgen synthesis and/or arrested follicle development.

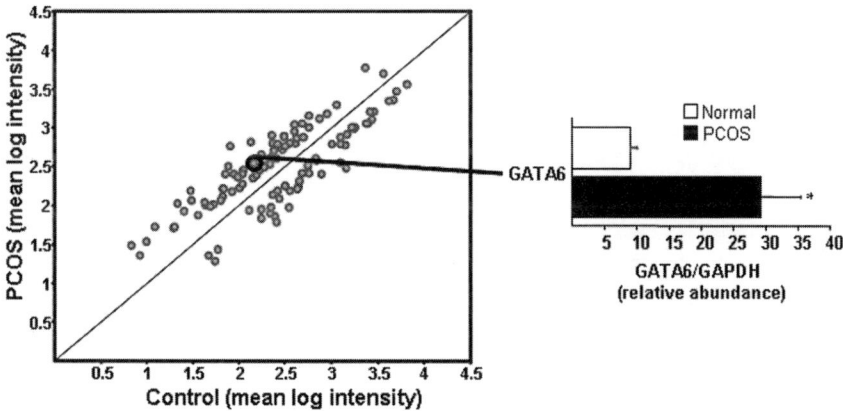

Fig. 3. Microarray analysis and QPCR demonstrate that GATA6 mRNA levels are increased in PCOS theca cells. Left panel: The mean log intensity of GATA6 in PCOS theca cell compared to the mean log intensity of GATA6 in normal theca cells is circled. Right panel: QPCR was carried out using cDNA from normal (*n* = 4; open bars) and PCOS (*n* = 4; closed bars) theca cells and gene-specific primers for GATA6. The mean ± S.E.M. normalized relative abundance for GATA6 was plotted and statistically significant differences in mRNA abundance between normal and PCOS theca cells was determined by Student's *t* -test (*, *P*-value < 0.05).

Table 4
GATA6 Action in Normal and PCOS Theca Cells

Functional Assay	Result	Reference(s)
Western blot	Increased levels of both the 52 kD and 64 kD GATA6 protein isoforms	(46)
Promoter activity	CYP17, CYP11A1, 3βHSD2, and StAR promoter activity is increased by both GATA6 protein isoforms in transiently transfected HeLa cells	(46, 65)
Genetic studies	No association between PCOS and three previously described SNPs	(65)
	No sequence variation in the GATA6 promoter, 5′ UTR, or 3′ UTR associated with PCOS	

On further analysis of the theca cell microarray data, 3-hydroxysteroid epimerase (HSD17B6, RoDH2) and aldehyde dehydrogenase 1A3 (ALDH1A3, ALDH6) which together regulate the conversion of vitamin A (retinol) to all-trans retinoic acid (atRA) *(66–69)* exhibit increased mRNA abundance in PCOS theca cells (Fig. 4) *(46)*. PCOS theca cells also show increased levels of an atRA-responsive gene, retinoic acid receptor responder 1 (RARRES1) (Fig. 4), and an enzymatic assay demonstrates increased conversion of retinol to retinaldehyde in the presence of PCOS compared to normal theca cell protein extract (Table 5) *(46)*.

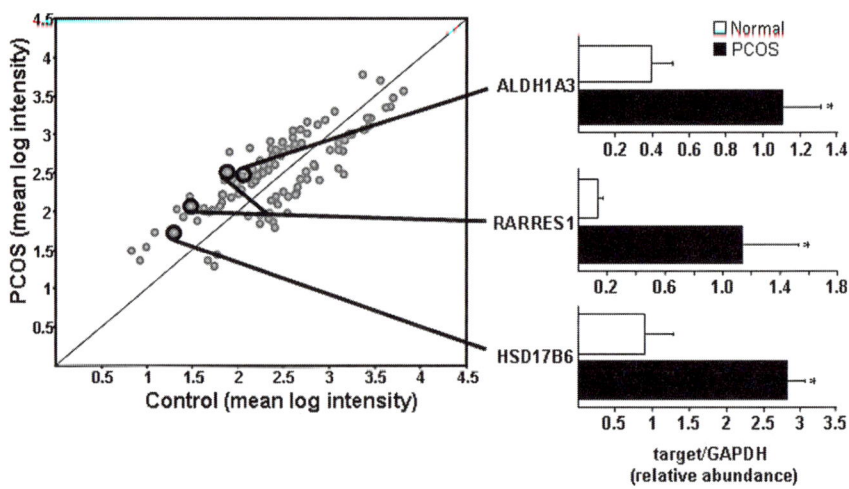

Fig. 4. Microarray analysis and QPCR demonstrate that genes involved in retinoic acid metabolism and action are differentially expressed in PCOS theca cells. Left panel: The mean log intensity of ALDH1A3, RARRES1, and HSD17B6 in PCOS compared to normal theca cells are indicated by circles. Right panel: QPCR was carried out using cDNA from normal ($n = 4$; open bars) and PCOS ($n = 4$; closed bars) theca cells and gene-specific primers for ALDH1A3, RARRES1, and HSD17B6. The mean ± S.E.M. normalized relative abundance for each target was plotted and statistically significant differences in mRNA abundance between normal and PCOS theca cells was determined by Student's t-test (*, P-value < 0.05).

In the adult testis, atRA regulates Leydig cell testosterone secretion and *CYP17* gene expression *(70)*. Likewise, in the ovarian granulosa cell, atRA inhibits FSH-dependent FSH- and LH-receptor expression *(71, 72)*. These studies suggest that atRA may play a role in increased steroid synthesis and arrested folliculogenesis in PCOS. Indeed, functional studies by Wickenheisser et al. confirm that retinoids including atRA regulate androgen biosynthesis and steroidogenic enzyme expression in normal and PCOS theca cells and likely contribute to the excessive theca-derived androgen production in PCOS (Table 5) *(73)*. Given that atRA also regulates cellular proliferation and differentiation *(74)*, it will be important to define the role of atRA in the process of folliculogenesis and determine if the increased atRA synthesis detected in the PCOS theca cell contributes in some way to the PCOS phenotype of arrested follicle development.

Genes Controlling PI3K/Akt Activity

Insulin acts to promote glucose uptake in muscle and fat. However, in the normal and PCOS ovary, insulin, acting as a co-gonadotropin, induces the expression of StAR,

Table 5
Retinoic Acid Action in Normal and PCOS Theca Cells

Functional assay	Result	Reference(s)
Vitamin A metabolism	Increased retinaldehyde production in PCOS theca cells	*(46)*
Sex steroid synthesis	Progesterone, 17α-hydroxyprogesterone, DHEA, and testosterone synthesis is increased by atRA in both normal and PCOS theca cells DHEA and testosterone synthesis is increased by vitamin A and 9-*cis*RA in PCOS but not normal theca cells	*(46, 73)*
Steroiodgenic enzyme mRNA abundance	CYP17 mRNA levels are increased by atRA and 9-*cis*RA in normal and PCOS theca cells and by vitamin A in PCOS but not normal theca cells CYP11A1 mRNA levels are increased by atRA and 9-*cis*RA in normal but not PCOS theca cells StAR mRNA levels are increased by atRA and 9-*cis*RA in normal and PCOS theca cells	*(46, 73)*
Promoter activity	CYP17 promoter activity is increased by atRA and 9-*cis*RA in normal and PCOS theca cells and by vitamin A in PCOS but not normal theca cells CYP11A1 promoter activity is increased by atRA and 9-*cis*RA in normal but not PCOS theca cells StAR promoter activity is increased by atRA and 9-*cis*RA in normal and PCOS theca cells	*(73)*

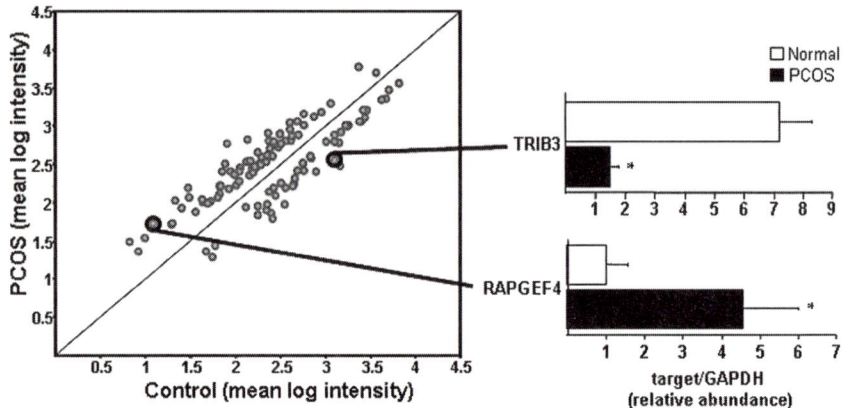

Fig. 5. Microarray analysis and QPCR demonstrate that genes involved in IP3K/Akt signaling are differentially expressed in PCOS theca cells. Left panel: The mean log intensity of TRIB3 and RAPGEF4 in PCOS compared to normal theca cells are indicated by circles. Right panel: QPCR was carried out using cDNA from normal ($n = 4$; open bars) and PCOS ($n = 4$; closed bars) theca cells and gene-specific primers for TRIB3 and RAPGEF4. The mean ± S.E.M. normalized relative abundance for each target was plotted and statistically significant differences in mRNA abundance between normal and PCOS theca cells was determined by Student's t-test (*, P-value < 0.05).

CYP11A1, and CYP17 and the production of ovarian steroids *(43, 57, 58)*. Given that PCOS women have higher circulating insulin levels compared to normal controls *(30)*, it has been suggested that this increased insulin promotes increased ovarian androgen synthesis. Indeed, several studies have demonstrated that PCOS women treated with insulin-sensitizing drugs, including troglitazones and metformin, exhibit a reversal of ovarian phenotypes *(4)*. Insulin-stimulated steroidogenesis is mediated through the PI3K/Akt pathway *(75, 76)*. Interestingly, tribbles homolog 3 (TRIB3) which binds to Akt/PKB and directly inhibits the transmission of the Akt/PKB signal *(77)* exhibits decreased mRNA abundance in PCOS theca cells. Conversely, the RAP guanine-nucleotide-exchange factor (RAPGEF4, cAMP-GEFII), which activates the small GTPase Rap 1 and subsequently augments PI3K activity *(78)*, has increased mRNA abundance in PCOS theca cells. The altered expression of TRIB3 and RAPGEF4 in PCOS theca cells has been correlated to increased, insulin-stimulated Akt phospho-rylation in PCOS theca cells (Fig. 5) *(45)*. These data suggest thatPCOS theca cells have an increased sensitivity to insulin signaling, a conclusion which implies that the insulin resistance in PCOS is organ or cell-type specific (i.e., there is a dichotomy of insulin signaling in PCOS women).

GENOMICS AND THE IDENTIFICATION OF NEW PCOS CANDIDATE GENES

The gene expression profiling experiments described herein demonstrate that the PCOS theca cell and whole ovary have a unique genomic signature compared to the normal theca cell and whole ovary. Furthermore, the functional studies demonstrate a link between altered mRNA levels and the biochemical phenotype of increased androgen synthesis. Current and future studies may elucidate the molecular mechanisms underlying PCOS and provide important information which will predict the response to interventions and reveal new therapeutic opportunities.

As shown in Table 2, most genetic studies of PCOS have focused on genes which directly impact androgen biosynthesis and/or glucose metabolism. Despite the long list of candidate gene studies investigated to date, there is no consensus as to the role of any of these loci with the possible exception of a locus on 19p13.2 *(79)*. The microarray data could provide a guide to other potential PCOS genes, although there has yet to be confirmation that a polymorphism or mutation in a gene identified through transcriptome profiling is linked and/or associated with PCOS. For example, single nucleotide polymorphism analysis of GATA6 *(65)* and RAPGEF4 (unpublished data) have been carried out. However, no genetic variants associated with PCOS in either of these gene loci were identified. Despite these preliminary results, the mining of the gene expression profiling data is still in progress and transcriptome profiling has yet to be married with linkage data from genome-wide scans and metabolomics, a multipronged approach which could refine the search for PCOS genes.

ACKNOWLEDGMENTS

The work described in the manuscript was supported by funding from the National Institutes of Health (U54-34449) and the Mellon Foundation (JRW).

REFERENCES

1. Stein, I. & Leventhal, M. (1935). Amenorrhea associated with bilateral polycystic ovaries. *Am J Obstet Gynecol* 29, 181–91.
2. Franks, S. (1995). Polycystic ovary syndrome [published erratum appears in *N Engl J Med* 1995 Nov 23; 333(21):1435] [see comments]. *N Engl J Med* 333, 853–61.
3. Ehrmann, D. A. (2005). Polycystic ovary syndrome. *N Engl J Med* 352, 1223–36.
4. Legro, R. S. (2001). Polycystic ovary syndrome: the new millenium. *Mol Cell Endocrinol* 184, 87–93.
5. Dunaif, A., Givens, J., Haseltine, F. & Merriam, G. (1992). *The Polycystic Ovary Syndrome*. Blackwell Scientific, Cambridge.
6. Moran, C., Knochenhauer, E., Boots, L. R. & Azziz, R. (1999). Adrenal androgen excess in hyperandrogenism: relation to age and body mass. *Fertil Steril* 71, 671–4.
7. Kumar, A., Woods, K. S., Bartolucci, A. A. & Azziz, R. (2005). Prevalence of adrenal androgen excess in patients with the polycystic ovary syndrome (PCOS). *Clin Endocrinol (Oxf)* 62, 644–9.
8. Legro, R. S., Spielman, R., Urbanek, M., Driscoll, D., Strauss, J. F. III & Dunaif, A. (1998). Phenotype and genotype in polycystic ovary syndrome. *Recent Prog Horm Res* 53, 217–56.
9. Gilling-Smith, C., Willis, D. S., Beard, R. W. & Franks, S. (1994). Hypersecretion of androstenedione by isolated thecal cells from polycystic ovaries. *J Clin Endocrinol Metab* 79, 1158–65.
10. Nelson, V. L., Legro, R. S., Strauss, J. F. III & McAllister, J. M. (1999). Augmented androgen production is a stable steroidogenic phenotype of propogated theca cells from polycystic ovaries. *Mol Endocrinol* 13, 946–57.
11. Nelson, V. L., Qin Kn, K. N., Rosenfield, R. L., Wood, J. R., Penning, T. M., Legro, R. S., Strauss, J. F. III & McAllister, J. M. (2001). The biochemical basis for increased testosterone production in theca cells propagated from patients with polycystic ovary syndrome. *J Clin Endocrinol Metab* 86, 5925–33.
12. Doi, S. A., Towers, P. A., Scott, C. J. & Al-Shoumer, K. A. (2005). PCOS: an ovarian disorder that leads to dysregulation in the hypothalamic-pituitary-adrenal axis? *Eur J Obstet Gynecol Reprod Biol* 118, 4–16.
13. Zeleznik, A. J. (2004). The physiology of follicle selection. *Reprod Biol Endocrinol* 2, 31.
14. van den Hurk, R. & Zhao, J. (2005). Formation of mammalian oocytes and their growth, differentiation and maturation within ovarian follicles. *Theriogenology* 63, 1717–51.

15. McGee, E. A. & Hsueh, A. J. (2000). Initial and cyclic recruitment of ovarian follicles. *Endocr Rev* 21, 200–14.

16. Knight, P. G. & Glister, C. (2001). Potential local regulatory functions of inhibins, activins and follistatin in the ovary. *Reproduction* 121, 503–12.

17. Hayashi, M., McGee, E. A., Min, G., Klein, C., Rose, U. M., van Duin, M. & Hsueh, A. J. (1999). Recombinant growth differentiation factor-9 (GDF-9) enhances growth and differentiation of cultured early ovarian follicles. *Endocrinology* 140, 1236–44.

18. Juengel, J. L. & McNatty, K. P. (2005). The role of proteins of the transforming growth factor-beta superfamily in the intraovarian regulation of follicular development. *Hum Reprod Update* 11, 143–60.

19. Franks, S., Mason, H. & Willis, D. (2000). Follicular dynamics in the polycystic ovary syndrome. *Mol Cell Endocrinol* 163, 49–52.

20. Ehrmann, D. A., Barnes, R. B. & Rosenfield, R. L. (1995). Polycystic ovary syndrome as a form of functional ovarian hyperandrogenism due to dysregulation of androgen secretion. *Endocr Rev* 16, 322–53.

21. Waldstreicher, J., Santoro, N. F., Hall, J. E., Filicori, M. & Crowley, W. F., Jr. (1988). Hyperfunction of the hypothalamic-pituitary axis in women with polycystic ovarian disease: indirect evidence for partial gonadotroph desensitization. *J Clin Endocrinol Metab* 66, 165–72.

22. McCartney, C. R., Eagleson, C. A. & Marshall, J. C. (2002). Regulation of gonadotropin secretion: implications for polycystic ovary syndrome. *Semin Reprod Med* 20, 317–26.

23. Ledee-Bataille, N., Lapree-Delage, G., Taupin, J. L., Dubanchet, S., Taieb, J., Moreau, J. F. & Chaouat, G. (2001). Follicular fluid concentration of leukaemia inhibitory factor is decreased among women with polycystic ovarian syndrome during assisted reproduction cycles. *Hum Reprod* 16, 2073–8.

24. Teixeira Filho, F. L., Baracat, E. C., Lee, T. H., Suh, C. S., Matsui, M., Chang, R. J., Shimasaki, S. & Erickson, G. F. (2002). Aberrant expression of growth differentiation factor-9 in oocytes of women with polycystic ovary syndrome. *J Clin Endocrinol Metab* 87, 1337–44.

25. Pigny, P., Merlen, E., Robert, Y., Cortet-Rudelli, C., Decanter, C., Jonard, S. & Dewailly, D. (2003). Elevated serum level of anti-mullerian hormone in patients with polycystic ovary syndrome: relationship to the ovarian follicle excess and to the follicular arrest. *J Clin Endocrinol Metab* 88, 5957–62.

26. Stubbs, S. A., Hardy, K., Da Silva-Buttkus, P., Stark, J., Webber, L. J., Flanagan, A. M., Themmen, A. P., Visser, J. A., Groome, N. P. & Franks, S. (2005). Anti-mullerian hormone protein expression is reduced during the initial stages of follicle development in human polycystic ovaries. *J Clin Endocrinol Metab* 90, 5536–43.

27. Welt, C. K., Taylor, A. E., Fox, J., Messerlian, G. M., Adams, J. M. & Schneyer, A. L. (2005). Follicular arrest in polycystic ovary syndrome is associated with deficient inhibin A and B biosynthesis. *J Clin Endocrinol Metab* 90, 5582–7.

28. Welt, C. K., Taylor, A. E., Martin, K. A. & Hall, J. E. (2002). Serum inhibin B in polycystic ovary syndrome: regulation by insulin and luteinizing hormone. *J Clin Endocrinol Metab* 87, 5559–65.

29. Fujiwara, T., Sidis, Y., Welt, C., Lambert-Messerlian, G., Fox, J., Taylor, A. & Schneyer, A. (2001). Dynamics of inhibin subunit and follistatin mRNA during development of normal and polycystic ovary syndrome follicles. *J Clin Endocrinol Metab* 86, 4206–15.

30. Dunaif, A. (1997). Insulin resistance and the polycystic ovary syndrome: mechanism and implications for pathogenesis. *Endocr Rev* 18, 774–800.

31. Dunaif, A. (1999). Insulin action in the polycystic ovary syndrome. *Endocrinol Metab Clin North Am* 28, 341–59.

32. Dunaif, A., Xia, J., Book, C. B., Schenker, E. & Tang, Z. (1995). Excessive insulin receptor serine phosphorylation in cultured fibroblasts and in skeletal muscle. A potential mechanism for insulin resistance in the polycystic ovary syndrome. *J Clin Invest* 96, 801–10.

33. Dunaif, A., Wu, X., Lee, A. & Diamanti-Kandarakis, E. (2001). Defects in insulin receptor signaling in vivo in the polycystic ovary syndrome (PCOS). *Am J Physiol Endocrinol Metab* 281, E392–9.

34. Corbould, A., Kim, Y. B., Youngren, J. F., Pender, C., Kahn, B. B., Lee, A. & Dunaif, A. (2005). Insulin resistance in the skeletal muscle of women with PCOS involves intrinsic and acquired defects in insulin signaling. *Am J Physiol Endocrinol Metab* 288, E1047–54.

35. Solomon, C. G. (1999). The epidemiology of polycystic ovary syndrome. Prevalence and associated disease risks. *Endocrinol Metab Clin North Am* 28, 247–63.

36. Legro, R. S., Driscoll, D., Strauss, J. F., III, Fox, J. & Dunaif, A. (1998). Evidence for a genetic basis for hyperandrogenemia in polycystic ovary syndrome. *Proc Natl Acad Sci U S A* 95, 14956–60.

37. Kahsar-Miller, M. D., Nixon, C., Boots, L. R., Go, R. C. & Azziz, R. (2001). Prevalence of polycystic ovary syndrome (PCOS) in first-degree relatives of patients with PCOS. *Fertil Steril* 75, 53–8.

38. Franks, S., Gharani, N. & McCarthy, M. (2001). Candidate genes in polycystic ovary syndrome. *Hum Reprod Update* 7, 405–10.

39. Legro, R. S. & Strauss, J. F., III (2002). Molecular progress in infertility: polycystic ovary syndrome. *Fertil Steril* 78, 569–76.

40. Roldan, B., San Millan, J. L. & Escobar-Morreale, H. F. (2004). Genetic basis of metabolic abnormalities in polycystic ovary syndrome: implications for therapy. *Am J Pharmacogenomics* 4, 93–107.

41. Rosenbaum, D., Haber, R. S. & Dunaif, A. (1993). Insulin resistance in polycystic ovary syndrome: decreased expression of GLUT-4 glucose transporters in adipocytes. *Am J Physiol* 264, E197–202.

42. Jakimiuk, A. J., Weitsman, S. R., Navab, A. & Magoffin, D. A. (2001). Luteinizing hormone receptor, steroidogenesis acute regulatory protein, and steroidogenic enzyme messenger ribonucleic acids are overexpressed in thecal and granulosa cells from polycystic ovaries. *J Clin Endocrinol Metab* 86, 1318–23.

43. Nestler, J. E., Jakubowicz, D. J., de Vargas, A. F., Brik, C., Quintero, N. & Medina, F. (1998). Insulin stimulates testosterone biosynthesis by human thecal cells from women with polycystic ovary syndrome by activating its own receptor and using inositolglycan mediators as the signal transduction system. *J Clin Endocrinol Metab* 83, 2001–5.

44. Wickenheisser, J. K., Quinn, P. G., Nelson, V. L., Legro, R. S., Strauss, J. F., III & McAllister, J. M. (2000). Differential activity of the cytochrome P450 17α-hydroxylase and steroidogenic acute regulatory protein gene promoters in normal and polycystic ovary syndrome theca cells. *J Clin Endocrinol Metab* 85, 2304–11.

45. Wood, J. R., Nelson-Degrave, V. L., Jansen, E., McAllister, J. M., Mosselman, S. & Strauss, J. F., III (2005). Valproate-induced alterations in human theca cell gene expression: clues to the association between valproate use and metabolic side effects. *Physiol Genomics* 20, 233–43.

46. Wood, J. R., Nelson, V. L., Ho, C., Jansen, E., Wang, C. Y., Urbanek, M., McAllister, J. M., Mosselman, S. & Strauss, J. F., III (2003). The molecular phenotype of polycystic ovary syndrome (PCOS) theca cells and new candidate PCOS genes defined by microarray analysis. *J Biol Chem* 278, 26380–90.

47. Jansen, E., Laven, J. S., Dommerholt, H. B., Polman, J., van Rijt, C., van den Hurk, C., Westland, J., Mosselman, S. & Fauser, B. C. (2004). Abnormal gene expression profiles in human ovaries from polycystic ovary syndrome patients. *Mol Endocrinol* 18, 3050–63.

48. Diao, F. Y., Xu, M., Hu, Y., Li, J., Xu, Z., Lin, M., Wang, L., Zhou, Y., Zhou, Z., Liu, J. & Sha, J. (2004). The molecular characteristics of polycystic ovary syndrome (PCOS) ovary defined by human ovary cDNA microarray. *J Mol Endocrinol* 33, 59–72.

49. Oksjoki, S., Soderstrom, M., Inki, P., Vuorio, E. & Anttila, L. (2005). Molecular profiling of polycystic ovaries for markers of cell invasion and matrix turnover. *Fertil Steril* 83, 937–44.

50. Chuaqui, R. F., Bonner, R. F., Best, C. J., Gillespie, J. W., Flaig, M. J., Hewitt, S. M., Phillips, J. L., Krizman, D. B., Tangrea, M. A., Ahram, M., Linehan, W. M., Knezevic, V. & Emmert-Buck, M. R. (2002). Post-analysis follow-up and validation of microarray experiments. *Nat Genet* 32 Suppl, 509–14.

51. Churchill, G. A. (2002). Fundamentals of experimental design for cDNA microarrays. *Nat Genet* 32 Suppl, 490–5.

52. Quackenbush, J. (2002). Microarray data normalization and transformation. *Nat Genet* 32 Suppl, 496–501.

53. Holloway, A. J., van Laar, R. K., Tothill, R. W. & Bowtell, D. D. (2002). Options available—from start to finish—for obtaining data from DNA microarrays II. *Nat Genet* 32 Suppl, 481–9.

54. Sharov, A. A., Dudekula, D. B. & Ko, M. S. (2005). A web-based tool for principal component and significance analysis of microarray data. *Bioinformatics* 21, 2548–9.

55. Attia, G. R., Dooley, C. A., Rainey, W. E. & Carr, B. R. (2000). Transforming growth factor beta inhibits steroidogenic acute regulatory (StAR) protein expression in human ovarian thecal cells. *Mol Cell Endocrinol* 170, 123–9.

56. Christenson, L. K., Johnson, P. F., McAllister, J. M. & Strauss, J. F., III (1999). CCAAT/enhancer-binding proteins regulate expression of the human steroidogenic acute regulatory protein (StAR) gene. *J Biol Chem* 274, 26591–8.

57. Devoto, L., Christenson, L. K., McAllister, J. M., Makrigiannakis, A. & Strauss, J. F., III (1999). Insulin and insulin-like growth factor-I and -II modulate human granulosa-lutein cell steroidogenesis: enhancement of steroidogenic acute regulatory protein (StAR) expression. *Mol Hum Reprod* 5, 1003–10.

58. Zhang, G., Garmey, J. C. & Veldhuis, J. D. (2000). Interactive stimulation by luteinizing hormone and insulin of the steroidogenic acute regulatory (StAR) protein and 17alpha- hydroxylase/17,20-lyase (CYP17) genes in porcine theca cells. *Endocrinology* 141, 2735–42.

59. Wood, J. R. & Strauss, J. F., III (2002). Multiple signal transduction pathways regulate ovarian steroidogenesis. *Rev Endocr Metab Disord* 3, 33–46.

60. Molkentin, J. D. (2000). The zinc finger-containing transcription factors GATA-4, -5, and -6. Ubiquitously expressed regulators of tissue-specific gene expression. *J Biol Chem* 275, 38949–52.

61. Tremblay, J. J. & Viger, R. S. (2003). Novel roles for GATA transcription factors in the regulation of steroidogenesis. *J Steroid Biochem Mol Biol* 85, 291–8.

62. Jimenez, P., Saner, K., Mayhew, B. & Rainey, W. E. (2003). GATA-6 is expressed in the human adrenal and regulates transcription of genes required for adrenal androgen biosynthesis. *Endocrinology* 144, 4285–8.

63. Saner, K. J., Suzuki, T., Sasano, H., Pizzey, J., Ho, C., Strauss, J. F., III, Carr, B. R. & Rainey, W. E. (2005). Steroid sulfotransferase 2A1 gene transcription is regulated by steroidogenic factor 1 and GATA-6 in the human adrenal. *Mol Endocrinol* 19, 184–97.

64. Fluck, C. E. & Miller, W. L. (2004). GATA-4 and GATA-6 modulate tissue-specific transcription of the human gene for P450c17 by direct interaction with Sp1. *Mol Endocrinol* 18, 1144–57.

65. Ho, C. K., Wood, J. R., Stewart, D. R., Ewens, K., Ankener, W., Wickenheisser, J., Nelson Degrave, V., Zhang, Z., Legro, R. S., Dunaif, A., McAllister, J. M., Spielman, R. & Strauss, J. F., III (2005). Increased transcription and increased mRNA stability contribute to increased GATA6 mRNA abundance in PCOS theca cells. *J Clin Endocrinol Metab* 90, 6596–6602.

66. Gottesman, M. E., Quadro, L. & Blaner, W. S. (2001). Studies of vitamin A metabolism in mouse model systems. *Bioessays* 23, 409–19.

67. Napoli, J. L. (1996). Biochemical pathways of retinoid transport, metabolism, and signal transduction. *Clin Immunol Immunopathol* 80, S52–62.

68. Chetyrkin, S. V., Belyaeva, O. V., Gough, W. H. & Kedishvili, N. Y. (2001). Characterization of a novel type of human microsomal 3alpha-hydroxysteroid dehydrogenase: unique tissue distribution and catalytic properties. *J Biol Chem* 276, 22278–86.

69. Rexer, B. N., Zheng, W. L. & Ong, D. E. (2001). Retinoic acid biosynthesis by normal human breast epithelium is via aldehyde dehydrogenase 6, absent in MCF-7 cells. *Cancer Res* 61, 7065–70.

70. Livera, G., Rouiller-Fabre, V., Pairault, C., Levacher, C. & Habert, R. (2002). Regulation and perturbation of testicular functions by vitamin A. *Reproduction* 124, 173–80.

71. Minegishi, T., Hirakawa, T., Kishi, H., Abe, K., Ibuki, Y. & Miyamoto, K. (2000). Retinoic acid (RA) represses follicle stimulating hormone (FSH)-induced luteinizing hormone (LH) receptor in rat granulosa cells. *Arch Biochem Biophys* 373, 203–10.

72. Minegishi, T., Hirakawa, T., Kishi, H., Abe, K., Tano, M., Abe, Y. & Miyamoto, K. (2000). The mechanisms of retinoic acid-induced regulation on the follicle-stimulating hormone receptor in rat granulosa cells. *Biochim Biophys Acta* 1495, 203–11.

73. Wickenheisser, J. K., Nelson-DeGrave, V. L., Hendricks, K. L., Legro, R. S., Strauss, J. F., III & McAllister, J. M. (2005). Retinoids and retinol differentially regulate steroid biosynthesis in ovarian theca cells isolated from normal cycling women and women with polycystic ovary syndrome. *J Clin Endocrinol Metab* 90, 4858–65.

74. Clagett-Dame, M. & DeLuca, H. F. (2002). The role of vitamin a in Mammalian reproduction and embryonic development. *Annu Rev Nutr* 22, 347–81.

75. Carvalho, C. R., Carvalheira, J. B., Lima, M. H., Zimmerman, S. F., Caperuto, L. C., Amanso, A., Gasparetti, A. L., Meneghetti, V., Zimmerman, L. F., Velloso, L. A. & Saad, M. J. (2003). Novel signal transduction pathway for luteinizing hormone and its interaction with insulin: activation of Janus kinase/signal transducer and activator of transcription and phosphoinositol 3-kinase/Akt pathways. *Endocrinology* 144, 638–47.

76. Munir, I., Yen, H. W., Geller, D. H., Torbati, D., Bierden, R. M., Weitsman, S. R., Agarwal, S. K. & Magoffin, D. A. (2004). Insulin augmentation of 17alpha-hydroxylase activity is mediated by phosphatidyl inositol 3-kinase but not extracellular signal-regulated kinase-1/2 in human ovarian theca cells. *Endocrinology* 145, 175–83.

77. Du, K., Herzig, S., Kulkarni, R. N. & Montminy, M. (2003). TRB3: a tribbles homolog that inhibits Akt/PKB activation by insulin in liver. *Science* 300, 1574–7.

78. Mei, F. C., Qiao, J., Tsygankova, O. M., Meinkoth, J. L., Quilliam, L. A. & Cheng, X. (2002). Differential signaling of cyclic AMP: opposing effects of exchange protein directly activated by cyclic AMP and cAMP-dependent protein kinase on protein kinase B activation. *J Biol Chem* 277, 11497–504.

79. Urbanek, M., Woodroffe, A., Ewens, K. G., Diamanti-Kandarakis, E., Legro, R. S., Strauss, J. F., III Dunaif, A. & Spielman, R. S. (2005). Candidate Gene Region for Polycystic Ovary Syndrome (PCOS) on Chromosome 19p13.2. *J Clin Endocrinol Metab* 90, 6623–6629.

13 Microarray Analysis of Alterations Induced by Obesity in White Adipose Tissue Gene Expression Profiling

Julien Tirard, Ricardo Moraes,
Danielle Naville, PhD,
and Martine Bégeot, PhD

CONTENTS

INTRODUCTION
ADIPOSE TISSUE: AN ACTIVE ENDOCRINE GLAND
APPLICATION OF MICROARRAY TECHNOLOGY TO ADIPOGENESIS
APPLICATION OF MICROARRAY TECHNOLOGY TO WHITE ADIPOSE
  TISSUE GENE EXPRESSION IN OBESITY
CONCLUSIONS
ACKNOWLEDGMENTS
REFERENCES

INTRODUCTION

The prevalence of overweight and obesity in many developed countries has increased over recent decades and represents a serious public health problem *(1, 2)*. Over 80% of patients with type 2 diabetes are obese and obesity is a very strong risk factor for the development of insulin resistance, cardiovascular diseases, and some types of cancer *(3)*. Obesity results from an increase in food intake and a decrease in energy expenditure leading to hyperplasia and hypertrophy of adipose tissue, fat deposits, and changes in gene expression profiles in this tissue. Adipose tissue was considered to be a passive tissue for storage of energy until it was clearly established that it is able to synthesize and secrete hormones, cytokines, and many other factors, and to express a great number of receptors *(4)*. Thus, it is now considered as an endocrine tissue with broad effects on body homeostasis at local, peripheral, and central levels, and adipocytes have been shown to express more genes than initially thought *(5)*. Oligonucleotide or cDNA microarrays have changed the study of the pattern of gene expression by permitting the simultaneous and global analysis of thousands of genes in a single experiment and the evaluation of their alterations in several conditions *(6)*.

From: *Contemporary Endocrinology: Genomics in Endocrinology:*
DNA Microarray Analysis in Endocrine Health and Disease
Edited by S. Handwerger and B. Aronow © Humana Press, Totowa, NJ

This technology has been widely applied to the discovery of sets of altered genes in diseases such as aging and caloric restriction *(7)*, cancer *(8–12)*, hypertension *(13)*, and several others *(14)*. Using both DNA and oligonucleotide microarrays, several studies have been performed on the well-known 3T3-L1 preadipocyte cell line in order to assess changes in gene expression profiles during adipocyte differentiation using a differentiation culture medium. This allows definition of different gene clusters as markers of the various stages of differentiation from preadipocytes to mature adipocytes *(15–19)*. The same types of studies were performed using human mesenchymal stem cells (hMSCs) from bone marrow stroma instead of 3T3-L1 since they are also able to differentiate into adipocytes under specific culture conditions *(20–22)*. In parallel, Soukas et al.*(23)* reported alterations in gene profiles between preadipocytes and adipocytes prepared from white adipose tissue in rodents and conducted a comparative analysis of adipogenesis profiles in vivo and in vitro. Human adipose tissue has not been investigated as much as rodent adipose tissue, in part due to a limited access to human tissues, but Urs et al. *(24)* for the first time presented lists of differentially expressed genes in adipocytes and preadipocytes from human adipose tissue. Beside this work, some other studies using human subcutaneous or omental white adipose tissues have established lists of the predominantly expressed genes in these tissues *(5, 25)*.

Therefore, DNA microarray technology constitutes a very interesting tool in identifying the molecular events involved in the development of obesity and defining the combinations of genes (known as well as unknown) that are altered *(26, 27)*. Many studies were performed in rodents using genetically obese *ob/ob* mice as a model of extreme obesity and diabetes *(28–31)* or mice fed on a high-fat diet as in our study *(32)*, a high-energy diet *(33, 34)* or even cafeteria obese rats *(35)*. The use of knockout mice for perilipin, a protein of the lipid droplet, was also very informative in understanding the biochemical pathways involved in obesity resistance *(36)*. Some studies using human white subcutaneous or omental adipose tissue measured the gene expression profiles of obese or diabetic patients in comparison with lean patients *(37–41)*.

ADIPOSE TISSUE: AN ACTIVE ENDOCRINE GLAND

Adipose tissue is no longer considered as a simple energy storage site for triglycerides but rather is identified as a major production site for several endocrine factors. The first secreted factor to be characterized was adipsin *(42)* and subsequently the discovery of leptin firmly established adipose tissue as an important endocrine organ *(43)*. Adipose tissue is composed of adipocytes surrounded by connective tissue matrix with collagen fibers, blood vessels with stromavascular cells and immune cells such as macrophages *(44, 45)*. Thus, many secreted proteins are derived from the non-adipocyte fraction of this tissue. The endocrine functions of white adipose tissue comprise various secreted proteins that have metabolic effects on other tissues and enzymes that are involved in the metabolism of steroid hormones. The secreted proteins belong to several families as illustrated in Fig. 1. The main family comprises cytokines and many of which are clearly associated with obesity and insulin resistance such as leptin, Interleukin 6 (IL6), and TNFα *(46, 47)*. Beside this group, another important one comprises prothrombin factors like plasminogen-activator inhibitor 1 (PAI-1), a serine protease

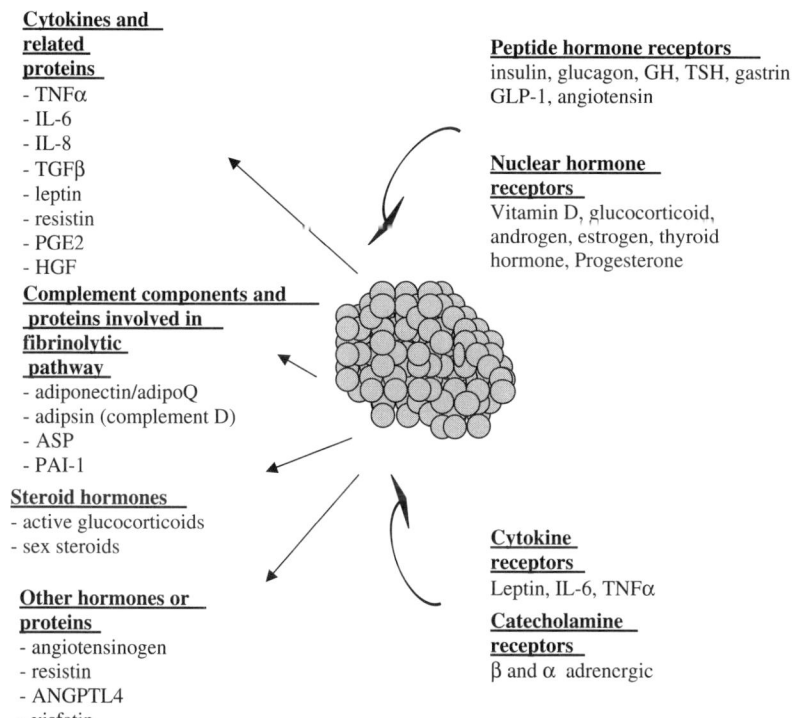

Cytokines and related proteins
- TNFα
- IL-6
- IL-8
- TGFβ
- leptin
- resistin
- PGE2
- HGF

Complement components and proteins involved in fibrinolytic pathway
- adiponectin/adipoQ
- adipsin (complement D)
- ASP
- PAI-1

Steroid hormones
- active glucocorticoids
- sex steroids

Other hormones or proteins
- angiotensinogen
- resistin
- ANGPTL4
- visfatin

Peptide hormone receptors
insulin, glucagon, GH, TSH, gastrin
GLP-1, angiotensin

Nuclear hormone receptors
Vitamin D, glucocorticoid,
androgen, estrogen, thyroid
hormone, Progesterone

Cytokine receptors
Leptin, IL-6, TNFα

Catecholamine receptors
β and α adrenergic

Fig. 1. Endocrine functions of white adipose tissue. One the left of the figure are listed known proteins or hormones that are secreted by white adipose tissue and on the right the different categories of receptors that have been characterized in this tissue indicating that the factors on the left could exert autocrine or paracrine effects (adapted from references 44, 45, 47, 48).

inhibitor whose plasma elevation in obesity is a risk factor for type 2 diabetes and cardiovascular disease *(48)*, and several components of the complement pathway such as adipsin (complement D). Several hormones or hormone precursors are also secreted by adipose tissue, in particular active corticoids, due to the presence of the enzyme 11β-hydroxysteroid dehydrogenase type 1, which catalyzes the conversion of inactive 11β-ketoglucocorticoid into hormonally active 11β-hydroxylated metabolites. All these factors could act through paracrine or autocrine mechanisms due to the presence of their specific receptors in adipose tissue (Fig. 1). However, for some of them like leptin or PAI-1, the plasma amounts are sufficient to exert their action on distant target tissues. In addition to these known factors, many of the expressed genes in adipose tissue need to be identified and characterized, and the link between increase or decrease in their level of expression and obesity has to be established.

APPLICATION OF MICROARRAY TECHNOLOGY TO ADIPOGENESIS

The 3T3-L1 Cell Line Model

The murine 3T3-L1 cell line from whole mouse embryos has been extensively used in *in vitro* studies of adipocyte differentiation *(49,50)*. Upon reaching confluence, these cells are in growth arrest but are able to differentiate into adipocytes by exposure to a culture medium containing insulin, dexamethasone, 3-isobutyl-1-methylxanthine in the presence of fetal calf serum (MDI). Within the first 24 hours following the hormonal stimulation, there is a re-entry into the cell cycle and clonal expansion. This is followed by permanent cell cycle withdrawal and differentiation to the terminal stage for two weeks or more, with accumulation of triglyceride lipid droplets in the cytoplasm in the mature adipocytes *(51)*. Over many years, a substantial number of genes that are coordinately induced or suppressed during 3T3-L1 adipogenesis have been identified. Some of them are transiently altered *(52)* and others such as transcription factors (e.g., CAAT-enhancer binding protein(s), peroxisome proliferator activated receptor are constitutively activated in distinct phases that contribute to the acquisition of the phenotype of mature adipocytes *(53)*. Despite the extensive analysis of these genes as more than 100 molecules have been identified as differentially expressed in adipogenesis, there are clearly many additional genes involved in the adipocyte differentiation process. In recent years, cDNA/oligonucleotide microarray technology has been used to investigate adipogenesis in the 3T3-L1 cell line model *(15–19, 23)*. In these studies, different types of microarrays have been used containing at least 10 000 *(17)* to 18 000 *(15)* cDNAs/EST. Among these genes or ESTs, the number of positive clones differentially expressed during adipogenesis varies among the different studies, since 119 *(17)*, 286 *(16)*, 754 *(15)*, 1259 *(23)* genes or EST were respectively identified depending on the type of arrays that have been used, the stringency (variation retained between threefold and tenfold) and type of data analysis that was performed and the time points of differentiation that were studied from 2 hours to 28 days. However, from different studies, it is clear that the number of genes whose expression is increased during adipogenesis is greater (about 70%) than the number whose expression is decreased *(15, 16)*. Table 1 shows a non-exhaustive list of the main families of genes that are up- or downregulated during differentiation. An early event in 3T3-L1 differentiation is re-entry into the cell cycle followed by clonal expansion during the first 24 hours of hormonal stimulation characterized by a decrease in the expression of genes involved in growth arrest, in contrast to genes involved in progression through the S phase. The role of Wnt signaling was studied by Ross et al.*(17)* and apparently the expression of 27 genes involved in cell cycle progression was blunted in Wnt-expressing cells compared to control cells through a reduction in E2F activity and thus inhibition of clonal expansion. Another characteristic of this early differentiation is the activation of some immediate early genes such as fos/jun in addition to the well-known C/EBPβ and δ as well as several signaling molecules. In the late differentiation stage, a great number of genes encoding transcription factors, nuclear hormone receptors, and many secreted molecules involved in specific functions of adipocytes are highly upregulated such as C/EBPα (at the same time the expression of C/EBPδ decreases), PPARγ2, SREBP-1, aP2, adiponectin, and other additional factors (X-box-binding protein, immunoglobulin/enhancer binding protein, for example) that were co-expressed but not previously characterized in adipogenesis. The genes known as targets for C/EBPα and SREBP-1 exhibited an increase in their expression profile starting from day 4 to peak expression later *(23)*. The genes encoding signaling molecules, such as insulin-like growth-factor-II, β-adrenergic receptor, or PEPCK known to have a PPAR-γ responsive element in their promoter, are maximally expressed in the mature adipocyte. Among the genes that are downregulated in late differentiation are cell cycle genes, genes associated with the cytoskeleton,

Table 1

Overview of Modified Gene Families during the Murine 3T3-L1 Adipogenesis Induced by MDI Medium (Methyl-3-isobutyl xanthine, Dexamethasone, Insulin) using Microarray Analyses [Adapted from *(15–19, 23)*]

Mitotic clonal expansion and early differentiation stage 0 to 24 Hours of MDI	*Late differentiation stage day 2 to day 4*	*Maturity stage (adipocytes) day 6 to day 28*
Upregulated genes		
Genes involved in re-entry in cell cycle (ex:E2F4,cdk2,cdc2, cyclins A, B1, B2...) and DNA replication (ex: helicase, DNA polymerase, topoisomerase...)	Genes encoding transcription factors or nuclear hormone receptors (C/EBPα, PPARγ2, SREBP-1/ADD1, RXR, LXRα(*although this factor is not adipogenic it increased GLUT-1 expression and glucose uptake and storage),* RARα,Myc...)	Same genes (with a higher expression level) for energy storage and lipogenesis than in late differentiation stage
Genes encoding transcription factors (ex: C/EBPδ, C/EBPβ, SREBP1, cfos/junB-AP1 complex, E2a forkhead box F2...), co-factors (Rb107) and signal molecules	Genes encoding additional factors (ex: X-box binding protein, oestrogen receptor related α...	Genes encoding regulatory and signal molecules, transcriptional co-regulators (ex: β3-adrenergic receptor, IGF2, IGF-BP2, STAT-1, cytochrome c oxidase VIII H, RIP 140...)
Nuclear hormone receptors encoding genes (ex: VDR, N10, thyroid receptor c-erbA-α2...)	Genes encoding specific proteins involved in energy storage, glycogen synthesis, oxidative phosphorylation, glycolysis, lipogenesis (ex: angiotensinogen, hormone sensitive lipase, aP2, adiponectin/Acrp30...) and signal molecules (*indicating that adipocytes are responsive to hormonal stimuli*)	Genes known as targets for C/EBP α and SREBP-1 and markers of differentiated adipocyte (ex: fatty acid binding proyein 4, fatty acid synthase, stearoyl-CoA desaturase 1 and 2, glycerophosphate dehydrogenase , phosphoenolpyruvate carboxykinase or PEPCK, phosphofructokinase I...) Other genes with unknown function in adipocyte (ex: Indian hedgehog Ihh, Cdc42 and Ack, integrin β...)and other signal molecules
Genes encoding proteins involved in the inhibition of apoptosis (ex: TIAP) Genes encoding metallo-proteases (ex: ADAM8, ...)		

(Continued)

Table 1
(Continued)

Mitotic clonal expansion and early differentiation stage 0 to 24 Hours of MDI	*Late differentiation stage day 2 to day 4*	*Maturity stage (adipocytes) day 6 to day 28*
Other genes unknown in adipogenesis (ex: wee1 kinase, vascular endothelial growth factor, gut-enriched Kruppel-lke factor, syndecan, anexin VIII, tolloid-like, latexin, TEA4, ...)		
Downregulated genes		
Genes involved in growth arrest (ex: growth arrest specific 1, GADD48, GADD 153, CHOP 10, Cdk inhibitors p21 and p 27, ...)	Gene encoding the transcription factor C/EBPδ	Genes involved in transcriptional repression (ex: AEBP1; Prx2 homeobox, Jun B, nuclear LIM interactor, Kruppel like factor, forkhead box F2, E2A, ...)
	Genes involved in cell cycle	
	Genes encoding markers of other cell types like cytoskeletal genes (ex: Myef2, Id2, Ihh, ...)	
	Genes associated with cytoskeleton	
	Genes encoding splicing factors and involved in protein turnover	
	Genes encoding DNA binding inhibitors (Id genes), high mobility group proteins (HMG genes)	

genes involved in protein turnover, and splicing factors encoding genes or those encoding markers of other cell lineages. An interesting point was the comparison made by Soukas et al.*(23)* between in vitro expression data and in vivo data obtained after preparation of preadipocytes (stromal fraction) and adipocytes from white adipose tissue of C57/Bl6J mice. In general, for many genes (up- or downregulated), adipocytes derived from the 3T3-L1 cell line exhibit a pattern of gene expression similar to that observed in the in vivo passage from preadipocytes to adipocytes. However, a certain number of genes (×68) are more highly expressed in adipocytes in vivo than in fully differentiated 3T3-L1 cells. A list of some of these genes is shown in Table 2, which indicates that additional factors are necessary to drive high levels of expression of these particular genes. Moreover, this study reveals some inconsistencies between these two gene expression profiles of adipogenesis. For instance, aP2, which is considered as a

Table 2
List of Genes Highly Expressed in Predipocytes Prepared in vivo (Stromal Fraction) versus
Confluent 3T3-L1 (Undifferentiated Stage or Preadipocytes) or in Adipocytes Prepared in vivo
versus Fully Differentiated (28 Days of Differentiation) 3T3-L1 in vitro (Mature Adipocytes)
from *(23)*

Preadipocytes

Genes encoding molecules with an immune function
  TNF
  Macrophage inflammatory protein 2
  IL-1β
  IL-6
  Several other chemokines (these molecules are also expressed at a lower level
  in adipocytes)

Mature adipocytes
Genes encoding metabolic enzymes
  ATP-citrate lyase (10 fold)
  PEPCK (2 fold)
  Acetyl-CoA synthetase (3fold)
Genes encoding hormones, receptors or signal molecules
  Leptin (63 fold)
  High molecular weight GH-receptor and TSH receptor (6fold)
  Muscle LIM protein FHL1 (18 fold)

marker of mature adipocytes in vitro, is also expressed at a high level in preadipocytes in vivo although C/EBPα and PPARγ2, which are considered as its inducers in vitro, are not expressed in preadipocytes in vivo. Two hundred ninety-three genes expressed in preadipocytes in vivo are not expressed in vitro (preadipocyte or adipocyte stages). Among these genes, many encode molecules that have immune functions such as cytokines, interleukins, TNFα, macrophage inflammatory protein 2, and correspond to immunologic cell markers of adipocyte lineage. Finally, several other groups of genes are highly expressed in the 3T3-L1 cell line, indicating that differences are noticeable between in vitro and in vivo differentiated adipocytes.

The Human Mesenchymal Stem Cell Model

Human mesenchymal stem cells prepared from bone marrow give rise to several cell types when cultured under defined culture conditions *(54)*. When cultured in the presence of dexamethasone, insulin and 3-isobutyl-1-methyl-xanthine, hMSCs, as 3T3-L1 cells, differentiate to the adipocyte lineage *(55)*. However, it seems that hMSCs are in a more undifferentiated state than 3T3-L1 cells at confluence since phenotypic changes appear later upon differentiation *(20)*. Several studies report the use of DNA/EST microarrays to follow the adipogenic differentiation of these cells *(20–22)*, and Table 3 presents an overview of the data that were obtained. During the early stage of differentiation up to day 3, some of the genes such as those involved in cell cycle progression or those encoding transcription factors like C/EBP(s) have an increased expression level and were also shown to be modified in the early phase of differentiation

Table 3
Overview of the Main Variations in Gene Families during Adipogenic Differentiation
of Human Mesenchymal Stem Cells from Bone Marrow Using Microarray Analyses
[Adapted from (20–22)].

Early stage of differentiation (day 1 to day 3)	*Late stage of differentiation (day 4 to day 14)*
Upregulated genes	
Genes encoding transcription factors and regulatory proteins (ex: C/EBPβ, C/EBPδ , SWI/SNF complex, zinc finger protein145, SLUG, and others, forkhead box O1A and B, Atpa1 regulatory element binding factor 6, PI3 kinase regulatory p85...)	Genes encoding transcription factors (ex: PPARγ2, C/EBPα, SREBP-1c, zinc finger E-box, E2F5, LXRα...)
Genes encoding different components of the extracellular matrix (ex : collagen type VI, alpha 3, metallothioneins 1R 1L 1B 1E 1H 1F 1G, laminin...)	Genes involved in lipid and carbohydrate metabolism (ex : PEPCK, hexokinase 2, acetyl CoA carboxylase, LPL, stearoyl-CoA desaturase, fatty acid synthase, FABP4, FABP5, hormone sensitive lipase, fatty acid translocase, aP2, glycero-3- phosphate dehydrogenase 1, 6-phosphofructo kinase, diacylglycerol kinase, glutathione peroxydase 1 and 3...)
genes encoding molecules of degradation of extracellular matrix (ex: metalloproteases ADAMTS1, ADAMTS4...)	
Genes involved in growth arrest (ex: GADD 45B, growth arrest-specific gene 1...)	Genes encoding specific or secreted proteins (ex: perilipin, apolipoprotein E, UCP4, angiotensinogen, retinol binding protein, adipose differentiation related protein, lipin 1, IGF-BP2 and 5, IL8...)
Genes encoding aldo/keto reductases AKR1C1 and C2	Genes encoding signal molecules, receptors, transporters (ex: STAT-5B, ERK3, IL8, TGFβ, 1 and 2, IGF-I and II, angiotensin receptor 1, ATP-binding cassette subfamily E,ATP binding cassette 1, transporter 1, β catenin, cytochrome b...)
	Genes involved in cell cycle progression (ex: CDC2-associated protein)
Downregulated genes	
Genes encoding cytoskeleton and components of the extracellular matrix (ex: tissue inhibitor of metalloproteinase 3, vascular cell adhesion 1...)	Genes involved in growth arrest (ex:GADD45, growth arrest specific 1,B-cell translocation gene 1, anti-proliferative cyclin dependent kinase inhibitor 1c...)
	Genes encoding extracellular matrix and cytoskeleton components or inflammatory molecules (ex: type XV collagen alpha1 and 3 chains, thrombospondin 2, actin γ2, keratin 19, tropomyosin 1α, PAI-1, IL-6...)

(Continued)

Table 3
(*Continued*)

Early stage of differentiation (day 1 to day 3)	Late stage of differentiation (day 4 to day 14)
	Genes encoding markers of other cell lineages (ex: osteoblast-specific factor-2os, osteoclastogenesis inhibitory factor, type IX collagen, IGF-BP3, L1 cell adhesion molecule, tubulin α3 and β tubulin, SM22-alpha homolog…)
	Genes encoding cell surface and adhesion molecules (ex :CD29/integrin b1,integrin α3, CD44 adhesion protein, CD105/endoglin…)

of 3T3-L1 cells. Several genes encoding some components of the extracellular matrix such as metallothioneins were also upregulated in these cells, probably due to stimulation by interleukin-6 *(21)*. Most of the differentially regulated genes (44%) are found during the late stage of differentiation. These include genes that have specific functions in adipocytes, such as genes encoding proteins or enzymes of lipid or carbohydrate metabolism and transcription factors. There is also increased expression of many genes encoding signaling molecules or receptors. Among the downregulated genes are genes encoding markers of other cell lineages but also several genes encoding cell adhesion molecules or components of the cytoskeleton. In conclusion, there are many similarities in expression profiles in particular for the late phases of differentiation between hMSC and 3T3-L1 adipogenesis, but hMSCs are probably an appropriate model for studying the very early stages of adipogenesis *(20)*.

Other Studies Using Human Adipose Tissue

A first report *(5)* describes a large-scale analysis of gene expression by using DNA array in human abdominal subcutaneous white adipose tissue in order to identify candidate genes in diseases like obesity. One hundred thirty-six genes with very high hybridization signals were classified into seven categories according to their functions and less than 50% of these genes had previously been reported to be expressed in adipose tissue. By using the same approach, Sjöholm et al. *(25)* searched for genes predominantly expressed in human omental tissue compared to 32 other human tissues and cells. Many genes predominantly expressed in this tissue were expected but the main conclusion of the report is that human omental tissue is a major site of expression of the acute-phase gene encoding serum amyloid A1, a known risk factor for coronary artery disease. Using the DNA array approach, a recent study of human subcutaneous abdominal adipose tissue has sought to define gene expression profiles in human preadipocytes and adipocytes *(24)*. As illustrated in Table 4 adapted from the report of Urs et al.*(24)*, most of the genes that are upregulated in adipocytes are involved directly or indirectly in lipid or glucose metabolism and include many genes encoding transcription factors, receptors and signaling or secreted proteins. Table 4 lists the genes that are overexpressed in preadipocytes, including genes encoding cytoskeletal and extracellular matrix components. During differentiation of preadipose cells into adipose cells, there is active synthesis of collagens as already observed for hMSCs,

Table 4

Overview of the Main Families of Genes whose Expression is Overexpressed up to 2.5-fold in Human Adipocytes (AD) versus Predipocytes (PA) on the Left and in Preadipocytes versus Adipocytes on the Right [Adapted from *(24)*]

Gene families	Mean ratio Log2 AD/PA	Gene families	Mean ratio Log2 PA/AD
Lipid and carbohydrate metabolism/antioxydative/ transport/ ECM		**Signaling molecules**	
Fatty acid binding protein 4	8	Protein tyrosine phosphatase receptor S	2.9
Lipoprotein lipase	6	Protein tyrosine phosphatase receptor Z1	4.9
Fatty acid CoA ligase	4.4	Protein tyrosine phosphatase N21	2.9
Adipose most abundant transcript 1	4.8	Adenosine receptor 2B	2.6
Perilipin	8	Angiotensin receptor L1	2.6
Stearoyl CoA desaturase	3.5	Src family associated phosphoprotein 1	3.4
Angiotensinogen	2.6	Interleukin 22 receptor	4.2
Uncoupling protein 4	4.3	Mitogen-activated protein kinase kinase	2.7
Glycerol 3 phosphate dehydrogenase	3.2	**Cellular matrix and cytoskeleton**	
Aldehyde dehydrogenase 6	2.6	Matrix metalloprotein 7	2.7
Aldehyde dehydrogenase	2.7	Extracellular matrix protein 2	3
Lipase hepatic	2.5	Dermatopontin	8.7
Lipase hormone sensitive	3.4		
Apolipoprotein B	3.1		
LDL receptor 8	2.7		
Diacylglycerol kinase γ	3.3		
ATPase aminophospholipid transporter	3.6		
ATPase-Ca2 transporting plasma membrane	3.3		
Cathepsin G	7.9		
Glutathione peroxidase 3	19.6		
Mannan-binding lectin serine protease 1	3.6		
Carbohydrate sulfotransferase 1	5.4		
Glutamine synthase	3.4		
Cytochrome b-5	3.2		
Crystalline αB	3.8		

(Continued)

Table 4
(**Continued**)

Gene families	Mean ratio Log2 AD/PA	Gene families	Mean ratio Log2 PA/AD
FXYD domain containing ion transport regulator 1	4.8		
Pleckstrin	3.6		
ATP binding cassette subfamily E	2.7		
Transcription factors/binding proteins			
Transcription factor CP2	3.1		
PPARγ	2.7		
Retinoid X receptor A	3.9		
Retinoid X receptor B	3.5		
E2F transcription factor 5	6.3		
Transcription activator of the c-fos promoter	2.6		
Zinc finger protein 336	2.9		
Signal transducer and activator of transcription 5B	3.6		
SWI/SNF complex	2.9		

which decrease during adipose conversion. Surprisingly, the gene encoding C/EBPα is also overexpressed which indicates that the preadipocytes from the stromal fraction of human adipose tissue are already committed to adipogenesis *(24)*.

APPLICATION OF MICROARRAY TECHNOLOGY TO WHITE ADIPOSE TISSUE GENE EXPRESSION IN OBESITY

As reported in a previous chapter, mature adipocytes express a number of genes whose products participate intimately in the regulation of energy homeostasis. This is the case for genes encoding transcription factors or signaling molecules essential for the activation of many genes involved in lipid metabolism and carbohydrate metabolism, both of which are upregulated in mature adipocytes. It is also clear that genes encoding molecules with an immune function are expressed in the stromal fraction of white adipose tissue (preadipocytes) and that genes encoding cellular matrix and cytoskeleton components are overexpressed in this fraction. Several studies have been designed for gene expression profiling of white adipose tissue in obesity and diabetes by using the microarray approach. Most of these studies have been performed in rodents (mouse or rat) but there are also few studies in human adipose tissue.

Studies in Rodents: The ob/ob and DIO Mice Models

Two studies using Affymetrix microarrays compared white adipose tissue gene expression in genetically obese *ob/ob* mice deficient in leptin and their lean counterparts. Epididymal fat tissue from male mice of two backgrounds of the same strain

C57Bl/6J *ob/ob* and BTBR *ob/ob* (28) or abdominal fat from obese C57Bl/6J ob/ob female mice (29) was compared with abdominal fat tissue removed from lean control mice of the same strain. Nadler et al.(28) analyzed more than 11 000 genes or EST and found 214 genes (2%) whose status was modified in obese mice after elimination of the strain background as a variable. Expression was increased in 136 genes and consistently decreased in 78. In their study of 6500 genes or EST, Soukas et al. (29) found 77 genes that were differentially expressed threefold or more in the *ob/ob* samples. In our study, we used male mice of the same C57Bl/6J strain fed either on a standard chow diet (control mice) or a fat-enriched diet from Harlan Teklad (DIO mice) (32). After 8 weeks of diet, the DIO mice exhibited an obese phenotype with a twofold increase in weight and fat mass over controls. Moreover, DIO mice displayed a sixfold increase in plasma leptin, were resistant to the infusion of leptin, and exhibited, respectively, twofold and 1.6-fold increases in plasma insulin and glucose levels. Abdominal fat tissue was removed and RNA was used for Affymetrix oligonu-cleotide microarray hybridization. Of 12 488 datasets a total of 472 genes or EST were differentially expressed (1.2-fold or more but at a sufficient level for detection in adipose tissue mRNA from control or DIO samples) and a total of 98 genes were differentially expressed threefold or more in the DIO samples. The majority of these genes were downregulated (about 70%) in DIO mice. As shown in Tables 5 and 6 that are a compilation of the data reported in the two papers (28, 29), the comparison of obese *ob/ob* and lean mice reveals a remarkable pattern of alteration in gene expression in white adipose tissue. Numerous genes that play a role in fatty acid or cholesterol biosynthesis, and whose expression was previously reported to increase during adipogenesis, have a significantly decreased expression level in the obese *ob/ob* mice (Table 5). These genes include lipogenic enzyme-encoding genes such as fatty acid synthase (fas), glycerol-3-phosphate dehydrogenase, ATP-citrate lyase, and squalene synthase, and specific genes encoding transcription factors such as SREBP-1/ADD1, which controls many lipogenic genes, as well as β3-adrenergic receptor, both of which exhibit increased expression during adipocyte differentiation. Leptin replacement normalized some but not all of these changes (29). A similar pattern of gene expression alteration was observed in our model of DIO mice for lipogenic genes as well as those encoding transcription factors or secreted proteins specific to mature adipocytes (32). However, all the genes that were strongly expressed in control mouse samples were moderately decreased in the DIO samples. These results suggest that leptin resistance occurring progressively in DIO is responsible for the moderate decrease in expression of these genes, while leptin deficiency in obese *ob/ob* mice induced a greater decrease in this expression. Another interesting feature was the absence of alteration of SREBP-1/ADD1 expression in DIO mouse samples, in contrast to obese *ob/ob* mice, although expression of the fas encoding gene that is a target of this transcription factor was also decreased. This would imply that the decrease in fas gene expression is not exclusively dependent on leptin, which normally controls its expression through the activation of SREBP-1 as already reported in primary cultured rat adipocytes (56). The decreased expression of adipogenic genes in obesity is surprising given that white adipose tissue mass in these different models (*ob/ob* or DIO) is greatly increased. This implies that adipocytes from obese mice in general have a decreased lipogenic capacity similar to preadipocytes. This reduces their capacity to synthesize fatty acids although they are lipid engorged, indicating that

Table 5

Overview of the Main Families of Genes whose Expression is Downregulated in White Adipose Tissue from Obese *ob/ob* Mice as Compared to their Lean Counterparts
[Adapted from *(28,29)*]

Gene families	Fold	Gene families	Fold
Lipid metabolism/energy metabolism/mitochondrial enzymes		**Antioxydant proteins**	
Phosphoenolpyruvate carboxykinase	−5.3	Glutathione transferase	−11.4
Glycerol phosphate dehydrogenase	−2.5 to −5	Glutathione peroxidise	−6.1
Spot 14	−4.6	o-class glutathione-S-transferase	−3.1 to −5.8
lactate dhydrogenase-B	−3.3	**Inflammatory/acute phase**	
LAF-1 transketolase	−3.2	IgK chain C- Region	−8.6
Fatty acid transport	−3.2	Kappa-immunoglobulin	−4.1
Fatty acid synthase	−3.1	Thrombomodulin	−3.6
Squalene synthase	−3.1	**Other factors**	
ATP-citrate-lyase	−2.9	B2-protein	−11.2
Stearoyl-CoA desaturase	−2.5	Keratin 19	−5.5
Aldolase A	−2.5	28 kD serine protease	−5.4
ATP synthase beta chain	−5.4	pcp4	−4.5
Fructose 2 6 bisphosphatase	−4.1	cdk inhibitor	−4.1
Alpha-amylase 1	−3.9	Asparagine synthetase	−4
Ubiquinol cytochrome c reductase core protein 2	−3.4	Laminin B1	−3.5
Aldehyde dehydrogenase	−3	−2-microglobulin	−3.2
Mitochondrial enoyl-CoA	−2.8	Uridine kinase	−2.7
Branched-chain aminoacid aminotransferase	−2.8	Ubiquitin	−2.7
Cytochrome c oxidase-subunit VIIIa	−2.7	Retinol-binding protein	−2.5
Transcription factors/hormones/signaling molecules/ secreted proteins		Glycogen phosphorylase	−2.5
β3 adrenergic receptor	−7.7 to 10.5		
Parathyroid hormone receptor	−7.7		
Albumin D-box binding protein	− 6.9		
Double LIM protein-1	−4.2		
AFBP-1	−3.7		
Interferon regulatory factor-7	−3.6		
Ras-related protein (DEXRAS1)	−2.9		
SREBP-1/ADD1	−2.7		
Adipsin	−8.3		
Angiotensinogen	−3 to 8.1		
Complement component C2	−3.2 to 6.8		
DAN protein	−4.5		
IGF-BP 5	−3		
Apolipoprotein E	−2.4		

Table 6
Overview of the Main Families of Genes whose Expression is Upregulated in White Adipose
Tissue from Obese *ob/ob* Mice as Compared to their Lean Counterparts
[Adapted from *(28,29)*]

Gene families	Fold	Gene families	Fold
Cytoskeleton/ECM/membrane trafficking		**Inflammatory/Acute phase**	
Myc basic motif homologue-1	+15	Macrophage metalloelastase	+31.5
Clathrin light chain B	+9.7	SV-40 induced 24p3	+10.7
Talin	+7.7	Heme oxydase	+ 9.8
Heparin binding protein 44	+4.8	Serum amyloid A3	+8.4
Filamin A	+4.5	PAI-1	+8.2
L-34 galactoside-binding lectin	+4.5	MRP8 (calcium binding)	+7.8
α actin	+3.8	Macrosialin	+6.6 to 3.7
Calveolin-1	+3.7	Macrophage specific cysteine-rich TM Glycoprotein	+5.7
Lysyloxidase	+3.3		
Sulfated glycoprotein (Sgp1)	+3.1	Serum amyloid A4	+5.2
Osteoblast specific factor (OSF-2)	+3.1	CD53	+4.9
Ion channel homolg RIC	+3	gp 49	+4.8
+-B2-crystallin	+2.8	Macrophage mannose receptor	+3.7
Metallothionein II	+2.6	CD18 beta subunit	+4
		LPS binding protein	+3.6
Secreted proteins/signaling and vesicle traffic molecules /transcription factors		Macrophage specific Nramp	+3.4
		pEL98	+3.4
Apolipoprotein D	+5.4	Lipocortin-1	+3.3
LIM protein 1	+4.9	**Protein synthesis/turnover/ tissue remodeling**	
MPS1	+4.9 to 5.4		
Low density lipoprotein receptor 2	+4.8	Cathepsin S or cathepsin S precursor	+3.9 to 7.7
5-lipoxygenase activating protein	+3.4 to 4.3	TI-225 polyubiquitin	+4.3
ADRP	+4.2	Preprocathepsin K	+4.5
14-3-3 eta (positive regulator of GR)	+4.2	Cathepsin Z precursor	+3.9
Protease-nexin 1	+3.7	Preprolegumain	+3.9

(Continued)

Table 6
(*Continued*)

Gene families	Fold	Gene families	Fold
15-hydroxyprostaglandin dehydrogenase	+3.3	Cathepsin B	+3.5
Leptin	+3 to 3.2	Spi2 protease inhibitor	+3.2
Immune /complement components		**Others**	
Fc receptor	+8.1	Mesodermal specific transcript	+8.8
γ-inducible lysosomal thiol reductase	+6	Superoxide dismutase SOD3	+5.9
Fc γ receptor	+4.8	D52	+5.2
C10 like chemokine	+3.3	Brain acid-soluble protein 1	+4.5
IFN-α induced protein	+2.8	Cystatin B	+4.4
P-selectin glycoprotein ligand	+2.8	UCP2	+4.3
Complement C1q βchain	+2.6	Hematopoietic specific protein 1	+4
		Schwannoma-associated protein	+3.9
		Retinoic acid-inducible E3 protein	+3.4
		Creatine kinase B	+3.4
		5-lipoxygenase- activating protein	+3.4

they have reached their peak storage capacity. Normally insulin and leptin promote lipogenesis in mature adipocytes and prevent excessive lipid deposition in other tissues such as liver. Obese mice are resistant to the action of these hormones and there is a shift to the liver, which explains the hepatic steatosis in these different models of obesity. This is a common feature with lipoatrophic diabetes. Thus, the lack of functional adipocytes but not the overabundance of fat cells is a main risk factor for diabetes *(27)*. Moreover, the decrease in lipogenic gene expression in adipose tissue became more pronounced with the development of diabetes in BTBR *ob/ob* diabetes susceptible mice *(30)*. Several other sets of genes displayed a strong inhibition in DIO mouse samples (Table 7). The first family is represented by genes encoding structural components of the vascular smooth muscle cells in association with the microfilament network corresponding to a disorganization of the vascular compartment in DIO adipose tissue, which has not been mentioned in the *ob/ob* model. Moreover, several genes encoding cell–cell or cell–matrix adhesion molecules are also highly downregulated in DIO samples (Table 7). A great number of genes encoding antioxidant proteins such as glutathione-S-transferase or μ-class glutathione transferase (Table 7) also had a decreased expression in DIO samples. This observation was also reported in the *ob/ob* model (Table 5). In DIO this is illustrated by the gene encoding the mouse vas deferens protein, a member of the aldoketoreductase gene superfamily whose expression has

Table 7
Genes with Modified Expression (up to threefold) in White Adipose Tissue in DIO versus
Control Mice from *(32)*

Genes with down-regulated expression	Function	Fold change
Smooth muscle calponin gene	Structural, actin-associated	–68.2
Androgen regulated vas deferens protein (MVDP)	Steroidogenesis, detoxication	–59.0
Purkinje cell protein 4 (pcp-4)	Apoptosis, signal.inhibition (b cells)	–44.4
Glucocorticoid-regulated inflammatory PG G/H synthase	Inflammatory	–41.1
Keratin complex 1	Structural	–35.2
GATA-binding protein 3	Transcription factor	–21.4
Smoothelin	Structural	–18.4
D-amino acid oxidase	Metabolic	–12.7
Myosine heavy chain 11	Structural	–11.6
Glutathione-S-transferase	Antioxidant, detoxication	–10.3
Alpha-actinin 2 associated LIM protein	Structural	–8.7
B6CBA Lisch7	Transcription factor	–7.5
Beta chemokine TCA4 gene	Inflammatory	–7.1
Placenta-specific ATP-binding cassette transp.(Abcg2)	Lipid transporter, detoxication	–7.1
4F2/CD98 light chain	Amino acid transporter	–6.5
Cytochrome P450 naphtalene hydroxylase	Detoxication	–6.2
Ankyrin 1	Structural	–6.2
Cadherin 16	Cell adhesion	–5.8
Cytokeratin endo A or keratin 8	Structural (intermediate filament)	–5.6
Cytokeratin (endo B) or keratin 18	Structural (intermediate filament)	–4.9
CPETR2 (Clostridium perfringens enterotoxin receptor)	Neurodevelopment	–4.7
Glycoprotein-associated amino acid transporter	Amino acid transporter	–4.7
Transition protein 2	Chromatin remodeling	–4.6
Actin, alpha 1	Structural	–4.5
Transgelin	Structural (actin-associated)	–4.4
Melanocyte specific gene (msg1)	Transactivator	–4.4
Matrix metalloproteinase 7 or matrilysin	Differentiation, remodeling	–4.2
Acidic epididymal glycoprotein 1	Detoxication	–4.0
AC133 antigen homolog	Unknown	–3.9
Thiosulfate sulfurtransferase, mitochondrial (rhodanese)	Detoxication	–3.8
Nitric oxide synthase 3	Metabolic	–3.7

(Continued)

Table 7
(*Continued*)

Genes with down-regulated expression	Function	Fold change
Thrombospondin type 1 domain or R-spondin	Cell adhesion, extracellular matrix	−3.2
Cystein rich protein (CRP-1)	Differentiation, cell growth	−3.2
Branched chain aminotransferase 2, mitoch. (Eca 40)	Proliferation	−3.1
Melanoma-inhibitory -activity protein	Unknown	−3.1
Cell death activator (CIDE-A)	Apoptosis	−3.0
Prolactin receptor	Signaling	−2.9
Protamine-3	Chromatin remodeling	−2.9
Seb4	Inflammatory	−2.9
Ets transcription factor (ELF3)	Transcription factor	−2.9
Transketolase	Metabolic	−2.9
Carbonyl reductase	Detoxication	−2.9

Genes with up-regulated expression	Function	Fold change
Macrophage metalloelastase or metalloproteinase (MME)	Inflammatory, elastolyse	+21.8
Plasminogen activator inhibitor (PAI-1)	Inflammatory	+11.7
Zinc finger protein (Peg3)	Growth, apoptosis	+8.3
Peg1/ MEST protein	Embryonic growth	+8.2
Alpha-1 protease inhibitor 3	Inflammatory, elastolyse	+4.8
Small inducible cytokine A2	Inflammatory	+4.8
Bcm-1 antigen	Inflammatory	+4.0
IIGP protein interferon gamma-induced GTPase	Growth	+3.8
Small inducible cytokine A6	Inflammatory	+3.7
Tenascin X	Cell adhesion	+3.7
Thrombospondin 1	Structural, cell adhesion	+3.6
Fibrillin 1	Structural, extracellular matrix	+3.5
Glycoprotein 49A	Inflammatory	+3.5
Neural precursor cell expressed developmentally downregulated Nedd 9	Apoptosis	+3.4
Ecotropic viral integration site 2	Proto-oncogene	+3.3
A disintegrin and metalloprotease (ADAM) 7	Structural, cell adhesion	+3.0
Complement component 3a receptor 1	Signaling	+3.0

never been reported in adipose tissue. We know from further studies that the expression of this gene is highly dependent on the localization of the selected adipose tissue mass (unpublished data). Many genes encoding mitochondrial enzymes such as aldehyde dehydrogenase *(32)* also have a decreased expression in DIO as reported for *ob/ob* mice *(28)*. An important point regarding obesity and its complications such as type 2 diabetes and cardiovascular diseases is the high upregulation of genes involved in inflammatory processes or acute-phase reaction and genes encoding lysosomal proteins. All these genes often displayed very low levels of expression in control mice. This is true for all the different models of obesity (*ob/ob* and DIO models) (Tables 6 and 7). Thus, leptin deficiency in obese *ob/ob* mice and leptin resistance in DIO mice produce a similar dysregulation in the expression of genes that function in response to inflammation in the adipocyte lineage. Among these genes are those encoding cathepsins or cathepsin precursors involved in matrix remodeling and turnover. Many of the genes involved in inflammation probably originate from macrophages, such as the macrophage metalloelastase gene, or are immune or complement factor-encoding genes. The high expression levels of genes encoding proteins involved in macrophage-mediated proteolysis or acute-phase reaction such as plasminogen activator inhibitor protein-1 (PAI-1) could widely contribute to the pathogenesis of type 2 diabetes and cardiovascular diseases. However, the elevated signals of inflammatory genes could be caused by an enrichment of macrophages in the white adipose tissue or could result from an elevated expression in the adipocytes. By using microarray analysis on perigonadal white adipose tissue of obese mice (Agouti Ay or ob/ob mice), Weisberg et al.*(31)* found that 30% of the genes that had an increased expression encoded proteins that are characteristic of macrophages and are positively correlated with body mass. By immunohistochemical analysis, they reported that the number of cells expressing the macrophage marker F4/80 was positively correlated with the degree of adiposity. A detailed comparison by means of microarray analysis was made in adipose tissue between the two backgrounds of the C57Bl/6J *ob/ob* diabetes-resistant and BTBR *ob/ob* diabetes-susceptible obese mice *(30)*. This study demonstrates that there is an increased expression of genes involved in inflammation in adipose tissue of the diabetic mice, suggesting that they have a role in this disease. In DIO mice, the progressive increase in the expression of these inflammatory genes when the high-fat diet is prolonged, as well as the decrease in the expression of lipogenic genes in adipose tissue, seems to be correlated with the appearance of insulin resistance and type 2 diabetes as indicated by our further studies (unpublished data). DIO mice are therefore a suitable model for further studies to establish a correlation between the alteration in gene expression profiles for defined gene families in response to a progressive development of obesity and appearance of metabolic disorders. Interestingly, Higami et al. *(57)* have shown by using microarray analysis (Affymetrix) that, in epididymal adipose tissue removed from male C57Bl/6J mice subjected to long-term caloric restriction (41%), the majority of the gene expression alterations are in the opposite direction than those displayed by obese (*ob/ob* or DIO) mice. In fact, the long-term reduction of caloric intake induced an increase in expression of genes involved in lipid, carbohydrate, and mitochondrial energy metabolism and a decrease in the expression of genes encoding proteins of the cytoskeleton, extracellular matrix, and mainly genes involved in inflammatory processes.

Studies in Rodents: The Obese Rat Model

Lopez et al. *(35)* performed a microarray analysis (Affymetrix) on epididymal white adipose tissue removed from male Wistar rats fed either on a standard diet (control group) or a fat-rich hypercaloric diet (cafeteria) containing pâté, chips, chocolate bacon, biscuits, and chow in appropriate proportion (obese group). After about 9 weeks, rats fed on the cafeteria diet exhibited a 1.5-fold increase in weight as compared to control rats, and a 1.7-fold increase in plasma leptin, but no change in plasma glucose. Of 12 000 genes or EST, 416 genes were differentially expressed twofold or more in obese rat samples. Many gene encoding transcription factors involved in adipogenesis or gene encoding enzymes involved in lipid metabolism, such as PPARγ, CEBP/α, FABP, glycerol-3-phosphate dehydrogenase, or squalene synthase, were upregulated as described during adipogenesis *(35)*. The expression level of β3 adrenergic receptor gene was also upregulated twofold and leptin gene 49-fold. These results are not in agreement with the observations made on obese mice (*ob/ob* or DIO) or in humans (see section "Studies in human obesity and insulin resistance"). Moreover, unlike obese mice or obese humans, there is no apparent increase in the expression of genes encoding inflammatory factors in the adipose tissue of the obese rats. Using microarray analyses (Affymetrix) on subcutaneous abdominal adipose tissue from DIO rats, Ramos et al.*(33)* showed that the gene encoding cytokine IL6 was increased 13-fold and that the Roux-en-Y gastric bypass operation reduced this level. Other cell structure genes are also upregulated such as B-crystallin, clathrin assembly protein, and microtubule-associated protein as in obese mice *(35)*. In contrast, several genes encoding antioxidant proteins involved in detoxication processes, such as glutathione-S-transferase and metallothionein, had a decreased expression level in obese rat samples as also reported in our model of DIO mice. Thus, there are some divergences between obese rat (cafeteria diet) and mouse (ob/ob and DIO) models, in particular, concerning the expression level of lipogenic or inflammatory genes. In rats, obesity development is apparently not associated with a reduction in lipogenic capacity of mature adipocytes from abdominal white adipose tissue, in contrast to obese mice or humans, which seems to indicate that they did not reach their maximal storage capacity. Moreover, obesity development is not necessarily associated with an increase in inflammatory proteins *(35)*. It could be that there are species differences or that a nine-week treatment with this particular cafeteria diet in rats is not sufficient to induce downregulation of the genes involved in lipogenesis in mature adipocytes from abdominal white adipose tissue. It could also be that new preadipocytes differentiate into mature adipocytes, which would explain the upregulation of the expression of genes belonging to this family.

Studies in Human Obesity and Insulin Resistance

A few microarray analyses have been performed on human omental or subcutaneous adipose tissue in cases of obesity or insulin resistance. Gomez-Ambrosi et al.*(38)* have compared by means of microarray analysis (Mergen Ltd, San Leandro, CA, USA) the pattern of gene expression in omental tissue biopsies of obese and lean volunteers. They found that 89 genes were upregulated while 64 were downregulated at least twofold in obese patients. The majority of the upregulated genes were involved in cell proliferation, such as several mitogen-activated protein kinases, or in the immune response, mainly receptors for Fc fragment of IgG. They found a general tendency to blunt signal transduction genes (insulin receptor substrate 4 or β-adrenergic receptors) and a global downregulation of genes encoding growth factors (IGF or FGF). Genes participating in the control of metabolism, such as those encoding adrenergic β receptor kinase 2, glycogen synthase kinase 3-α, were either upregulated or decreased, such as those coding for low-density lipoprotein receptor, prostaglandin E receptor 3. The expression of genes encoding several proteins involved in food intake, like the neuropeptide Y receptors Y1 and Y5, was also increased in obese patients. Another report *(58)* confirms by DNA microarray analysis (Affymetrix) the importance of immune markers in obesity and its metabolic consequences. The authors reported that complement components C2, C3, C4, C7, and factor B had a higher expression level in omental as compared to subcutaneous adipose tissue from biopsies of obese men, and the development of visceral adiposity widely contributes to the metabolic complications. By using a subtractive hybridization strategy on visceral and subcutaneous adipose tissue of obese patients (male and female), Linder et al.*(59)* identified a set of differentially expressed transcripts that they further spotted to perform a microarray analysis. They identified five differentially expressed genes in both types of adipose tissues. Ras homolog gene family member G and the gene encoding phospholipid transfer protein were upregulated in omental tissue, and three genes were upregulated in subcutaneous tissue; those encoding calcyclin and adipsin in both sexes, and the third one corresponding to an unidentified clone (possibly related to a calcium channel α2-δ3 encoding gene) was only detected in male subcutaneous tissue, which indicates that sex differences exist in the adipose tissue gene expression patterns. In an interesting report, Clément et al. *(37)* examined the gene expression profile of subcutaneous adipose tissue obtained from a large number of obese patients (×29) on very low-calorie diet (VLCD) as compared to non-obese subjects (×17) by using DNA microarray analysis (Stanford University Microarray). They reported that in obese patients a great number (×100) of inflammation-related transcripts were downregulated during a 28-day VLCD and the gene expression profile was closer to that of lean subjects than to the pattern of gene expression in obese subjects before VLCD. Thus, it appears that weight loss in obese patients improves inflammatory status in adipose tissue through a decrease in the expression of proinflammatory molecules originating from the macrophage population and an increase in antiinflammatory molecules. Two studies have used microarray analysis to characterize the gene expression profile of adipose tissue and skeletal muscle in diabetic and non-diabetic insulin-resistant or insulin-sensitive subjects *(39, 40)*. In subcutaneous adipose tissue, genes encoding proteins involved in the cell cycle and cytoskeleton reorganization were upregulated in diabetic patients. In contrast, adipogenic transcription factors, such as C/EBPα and β, PPARγ, and SREBP-1 as well as Wnt signaling genes such as WNT1, FZD1, GSK3, and β-catenin, have decreased expression in diabetic patients. There is also reduced expression of adipose-specific secreted proteins such as adiponectin and aP2, which are involved in adipogenesis. All these data suggest that insulin resistance is associated with impaired adipogenesis, as already reported in obese diabetic mice (see section "Studies in rodents: the ob/ob and DIO mice models").

CONCLUSIONS

Over the last five years, several significant advances have been made in adipocyte gene expression profiling during adipogenesis through the application of DNA microarrays. In general, there is a remarkable overlap of studies in the identification of genes. During the same period, a few microarray studies investigated molecular changes in adipose tissue gene expression profiles that occur in obesity and insulin resistance by comparison of obese and lean subjects. Several rodent models of obesity were used, but few reports concerned human adipose tissue in obese patients. This has provided valuable new insights into the pathology of obesity since a good many genes were found to be changed. In general, it appears that the enlarged white adipose tissue in obese subjects shows a globally reduced expression of lipogenic genes, indicating a decrease in lipogenic capacity of adipocytes as compared to normal mature adipocytes. Other important findings are the demonstration of decreased expression of genes encoding antioxidant proteins associated with increased expression of genes encoding immune and inflammatory proteins, which could be responsible for the development of insulin resistance and type 2 diabetes in obese subjects. However, further microarray studies are needed to identify the early events leading to white adipose tissue dysfunction during the development of obesity as well as the subsequent cascade of events.

ACKNOWLEDGMENTS

This work was supported by Institut National de le Santé et la Recherche Médicale (INSERM, Paris), Institut National de la Recherche Agronomique (INRA, Paris), Université Claude Bernard Lyon 1 (Lyon), and Institut Fédératif de Recherche Laennec (IFR 62, Lyon). We thank David Marsh for the English reviewing.

REFERENCES

1. Mokdad, A.H., Serdula, M.K., Dietz, W.H., Bowman, B.A., Marks, J.S. and Koplan, J.P. (1999) The spread of the obesity epidemic in the United States, 1991–1998. *JAMA* **282**, 1519–1522.
2. Rosenbaum, M., Leibel, R.L. and Hirsch, J. (1997) Obesity. *N. Engl. J. Med.* **337**, 396–407.
3. Calle, E.E., Rodriguez, C., Walker-Thurmond, K. and Thun, M.J. (2003) Overweight, obesity, and mortality form cancer in a prospectively studied cohort of U.S. adults. *N. Engl. J. Med.* **348**, 1625–1638.
4. Frühbeck, G., Gomez-Ambrosi, J., Muruzabal, F.J. and Burrell, M.A. (2001) The adipocyte: a model for integration of endocrine and metabolic signaling in energy metabolism regulation. *Am. J. Physiol. Endocrinol. Metab.* **280**, E827–E847.
5. Gabrielsson, B.L., Carlsson, B. and Carlsson, L.M.S. (2000) Partial genome scale analysis of gene expression in human adipose tissue using DNA array. *Ob. Res.* **8**, 374–384.
6. Yang, G.P., Ross, D.T., Kuang, W.W., Brown, P.O. and Weigel, R.J. (1999) Combining SSH and cDNA microarrays for rapid identification of differentially expressed genes. *Nucl. Ac. Res.* **26**, 1517–1523.
7. Lee, C., Klopp, R., Weindruch, R. and Prolla, T. (1999) Gene expression profile of aging and its retardation by caloric restriction. *Science* **285**, 1390–1393.
8. DeRisi, J.L., Penland, J.L., Brown P.O., Bittner, M.L., Meltzer, P.S., Ray, M, Chen, Y., Su, Y.A. and Trent, J.M. (1996) Use of a cDNA microarray to analyse gene expression patterns in human cancer. *Nat. Genet.* **14**, 457–460.
9. Kononen, J., Bubendorf, L., Kallioniemi, A., Barlund, M., Schraml, P., Leighton, S., Torhorst, J., Mihatsch, M.J., Sauter, G. and Kallioniemi, O.P. (1998) Tissue microarrays for high-throughput molecular profiling of tumour specimens. *Nat. Med.* **4**, 844–847.
10. Cole, K.A., Krizman, D.B. and Emmert-Buck, M.R. (1999) The genetics of cancer: a 3D model. *Nat. Genet.* **21**, 38–41.
11. Golub, T., Slonim, D., Tamayo, P., Huard, C., Gaasenbeek, M., Mesirov, P., Coller, H., Loh, M., Downing, J., Caligiuri, M.A., Bloomfield, C.D. and Lander, E.S. (1999) Molecular classification of cancer: class discovery and class prediction by gene expression monitoring. *Science* **286**, 531–537.

12. Perou, C., Jeffrey, S., Rijn, M.V.D., Rees, D., Eisen, M., Ross, D., Pergamenschikov, A., Williams, C., Zhu, S., Lee, J.C., Lashkari, D., Shalon, D., Brown, P.O. and Botstein, D. (1999) Distinctive gene expression patterns in human mammary epithelial cells and breast cancers. *Proc. Natl. Acad. Sci. U.S.A.* **95**, 9212–9217.

13. Lee, W.K., Padmanabha, S. and Dominiczak, A.F. (2000) Genetics of hypertension: from experimental models to clinical applications. *J. Hum. Hypertens.* **14**, 631–647.

14. Kaminski, N., Allard, J., Pittet, J., Zuo, F., Griffiths, M., Morris, D., Huang, X., Sheppard, D. and Heller, R.A. (2000) Global analysis of gene expression in pulmonary fibrosis reveals distinct programs regulating lung inflammation and fibrosis. *Proc. Natl. Acad. Sci. U.S.A.* **97**, 1778–1783.

15. Guo, X. and Liao, K. (2000) Analysis of gene expression profile during 3T3-L1 preadipocyte differentiation. *Gene* **251**, 45–53.

16. Burton, G.R., Guan, Y., Nagarajan, R. and McGehee, Jr R.E. (2002) Microarray analysis of gene expression during early adipocyte differentiation. *Gene* **293**, 21–31.

17. Ross S.E., Erickson, R.L., Gerin, I., DeRose, P.M., Bajnok, L., Longo, K.A., Misek, D.E., Kuick, R., Hanash, S.M., Atkins, K.B., Andresen, S.M., Nebb, H.I., Madsen, L., Kristiansen, K. and MacDougald, O.A. (2002) Microarray analyses during adipogenesis: understanding the effects of Wnt signaling on adipogenesis and the roles of liver X receptor α in adipocyte metabolism. *Mol. Cell. Biol.* **22**, 5989–5999.

18. Burton, G.R. and McGehee, Jr R.E. (2004) Identification of candidate genes involved in the regulation of adipocyte differentiation using microarray-based gene expression profiling. *Nutrition* **20**, 109–114.

19. Burton, G.R., Nagarajan, R., Peterson, C.A. and McGehee, Jr R.E. (2004) Microarray analysis of differentiation-specific gene expression during 3T3-L1 adipogenesis. *Gene* **329**, 167–185.

20. Nakamura, T., Shiojima, S., Hirai, Y., Iwama, T., Tsuruzoe, N., Hirasawa, A., Katsuma, S. and Tsujimoto, G. (2003) Temporal gene changes during adipogenesis in human mesenchymal stem cells. *Biochem. Biophys. Res. Commun.* **303**, 306–312.

21. Hung, S.C., Chang, C.F., Ma, H.L., Chen, T.H. and Ho, L.L.T. (2004) Gene expression profiles of early adipogenesis in human mesenchymal stem cells. *Gene* **340**, 141–150.

22. Sekiya, I., Larson, B.L., Vuoristo, J.T., Cui, J.G. and Prockop, D.J. (2004) Adipogenic differentiation of human adult stem cells from bone marrow stroma (MSCs). *J. Bone Miner. Res.* **19**, 256–264.

23. Soukas, A., Socci, N.D., Saatkamp, B.D., Novelli, S. and Friedman, J.M. (2001) Distinct transcriptional profiles of adipogenesis in vivo and in vitro. *J. Biol. Chem.* **276**, 34167–34174.

24. Urs, S., Smith, C., Campbell, B., Saxton, A.M., Taylor, J., Zhang, B., Snoddy, J., Voy, B.J. and Moustaid-Moussa, N. (2004) Gene expression profiling in human preadipocytes and adipocytes by microarray analysis. *J. Nut.* **134**, 762–770.

25. Sjöholm, K., Palming, J., Olofsson, L.E., Gummesson, A., Svensson, P.A., Lystig, T.C., Jennische, E., Brandberg, J., Torgerson, J.S., Carlsson, B. and Carlsson, M.L.S. (2004) A microarray search for genes predominantly expressed in human omental adipocytes: adipose tissue as a major production site of serum amyloid A. *J. Clin. Endocrinol. Metab.* **90**, 2233–2239.

26. Moreno-Aliaga, M.J., Marti, A., Garcia-Foncillas, J. and Martinez, J.A. (2001) DNA hybridization arrays: a powerful technology for nutritional and obesity research. *Brit. J. Nutrition* **86**, 119–122.

27. Nadler, S.T. and Attie, A.D. (2001) Please pass the chips: genomic insights into obesity and diabetes. *J. Nutr.* **131**, 2078–2081.

28. Nadler, S.T., Stoeh, J.P., Schueler, K.L., Tanimoto, G., Yandell, B.S. and Attie, A.D. (2000) The expression of adipogenic genes is decreased in obesity and diabetes mellitus. *Proc. Natl. Acad. Sci. U.S.A.* **97**, 11371–11376.

29. Soukas, A., Cohen, P., Socci, N.D. and Friedman, J.M. (2000) Leptin-specific patterns of gene expression in white adipose tissue. *Genes Dev.* **14**, 963–980.

30. Lan, H., Rabaglia, M.E., Stoehr, J.P., Nadler, S.T., Schuele, R.K.L., Zou, F., Yandell, B.S. and Attie, A.D. (2003) Gene expression profiles of nondiabetic and diabetic obese mice suggest a role of hepatic lipogenic capacity in diabetes susceptibility. *Diabetes* **52**, 688–700.

31. Weisberg, S.P., McCann, D., Desai, M., Rosenbaum, M., Leibel, R.L. and Ferrante, Jr A.W. (2003) Obesity is associated with macrophage accumulation in adipose tissue. *J. Clin. Invest.* **112**, 1796–1808.

32. Moraes, R.C., Blondet, A., Birkenkamp-Demtroder, K., Tirard, J., Orntoft, T.F., Gertler, A., Durand, P. Naville, D. and Bégeot, M. (2003) Study of the alteration of gene expression in adipose tissue of diet-induced obese mice by microarray and reverse transcription-polymerase chain reaction analyses. *Endocrinology* **144**, 4773–4782.

33. Ramos, E.J.B., Xu, Y., Romanova, I., Middleton, F., Chen, C., Quinn, R., Inui, A., Das, U. and Meguid, M.M. (2003) Is obesity an inflammatory disease? *Surgery* **134**, 329–335.

34. Middleton, F.A., Ramos, E.J.B., Xu, Y., Diab, H., Zhao, X., Das, U.N. and Meguid, M. (2004) Application of genomic technologies: DNA microarrays and metabolic profiling of obesity in the hypothalamus and in subcutaneous fat. *Nutrition* **20**, 14–25.

35. Lopez, I.P., Marti, A., Milagro, F.I., Zulet, M.L.A., Moreno-Aliaga, M.J., Martinez, J.A. and De Miguel, C. (2003) DNA microarray analysis of genes differentially expressed in diet-induced (cafeteria) obese rats. *Ob. Res.* **11**, 188–194.

36. Castro-Chavez, F., Yechoor, V.K., Saha, P.K., Martinez-Botas, J., Wooten, E.C., Sharma, S., O'Connell, P., Taegtmeyer, H. and Chan L. (2003) Coordinated upregulation of oxidative pathways and downregulation of lipid biosynthesis underlie obesity resistance in perilipin knockout mice: a microarray gene expression profile. *Diabetes* **52**, 2666–2674.

37. Clément, K., Viguerie, N., Poitou, C., Carette, C., Pelloux, V., Curat, C.A., Sicard, A., Rome, S., Benis, A., Zucker, J.D., Vidal, H., Laville, M., Barsh, G.S., Basdevant A., Stich, V., Cancello, R. and Langin, D. (2004) Weight loss regulates inflammation-related genes in white adipose tissue of obese subjects. *FASEB J.* **18**, 1657–1669.

38. Gomez-Ambrosi, J., Catalan, V., Diez-Caballero, A., Martinez-Cruz, M.J.G., Garcia-Foncillas, J., Cienfugos, J.A., Mato, J.M. and Frühbeck, G. (2004) Gene expression profile of omental adipose tissue in human obesity. *FASEB J.* **18**, 215–217.

39. Permana, P.A., Del Parigi, A. and Tataranni, P.A. (2004) Microarray gene expression profiling in obesity and insulin resistance. *Nutrition* **20**, 134–138.

40. Yang, X., Jansson, P.A., Nagaev, I., Jack, M.M., Carvalho, E., Stitbrant-Sunnerhagen, K., Cam, M.C., Cushman, S.W. and Smith, U. (2004) Evidence of impaired adipogenesis in insulin resistance. *Biochem. Biophys. Res. Commun.* **317**, 1045–1051.

41. Viguerie, N., Poitou, C., Cancello, R., Stich, V., Clément, K. and Langin, D. (2005) Transcriptomics applied to obesity and caloric restriction. *Biochimie* **87**, 117–123.

42. Flier, J.S., Cook, K.S., Usher, P. and Spiegelman, B.M. (1987) Severely impaired adipsin expression in genetic and acquired obesity. *Science* **237**, 405–408.

43. Zhang, Y., Proenca, R., Maffei, M., Baronne, M., Leopold, L. and Friedman, J.M. (1994) Positional cloning of the mouse obese gene and its human homologue. *Nature* **372**, 425–432.

44. Ahima, R.S. and Flier, J.S. (2000) Adipose tissue as an endocrine organ. *Trends Endocrinol. Metab.* **11**, 327–332.

45. Gimeno, R.E. and Klaman, L.D. (2005) Adipose tissue as an active endocrine organ: recent advances. *Curr. Opin. Pharmacol.* **5**, 122–128.

46. Mohamed-Ali, V., Pinkney, J.H. and Coppack, S.W. (1998) Adipose tissue as an endocrine and paracrine organ. *Int. J. Obes. Relat. Metab. Disord.* **22**, 1145–1158.

47. Fain, J.N., Madan, A.K., Hiler, M.L., Cheema, P. and Batouth, S.W. (2004) Comparison of the release of adipokines by adipose tissue, adipose tissue matrix, and adipocytes from visceral and subcutaneous abdominal adipose tissues of obese humans. *Endocrinology* **145**, 2273–2282.

48. Kershaw, E.E. and Flier, J.S. (2004) Adipose tissue as an endocrine organ. *J. Clin. Endocrinol. Metab.* **89**, 2548–2556.

49. MacDougald, O.M. and Lane, M.D. (1995) Transcriptional regulation of gene expression during adipocyte differentiation. *Annu. Rev. Biochem.* **64**, 345–373.

50. Gregoire, F.M., Smas, C.M. and Sul, H.S. (1998) Understanding adipocyte differentiation. *Physiol. Rev.* **78**, 783–809.

51. Green, H. and Kehinde, O. (1975) An established preadipose cell line and its differentiation in culture. II. Factors affecting the adipose conversion. *Cell* **5**, 19–27.

52. Richon, V.M., Lyle, R.E. and McGehee, Jr R.E. (1997) Regulation and expression of retinoblastoma proteins p107 and p130 during 3T3-L1 adipocyte differentiation. *J. Biol. Chem.* **272**, 10117–10124.

53. Cowherd, R.M., Lyle, R.E. and McGehee, Jr R.E. (1999) Molecular regulation of adipocyte differentiation. *Cell. Dev. Biol.* **10**, 3–10.

54. Jaiswal, R.K., Jaiswal, N., Bruder, S.P., Mbalaviele, G., Marshak, D.R. and Pittenger, M.F. (2000) Adult human mesenchymal stem cell differentiation to the osteogenic or adipogenic lineage is regulated by mitogen-activated protein kinase. *J. Biol. Chem.* **275**, 9645–9652.

55. Pittenger, M.F., Mackay, A.M., Beck, S.C., Jaiswal, R.K., Douglas, R., Mosca, J.D., Moorman, M.A., Simonetti, D.W., Craig, S. and Marshak, D.R. (1999) Multilineage potential of adult human mesenchymal stem cells. *Science* **284**, 143–147.

56. Fukuda, H., Iritani, N., Sugimoto, T. and Ikeda, H. (1999) Transcriptional regulation of fatty acid synthase gene by insulin/glucose, polyinsatured fatty acid and leptin in hepatocytes and adipocytes in normal and genetically obese rats. *Eur. J. Biochem.* **260**, 505–511.

57. Higami, Y., Pugh, T.D., Page, G.P., Allison, D.B., Prolla, T.A. and Weindruch, R. (2004) Adipose tissue energy metabolism: altered gene expression profile of mice subjected to long-term caloric restriction. *FASEB J.* **18**, 415–417.

58. Gabrielson, B.G., Johansson, J.M., Lönn, M., Jernas, M., Olbers, T., Peltonen, M., Larsson, I., Lönn, L., Sjöström, L., Carlsson, B., and Carlsson, L.M.S. (2003) High expression of complement components in omental adipose tissue in obese men. *Obes. Res.* **11**, 699–708.

59. Linder, K., Arner, P., Flores-Morales, A., Tollet-Egnell, P. and Norstedt, G. (2004) Differentially expressed genes in visceral or subcutaneous adipose tissue of obese men and women. *J. Lipid. Res.* **45**, 148–154.

14

Novel Molecular Signaling and Classification of Human Clinically Nonfunctioning Pituitary Adenomas Identified by Microarray and Reverse Transcription-Quantitative Polymerase Chain Reaction

Chheng-Orn Evans, MS,
Carlos S. Moreno, PhD,
and Nelson M. Oyesiku, MD, PhD, FACS

Contents

INTRODUCTION
MATERIALS
METHODS
NOTES
REFERENCES

Abstract

Pituitary adenomas comprise 10% of intracranial tumors and occur in about 20% of the population. Nonfunctioning (NF) pituitary adenomas do not cause clinical hormone hypersecretion, they account for approximately 30% of pituitary tumors. Their molecular pathogenesis is unclear. To elucidate the molecular changes that contribute to the development of these tumors and classify them by their molecular characteristics, we investigated 11 NF pituitary adenomas and eight normal pituitary glands, using 33 oligonucleotide GeneChip microarrays. We validated microarray results with the reverse transcription-real time quantitative polymerase chain reaction (RT-qPCR) using 23 NF adenomas and eight normal pituitary glands.

Microarray analysis identified significant increases in the expression of 115 genes and decreases in 169 genes. We observed changes in expression of SFRP1, TLE2, PITX2, Notch3, and delta-like 1, suggesting that the developmental Wnt and Notch pathways are activated and important for the progression of NF pituitary adenomas. We further analyzed gene expression profiles of all NF pituitary subtypes compared to each other and identified genes that were uniquely altered in each subtype.

The results provide new insight into the pathogenesis and molecular classification of NF pituitary adenomas, and suggest that therapeutic targeting of the Notch pathway could be effective for these tumors.

From: *Contemporary Endocrinology: Genomics in Endocrinology:*
DNA Microarray Analysis in Endocrine Health and Disease
Edited by S. Handwerger and B. Aronow © Humana Press, Totowa, NJ

Key Words: Nonfunctioning pituitary adenomas; microarrays; RT-qPCR; genes expression; molecular classification

INTRODUCTION

Pituitary adenomas cause significant morbidity by compression of regional structures and the inappropriate expression of pituitary hormones *(1, 2)*. Nonfunctioning (NF) pituitary adenomas, so-called because they do not cause clinical hormone hypersecretion *(2–5)*, account for approximately 30% of pituitary tumors *(3)*. The NF tumors are uniquely heterogenous, typically quite large, and cause hypopituitarism or blindness from regional compression *(1)*.

Despite the lack of clinical hormone hypersecretion, immunocytochemical staining for hormones reveals evidence for hormone expression in up to 79% of these tumors, which we will refer to as immunohistochemically positive NF (NF+) tumors. The remainder are negative for hormone expression *(2,5)* and will be referred to as immunohistochemically negative NF (NF–) tumors. However, current pathological classification of these tumors has no molecular basis, and surprisingly is based only on anterior pituitary hormone histochemistry for hormones.

Unlike the functional pituitary tumors, there is no available effective medical therapy for the NF tumors, and only a better understanding of the molecular biology of these tumors will provide needed medical treatment options.

Recently, new insights into neoplasia have emerged as technical advances have permitted large-scale analysis of eukaryotic gene expression. Microarray analysis has successfully identified unexpected gene expression patterns and allowed researchers to define clinically relevant phenotypic differences in several types of human tumors that were indistinguishable by traditional histopathological examination *(6–10)*.

To elucidate the molecular changes that contribute to the development of these NF pituitary tumors and reclassify them according to their molecular fingerprints, we used microarray analysis to elucidate the gene expression profile of 11 NF pituitary adenomas compared to eight normal pituitary glands. We verified the gene expression changes of four genes that were detected by microarray analysis in 23 NF pituitary tumors and eight normal pituitary glands by reverse transcription-real time quantitative polymerase chain reaction (RT-qPCR) *(11)*.

MATERIALS

Patients and Tumor Characterization

Sporadic pituitary adenomas were obtained from patients at Emory University Hospital during transsphenoidal surgery. Informed consent for inclusion in this study was obtained. Pituitary adenomas are anatomically and pathologically distinct from the normal anterior lobe making them easy to dissect under the surgical microscope. All tumors were microdissected and removed using the surgical microscope, rinsed in sterile saline, snap-frozen in liquid nitrogen, and stored (–80°C) for molecular analysis. Each tumor fragment was then confirmed independently by a neuropathologist as being homogenous and unadulterated by histology and immunohistochemistry prior to molecular analysis. Eight normal pituitary glands obtained from the National Hormone and Pituitary Program, National Institute of Diabetes and Digestive and

Kidney Diseases (Bethesda, MD, USA; $n = 3$) and from National Disease Research Interchange (Philadephia, PA, USA; $n = 5$) were used as controls for microarray and RT-qPCR analyses.

Reagents

Oligonucleotide Human Genome U95Av2 (HG-U95Av2) arrays were purchased from Affymetrix (www.affymetrix.com/products/arrays/specific/hgu95.affx; Santa Clara, CA, USA). The hybridization, washing, scanning, and data analysis were performed at the Emory/VA Medical Center DNA Core Facility in Atlanta, GA, USA, and the Moffit Comprehensive Cancer Center, University of South Florida, Tampa, FL, USA. TRIzol reagent (Cat. No.15596-026), DNase I Amplification Grade (Cat. No.18068-015), SuperScript™ Double-Stranded cDNA Synthesis Kit (Cat. No.11917-010), and SuperScript First Strand Synthesis System for RT-PCR (Cat. No 11904-018) were from Invitrogen (www.invitrogen.com, Carlsbad, CA, USA). Diethyl Pyrocarbonate (Cat. No. D-5758) was from Sigma-Aldrich, St. Louis, MO, USA. Plastic, disposable rotor stator generator probes OMNI-Tips and OMNI TH homogeniger (S/N TH-2502) were from OMNI International (www.omni-inc.com, Marietta, GA, USA). RNeasy Mini Kit was from Qiagen (Cat.No.74104, QIAGEN Inc., Valencia, CA, USA). Enzo BioArray™ High Yield™ RNA Transcription Labeling Kit was from Enzo Diagnostics (Cat No.900182, Enzo diagnostics Inc, Farmingdale, NY, USA). 2X SYBR Green I dye PCR Master Mix (Cat No. 4309155), 96-well Optical Reaction Plate (Part No. 4306737), Optical Caps (Part No. N801-0935), Primer Express software version 2.0, and 18S primers were from AB [Applied Biosystems (www.appliedbiosystems.com, Foster City, CA, USA)].

METHODS
Total RNA Extraction and Synthesis of Biotin-labeled cRNA

Total RNA was extracted from normal pituitaries (80–120 mg) or pituitary adenomas (60–100 mg). Tissue was washed once in ice cold phosphate-buffered saline to remove contaminating blood. Then, it was ground in liquid nitrogen, and homogenized in TRIzol according to the manufacturer recommendation. Total RNA (100 μg) was further purified, using the RNeasy Mini Kit with a few modifications from the manufacturer's instruction (see section "Notes"). Spectrophotometry (OD 260/280 nm) and denaturing agarose gels were performed to determine the integrity of total RNA samples. The first strand synthesis was performed from 25 μg of total RNA (5–8 μl), 1 μl of T7-(dT) 24 oligomer, 1 μl of SuperScript™ Double-Stranded cDNA Synthesis (Cat. No. 11917-010), 4 μl of 5X First Strand Reaction buffer, 1 μl of 10 mM dNTP mix, 2 μl of 0.1 M DTT and 2–4 μl of DEPC-treated H_2O. The reaction was incubated at 45°C for 1 h 30 min. T7-(dT) 24 oligomer and phase-lock gel were provided by the Emory/VA Medical Center DNA Core Facility. The second strand DNA synthesis was performed by adding to the first strand reaction tube 91 μl of DEPC-treated H_2O, 30 μl of 5X Second Strand Reaction buffer, 3 μl of 10 mM dNTP mix, 1 μl of 10 U/μl DNA Ligase, 4 μl of 10 U/μl DNA polymerase I, and 1 μl of 2 U/μl RNase H. The reaction was incubated at 16°C for 2 h. Then, 2 μl of 10 U/μl T4 DNA polymerase was added to the reaction which continued incubating for 5 min at 16°C. Ten microliters of 0.5 M ETDA was added to stop the reaction. The reaction product was then subjected to phenol/chloroform extraction by phase-lock gel and ethanol precipitation. To produce biotin-labeled cRNA with Enzo BioArray High Yield RNA Transcription Labeling Kit, in vitro transcription was performed as described by the manufacturer by adding all the following: 22 µl of cDNA, 4 µl of 10X HY reaction buffer, 4 µl of biotin-labeled ribonucleotides, 4 µl of 10X DTT, 4 µl of 10X RNA inhibitor, and 2 µl of 20X T7 RNA polymerase. The reaction was incubated at 37°C for 5 h by gently mixing the contents of the tube every 30–45 min during the incubation. The reaction product was then purified by using RNeasy column and quantified by OD 260/280 nm. The biotinylated cRNA was fragmented to 50–200 nucleotides by heating (94°C for 35 min), and chilled on ice.

Microarray Hybridization

The Affymetrix HG-U95Av2 arrays contain approximately 12,600 full-length gene sequences previously characterized in terms of function or disease association. The hybridization was carried out (45°C for 16 h) with mixing (rotisserie at 60 rpm) with a hybridization buffer that consisted of appropriate oligonucleotide controls and the fragmented biotin-labeled cRNA. The hybridization buffer consisted of the following: 40 µl of 25 µg of labeled cRNA and 5X fragmentation buffer, 5 µl of control oligo B2, 5 µl of 100X Eukaryotic hybridization control, 3 µl of Herring Sperm DNA, 3 µl of acetylated BSA, 150 µl of 2X hybridization buffer, and 94 µl of DEPC-treated H_2O. The fragmentation buffer, the control oligo B2, the 100X Eukaryotic hybridization control, the Herring Sperm DNA, the acetylated BSA, and the 2X hybridization buffer were provided by the Emory/VA Medical Center DNA Core Facility. After washing and staining with streptavidin-phycoerythrin (Molecular Probe, Eugene, OR, USA), the arrays were scanned using a GeneChip Fluidics station and scanner. Affymetrix Microarray Suite Version 5.0, GeneTraffic (Iobion, La Jolla, CA, USA), Spotfire DecisionSite 8.1, Significance Analysis of Microarrays (SAM), and GenePattern software packages were used to analyze the results (see section "Data analysis").

For microarray analysis, three normal pituitary glands and 11 NF pituitary adenomas were analyzed using HG-U95Av2 GeneChips (Affymetrix) at the Emory/VA Medical Center. All of these samples were analyzed in duplicate, starting from the extraction of total RNA, GeneChip hybridization, washing, scanning, and data analysis. Five additional normal pituitary glands were analyzed once using the same chips, HG-U95Av2 GeneChips at the Moffit Comprehensive Cancer Center, University of South Florida, Tampa, FL.

Reverse Transcriptase-real time Quantitative Polymerase Chain Reaction (RT-qPCR)

RT-qPCR was performed as described in Evans et al. *(12, 13)* on four gene transcripts in 23 NF pituitary adenomas and eight normal pituitary glands in a blinded fashion. To validate the findings of gene transcript measurements by the microarray approach, we performed RT-qPCR of selected gene transcripts. We measured the expression levels of four genes in eight normal pituitary glands and 23 NF pituitary adenomas which composed of 6 NF+ expressing FSH+, 6 NF+ expressing LH+, 6 NF+

expressing FSH+ and LH+, and 5 NF– null cell. The four genes analyzed were isocitrate dehydrogenase 1 (NADP-dependent isocitrate dehydrogenase, IDH1, GenBank AF020038), paired-like homeodomain transcription (PITX2, GenBank AF048722), Notch homolog 3 (NOTCH3, GenBank U97669), and delta-like 1 homolog (DLK1, GenBank U15979). Primers were selected using Primer Express software version 2.0 (PE Applied Biosytems), BLASTed against all *H. sapiens* gene sequences in GenBank for selectivity, and synthesized by the Microchemical Facility at Emory University. The primers of these genes were as the following:

	Forward primer	Reverse primer
IDH1	CACTACCGCATGTACCAGAAAGG	TCTGGTCCAGGCAAAAATGG
PITX2	GCCGGGATCGTAGGACCTT	GTGCCCACGACCTTCTAGCA
NOTCH3	TCTCAGACTGGTCCGAATCCAC	CCAAGATCTAAGAACTGACGAGCG
DLK1	CACATGCTGCGGAAGAAGAAGAAC	ACCGCGTATAGTAAGCTCTGAGGA

Ribosomal RNA (18S rRNA) was used as internal control. After purifying using RNeasy Mini Kit, total RNA (3 µg) of each sample was DNAse digested for 12 min at room temperature and reverse transcribed in 20 µl using 150 ng of random prime hexamers, 0.5 mM of deoxynucleotide triphosphate, and 50 units of SuperScript reverse transcriptase as recommended by the manufacturer (SuperScript First Strand Synthesis System, Cat. No 11904-018). The Reverse Transcription products were then diluted in water (5-fold for candidate genes and 1,000-fold for 18S rRNA) and subjected to PCR according to Applied Biosystems recommendations with few modifications. The PCR was performed in a 25-µl reaction using 2.5 µl of the diluted first strand cDNA template, optimized amount of primers, water and 12.5 µl of the 2X SYBR Green I dye PCR Master Mix. All PCRs were performed at least in duplicate and were cycled in the GeneAmp 5700 Sequence Detection System at 50°C for 2 min, 95°C for 10 min, and 40 cycles of 95°C for 15 s and 60°C for 1 min. The specificity of PCRs was determined from the dissociation curve analysis. Standard curve of each gene and 18S rRNA were performed each time the genes were analyzed. All PCR products were in the linear range of the exponential phase of PCR amplification. The quantity of the specific genes obtained from standard curves was normalized to that of the 18S rRNA of the same sample. Fold change of each gene was calculated as the ratio of the mean of the normalized mRNA of NF compared to the mean of the normalized mRNA of the normal pituitary controls.

Data Analysis

Gene expression data from 12,625 probe sets on the HG-U95Av2 GeneChips were processed into.CEL files using Affymetrix MAS 5.0 and uploaded to GeneTraffic software (Iobion) where they were normalized using the GCRMA method *(14)*. After data normalization, genes with uniformly low expression (maximum signal in any sample less than 32) were removed from consideration, leaving 7241 probe sets for analysis using SAM software *(15)*. Relevant parameters for the SAM analysis were— Imputation engine: 10-Nearest Neighbor; Number of Permutations: 500; RNG Seed:

1234567; Delta: 1.063; Fold-Change: 2.0. Normalized expression data from the 297 significant probe sets were analyzed by a two-dimensional hierarchical clustering, using Spotfire DecisionSite 8.1 software. Data were clustered using unweighted averages and ordered using average Euclidian distance.

For K-nearest neighbor (KNN) prediction, the normalized RT-qPCR data were analyzed with GenePattern software (http://www.broad.mit.edu/cancer/software/genepattern/) and both the KNN cross-validation and class prediction modules were used (KNN = 3). For these analyses, the four genes (or features) that were included were NADP-dependent isocitrate dehydrogenase (IDH1), paired-like homeodomain transcription factor 2 (PITX2), Notch homolog 3 (NOTCH3), and delta-like 1 homolog (Drosophila) (DLK1).

To identify genes uniquely altered in tumor subtypes, SAM analysis was performed with both the normal samples and other tumor subtypes as the control group, and the subtype of interest as the experimental group. Analyses were performed with 500 permutations, fold-change of 2.0, and FDR < 1%.

NOTES

For cleanup total RNA using RNeasy Mini Kit column, follow the QIAGEN protocol. However, to concentrate and maximize the yield of total RNA, the column should be warmed to 65°C for 5 min before eluting total RNA and the elution should be performed twice using the same eluant from the first elution.

There are two variants of the microarray technology. (a) cDNA microarray: the probe cDNA (500–5,000 bases long) is immobilized to a solid surface such as glass using robot spotting and exposed to a set of targets either separately or in a mixture. This method, "traditionally" called DNA microarray, is widely considered as developed at Stanford University. We performed our previous study using this cDNA microarray *(12)*. (b) An array of oligonucleotide (20–80-mer oligos) or peptide nucleic acid (PNA) probes is synthesized either in situ (on-chip) or by conventional synthesis followed by on-chip immobilization. The array is exposed to labeled sample DNA, hybridized, and the identity/abundance of complementary sequences is determined. This method, "historically" called DNA chip, was developed at Affymetrix, Inc., which sells its photolithographically fabricated products under the GeneChip trademark. In this study, we used the oligonucleotide array from Affymetrix.

We used HG-U95Av2 array because these arrays were available by then for human genome. The Affymetrix HG-U95Av2 arrays contain 12,625 full-length gene sequences previously characterized in terms of function or disease association. Subsequently, HG-U133 set, HG U133A 2.0 array, HG U133 Plus 2.0 array, Human Cancer G110 array, HuGeneFL Genome array are now available.

How does one know that the result is right? Affymetrix designed several controls: B2 Oligo serves as a positive hybridization control and is used by the software to place a grid over the image. Variation in B2 hybridization intensities across the array is normal and does not indicate variation in hybridization efficiency. Hybridization controls, bioB, bioC, bioD and cre, are spiked into the hybridization cocktail, independent of RNA sample preparation, and are thus used to evaluate sample hybridization efficiency to gene expression arrays. Internal control genes, β-actin and GAPDH, are used to assess RNA sample and assay quality. Specifically, the signal values of the 3′ probe set to the 5′ probe set should be no more than 3. Since the gene expression assay has inherent 3′ bias, a high 3′ to 5′ ratio may indicate degraded RNA or inefficient transcription of ds cDNA or biotinylated cRNA. Ratios 3′ to 5′ for internal controls are displayed in the expression Report (.rpt) file. All of our Report files showed that the ratio of 3′ to 5′ of β-actin and GAPDH were less than 3.

Scaling and normalization factors: for the majority of experiments where a relatively small subset of transcripts is changing, the global method of scaling/normalization is recommended. Differences in overall intensity are most likely due to assay variables including pipetting error, hybridization, washing and staining efficiencies, which are all independent of relative transcript concentration. Applying the global method, corrects for these variables. For global scaling, it is imperative that the same target intensity value is applied to all arrays being compared. In this study, we used global scaling to analyze the data. For some experiments, where a relatively large subset of transcripts is affected, the "Selected Probe Sets" method of scaling/normalization is recommended. Differences in overall intensity are due to biological and /or environmental conditions.

We used Spotfire DecisionSite 8.1 software for analyzing a two-dimensional hierarchical clustering. For KNN prediction of the blinded samples for RT-qPCR, we used GenePattern software to analyze the normalized RT-qPCR data.

For quantitative PCR, optimization of the primers should be performed first. The cDNA obtained from reverse transcription should be diluted 5- to 10-fold for the candidate genes and it should be diluted 1,000- to 10,000-fold for 18S before the PCR should be performed.

REFERENCES

1. Greenman, Y. and Melmed, S. (1996) Diagnosis and management of nonfunctioning pituitary tumors. *Annu Rev Med* **47**, 95–106.
2. Asa, S. L. and Ezzat, S. (1998) The cytogenesis and pathogenesis of pituitary adenomas. *Endocr Rev* **19**, 798–827.
3. Asa, S. L. and Kovacs, K. (1992) Clinically non-functioning human pituitary adenomas. *Can J Neurol Sci* **19**, 228–235.
4. Black, P. M., Hsu, D. W., Klibanski, A., Kliman, B., Jameson, J. L., Ridgway, E. C., Hedley-Whyte, E. T. and Zervas, N. T. (1987) Hormone production in clinically nonfunctioning pituitary adenomas. *J Neurosurg* **66**, 244–250.
5. Katznelson, L., Alexander, J. M. and Klibanski, A. (1993) Clinical review 45: clinically nonfunctioning pituitary adenomas. *J Clin Endocrinol Metab* **76**, 1089–1094.
6. DeRisi, J., Penland, L., Brown, P. O., Bittner, M. L., Meltzer, P. S., Ray, M., Chen, Y., Su, Y. A. and Trent, J. M. (1996) Use of a cDNA microarray to analyse gene expression patterns in human cancer [see comments]. *Nat Genet* **14**, 457–460.
7. Moch, H., Schraml, P., Bubendorf, L., Mirlacher, M., Kononen, J., Gasser, T., Mihatsch, M. J., Kallioniemi, O. P. and Sauter, G. (1999) High-throughput tissue microarray analysis to evaluate genes uncovered by cDNA microarray screening in renal cell carcinoma [see comments]. *Am J Pathol* **154**, 981–986.
8. Schena, M., Shalon, D., Davis, R. W. and Brown, P. O. (1995) Quantitative monitoring of gene expression patterns with a complementary DNA microarray [see comments]. *Science* **270**, 467–470.
9. Wang, K., Gan, L., Jeffery, E., Gayle, M., Gown, A. M., Skelly, M., Nelson, P. S., Ng, W. V., Schummer, M., Hood, L. and Mulligan, J. (1999) Monitoring gene expression profile changes in ovarian carcinomas using cDNA microarray. *Gene* **229**, 101–108.
10. Wang, T., Hopkins, D., Schmidt, C., Silva, S., Houghton, R., Takita, H., Repasky, E. and Reed, S. G. (2000) Identification of genes differentially over-expressed in lung squamous cell carcinoma using combination of cDNA subtraction and microarray analysis. *Oncogene* **19**, 1519–1528.

11. Moreno, C. S., Evans, C. O., Zhan, X., Okor, M., Desiderio, D. M. and Oyesiku, N. M. (2005) Novel molecular signaling and classification of human clinically nonfunctioning pituitary adenomas identified by gene expression profiling and proteomic analyses. *Cancer Res* **65**, 10214–10222.

12. Evans, C. O., Young, A. N., Brown, M. R., Brat, D. J., Parks, J. S., Neish, A. S. and Oyesiku, N. M. (2001) Novel patterns of gene expression in pituitary adenomas identified by complementary deoxyribonucleic acid microarrays and quantitative reverse transcription-polymerase chain reaction. *J Clin Endocrinol Metab* **86**, 3097–3107.

13. Evans, C. O., Reddy, P., Brat, D. J., O'Neill, E. B., Craige, B., Stevens, V. L. and Oyesiku, N. M. (2003) Differential expression of folate receptor in pituitary adenomas. *Cancer Res* **63**, 4218–4224.

14. Irizarry, R. A., Hobbs, B., Collin, F., Beazer-Barclay, Y. D., Antonellis, K. J., Scherf, U. and Speed, T. P. (2003) Exploration, normalization, and summaries of high density oligonucleotide array probe level data. *Biostatistics* **4**, 249–264.

15. Tusher, V. G., Tibshirani, R. and Chu, G. (2001) Significance analysis of microarrays applied to the ionizing radiation response. *Pro Natl Acad Sci U S A* **98**, 5116–5121.

15

Gene Expression Studies of Prostate Hyperplasia in Prolactin Transgenic Mice

Karin Dillner, PhD, Jon Kindblom, MD, PhD, Amilcar Flores-Morales, PhD, and Håkan Wennbo, MD, PhD

CONTENTS

BENIGN PROSTATE HYPERPLASIA
PROLACTIN ACTION ON THE PROSTATE
PROSTATE HYPERPLASIA MOUSE MODELS OVEREXPRESSING
  THE PRL GENE
FUNCTIONAL GENOMICS IN THE STUDY OF THE PROSTATE GLAND
GENE EXPRESSION STUDIES IN PROSTATE HYPERPLASIA
  OF PRL-TRANSGENIC MICE
DIFFERENTIALLY EXPRESSED GENES IN THE ENLARGED PROSTATES
  OF PB-PRL TRANSGENIC MICE AS COMPARED TO CONTROLS
CONCLUSION
REFERENCES

Abstract

Benign prostatic hyperplasia (BPH) and prostate cancer are age-related diseases, affecting a majority of elderly men in the western world, and are known to be influenced by several different hormones, including sex hormones. Although the hormone prolactin (PRL) is well known to exert trophic effects on prostate cells, its involvement in pathophysiological conditions is still poorly characterized. In order to evaluate the potential role of PRL in promoting prostate growth, we have used a recently developed transgenic mouse model that overexpresses the PRL gene specifically in the prostate (Pb-PRL transgenic mice). The PRL transgenic mice develop a significant prostate hyperplasia which increases with age. The prostates of the Pb-PRL transgenic mice display a prominent stromal hyperplasia with mild epithelial dysplastic features, leading to an increased stromal/epithelial ratio. Accumulation of secretory material is also a major characteristic. By using cDNA microarray analysis we have obtained interesting insights into the molecular mechanisms involved in the prostate hyperplasia. Of particular interest is the significance of reduced apoptosis for the development/progression of the prostate phenotype. This finding was further confirmed by immunohistochemical analysis using two different apoptosis markers. Moreover, in line with the prominent expansion of the stromal compartment were the identified changes in gene expression seen

From: *Contemporary Endocrinology: Genomics in Endocrinology:*
DNA Microarray Analysis in Endocrine Health and Disease
Edited by S. Handwerger and B. Aronow © Humana Press, Totowa, NJ

in the PRL transgenic prostate, suggesting that activation of the stroma is important for the development of the prostate hyperplasia.

Overall, we demonstrate histological and molecular similarities between the prostate hyperplasia of Pb-PRL transgenic mice and human prostate pathology, including both BPH and prostate cancer.

BENIGN PROSTATE HYPERPLASIA

Benign prostatic hyperplasia (BPH) is one of the most common age-associated patho-physiological conditions in the male population (1). BPH is characterized as a slow, progressive enlargement of the prostate gland, which eventually causes obstruction and subsequent problems with urination. Despite of its obvious importance as a major health problem, little is known in terms of the biological processes that contribute to the pathogenesis of BPH. The BPH progression is characterized by hyperplasia of both the stromal and epithelial compartments. Although, the exact etiology of BPH is unknown, it is thought to arise as a result of epithelial–stromal interactions in a certain hormonal milieu (2). Tissue growth depends upon a complex balance between the rates of cell proliferation and cell death (apoptosis). Alterations in the molecular mechanisms regulating these two processes may underlie the abnormal growth of the gland, leading to BPH and even prostate carcinoma. Quantitative analyses, comparing BPH and normal prostatic tissues, have revealed that the total increase of both stromal and epithelial cells is a result of reduced apoptotic activity in parallel with increased cell proliferation (3,4). However, other studies, comparing normal and BPH epithelium only observed an increased cell proliferation but similar levels of apoptosis (5). When calculating the ratio of stromal to epithelial compartments in human BPH tissue, clinical reports have firmly established a dominance of the stromal component in BPH tissue, which is in contrast to the predominance of the epithelial compartments in normal prostate (6–9). Furthermore, in symptomatic BPH patients the stromal/epithelial ratio has been reported to be significantly higher than in asymptomatic patients (7). Together, these observations support the long held contention that BPH arises from changes in the fibromuscular stroma (10). This proposition was based on BPH histopathology features from which McNeal concluded that the prostate stroma undergoes an "embryonic reawakening," resulting in inductive effects of the local stroma, which in turn induces hyperplastic changes in the epithelium through stromal–epithelial interactions.

Androgens are known to be of major importance for the growth, differentiation, and apoptosis of prostatic cells. The development of BPH in man is an androgen-dependent process, as evidenced by the lack of prostate disease in prepubertally castrated males (11). However, nonandrogenic hormones and growth factors are also considered necessary for normal prostatic development and function (12). These factors are also believed to be involved in the development of BPH and prostate carcinoma.

PROLACTIN ACTION ON THE PROSTATE

Prolactin (PRL) is one well-studied factor which, alone or synergistically with androgens, exerts trophic effects and regulatory actions on the secretory and metabolic functions of the prostatic cells, both during glandular development and in the mature gland in rodent and human (12, 13). The PRL receptor (PRLR) is expressed in both human and rodent prostate (14, 15). Interestingly, expression of the PRL ligand has also been demonstrated locally in both human and rodent prostate (15, 16). Moreover, clinical reports have shown that PRL also may be elevated locally in BPH tissue, without any significant increase in serum PRL levels *(17)*.

The role of PRL action in promoting pathological conditions of the prostate has been extensively investigated. Several in vivo studies, in rodents, have demonstrated the growth-promoting effects of PRL on the prostate *(18–20)*. Moreover, PRL has been shown to stimulate growth and significantly increase the cell proliferation rate in human BPH organ cultures, human primary prostate epithelium, and in the androgen refractory human prostate cancer cell lines PC-3 and DU145 *(15, 21–23)*. In parallel to inducing proliferation, PRL has been shown to decrease epithelial apoptosis induced by androgen deprivation, thereby acting as a survival factor *(24, 25)*.

PROSTATE HYPERPLASIA MOUSE MODELS OVEREXPRESSING THE PRL GENE

Male mice ubiquitously overexpressing the rat PRL gene using the metallothionein (Mt) promoter, *Mt-PRL transgenic mice*, develop a dramatic enlargement of the prostate gland with parallel chronic hyperprolactinemia and elevated serum androgen levels *(26)*. However, further studies showed that the elevated levels of circulating androgen in the adult Mt-PRL male mice were not necessary for the abnormal prostate growth *(27)*.

In order to evaluate the role of local PRL action in the prostate, under physiological systemic androgen levels, a new transgenic mouse model was recently generated. In these transgenic mice, *Pb-PRL transgenic mice*, the rat PRL gene is driven by the prostate-specific rat probasin (Pb) promoter. The male mice expressing the transgene developed a significant enlargement of both the dorso-lateral prostate lobe (DLP) and ventral prostate lobe (VP), evident from 10 weeks of age and the prostate enlargement progresses throughout adulthood *(28)*. Histologically, the Pb-PRL prostate displays a significant stromal hyperplasia, ductal dilation, and focal areas of epithelial dysplasia, resulting in a major increase in stromal/epithelial ratio similar to observations made in human BPH tissue *(6–9)*. The glandular dysplastic lesions demonstrated in the Pb-PRL prostate resemble the low-grade prostatic intraepithelial neoplasia (PIN) described in the recently proposed classification of PIN in genetically engineered mice *(29)* and those observations may indicate a potential for malignant disease. However, no progression to high-grade PIN or prostate tumor formation has yet been detected in Pb-PRL transgenic prostate. From this study, it could be concluded that abnormal growth of the Pb-PRL prostate gland occurs primarily at the postpubertal stage and in a setting of normal serum androgen levels, thereby resembling the situation in the adult human prostate.

FUNCTIONAL GENOMICS IN THE STUDY OF THE PROSTATE GLAND

The network of action of different hormones and growth factors on the prostate gland and their involvement in prostate pathophysiology are unquestionably complex. The recent completion of the human *(30)*, and the mouse *(31)* genome sequence, together with the improvement of high-throughput technologies, such as gene expression profiling, will hopefully provide a basis for rational determination of which pathways and molecular targets are appropriate to study further. The unveiling of a detailed genetic map of the main species and models of prostate research promise to dramatically increase our understanding of the genetic basis of prostate disorders together with the basic mechanism of action and involvement of hormones and growth factors for the induction of prostate disease.

GENE EXPRESSION STUDIES IN PROSTATE HYPERPLASIA OF PRL-TRANSGENIC MICE

In order to examine the molecular mechanisms involved in the prostate hyperplasia of Pb-PRL transgenic mice over-expressing the PRL gene specifically in the prostate, cDNA microarray analysis was performed *(32)*. Briefly, a pool of total RNA samples, isolated from prostates of 6-months-old transgenic (*n*= 4) and control mice (*n* = 5), were used to analyze the gene expression by using in-house printed cDNA microarray chip containing about 6250 cDNAs of rat or mouse origin. Three independent experiments were performed for DLP and VP, respectively. To identify the significant differentially expressed transcripts, the data from the three repeated experiments were statistically analyzed using SAM algorithm *(33)*. To the statistically based criteria we added a further requirement based on the absolute changes in expression ratios. Only genes where expression ratios could be measured in all replicates were used in the analysis. Moreover, only genes with average changes of more than 70% were counted as differentially expressed (correspond to a fold change of 1.7). In the present study, 266 genes were found to be differentially expressed (175 upregulated and 91 downregulated) in the enlarged prostate of the Pb-PRL transgenic mice as compared to controls in at least one of the lobe types analyzed *(32)*. An overview of the gene expression study can be viewed in Fig. 1. Of those, 117 genes were commonly differentially expressed in both DLP and VP (95 up and 22 down). The differences between the gene expression of VP and DLP can reflect biological differences and/or being a consequence of the limited detection sensitivity associated with the technique of cDNA microarray. Consequently, no specific attention was paid to these differences.

The differentially expressed genes were categorized based on their known functions. Effects were seen in virtually all cellular processes, including genes involved in cell tissue structure, metabolism, signal transduction, protein modification and protease inhibitors, transport and binding proteins, and immune responses. The two largest functionally categorized groups, of both the DLP and VP, were those of signal transduction and cell tissue structure. The majority of the genes in these two categories were highly expressed in the hyperplastic relative to the control prostate lobes, suggesting that prostate hyperplasia is closely associated with upregulation of genes in these two categories. Moreover, several immune system-associated genes were found to be upregulated, which is in line with previous observations of areas of mild to moderate chronic inflammation, exhibiting stromal mononuclear (primarily lymphocytes and macrophage) infiltrate, in both ventral and lateral Pb-PRL prostate lobes *(28)*. In accordance with these findings, chronic prostatitis has previously been observed in hyperprolactinemic conditions in rodent models *(34, 35)*.

Ten transcripts found to be differentially expressed by using cDNA microarray analysis were subjected to RT-PCR verification using mouse specific primers for the orthologous mouse genes. The same pool of total RNA that was used in the cDNA

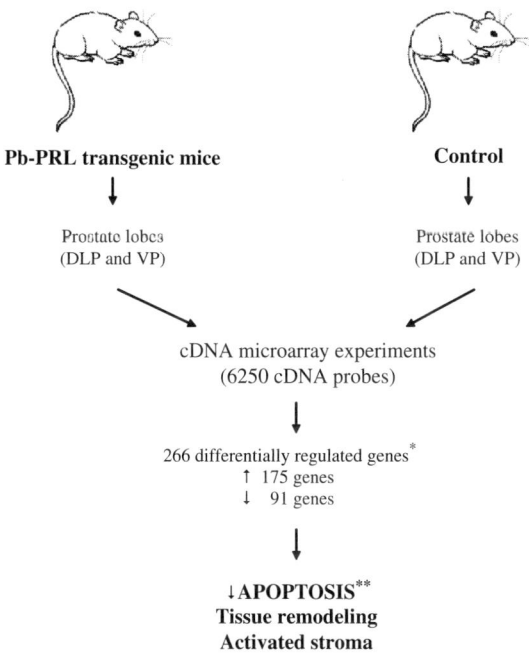

Fig. 1. An overview of the gene expression study used to identify differentially regulated genes in the prostate hyperplasia in Pb-PRL transgenic mice as compared to controls performed by Dillner et al. *(32)*. The analysis was performed in dorso-lateral prostate (DLP) and ventral prostate (VP) and analyzed, separately. The gene expression analysis was performed using in-house printed cDNA microarray chips consisting of 6250 cDNAs. The hyperplastic prostate of the Pb-PRL transgenic mice showed altered expression of more than 250 genes and provide evidence for reduced apoptosis, increased tissue remodeling, and activation of the stroma.

microarray experiments was used in the RT-PCR reactions. As in the cDNA microarray experiments, genes were denoted as differentially regulated if their level of expression was changed by 70% or more. The analysis was performed in triplicates at the lobe level. All the ten transcripts [TIMP-1, SARP-1, Cathepsin B, Cathepsin D, Clusterin, Decorin, Nuclear protein 1 (p8), BOK, CIPAR-1, and EGF] were successfully verified using the RT-PCR to be regulated in the same direction and at similar degree, as the fold changes obtained from the cDNA microarray experiments. This verification indicates the validity of using cDNA microarray technology in combination with statistical methods for identifying differentially regulated genes. Moreover, the verification using mouse-specific primers for the orthologous mouse genes to the cDNA clone of rat origin on the cDNA microarray further supports the use of rat probes to measure orthologous mouse genes.

DIFFERENTIALLY EXPRESSED GENES IN THE ENLARGED PROSTATES OF Pᴃ-PRL TRANSGENIC MICE AS COMPARED TO CONTROLS

Reduced Apoptotic Activity

Interestingly, a number of the identified differentially expressed genes in Pb-PRL transgenic as compared to control prostate gave indication of possible molecular mechanisms involved in the development/progression of the prostate hyperplasia. Among others, a number of apoptosis-related genes were differentially expressed. To further validate these results, immunohistochemical analysis for detection of two established apoptosis markers, activated caspase-3 and presence of ssDNA, were performed. In the Pb-PRL transgenic prostate, there were no immuno detectable levels of either activated caspase-3 or ssDNA. In contrast, distinct clusters of apoptotic epithelial cells were occasionally detected, using either of the two apoptotic markers, in distal regions of the ductal system in all prostate lobes of control samples. These results clearly indicated an overall diminished apoptotic activity in all prostate lobes of the Pb-PRL transgenic mice as compared to controls. In addition, a previously performed molecular characterization of the Mt-PRL transgenic mice, using cDNA representational difference analysis (RDA), did also find differential regulation of apoptosis-associated transcripts *(36)*.

In the list of apoptosis-related genes, that were differentially expressed in the Pb-PRL transgenic hyperplastic prostate as compared to controls, were down-regulation of pro-apoptotic genes [Bok (Bcl-2-related ovarian killer protein) *(37)*, CIPAR-1 (castration induced prostatic apoptosis related protein-1) *(38)* and Nuclear protein 1 *(4)*] in parallel with up-regulation of some genes with anti-apoptotic activity [SARP-1 (Secreted apoptosis-related protein 1) *(39)* and clusterin (also known as testosterone repressed prostate message-2 or sulfated glycoprotein-2)]. Interestingly, the ubiquitous protein clusterin has previously been shown to be upregulated in prostate hyperplasia *(40)*. In this study, Ouyang et al. showed that, as compared to normal prostate, the clusterin gene expression was highly upregulated in both hyperplastic and malignant rat prostate following sex hormone treatment. Moreover, a powerful survival/antiapoptotic activity of clusterin has been noted in vitro in human prostate cancer cells *(41)*.

These results correlate well with the proposed theories behind the etiology of BPH, suggesting an involvement of reduced rate of apoptosis *(42)*, based on the observations of reduced apoptotic activity in BPH tissue as compared to control *(3, 4, 43)*. Other studies have suggested a potential role for the antiapoptotic gene Bcl-2 in BPH. In benign prostatic tissues, Bcl-2 expression is predominantly seen in basal epithelial cells and has been associated with resistance to androgen ablation in BPH epithelium *(44)*.

The importance of PRL in regulation of apoptotic activity in the prostate is yet to be determined. It is possible that PRL acts preferentially as a survival factor rather than a growth factor. Earlier work has demonstrated PRL-induced delay in castration-induced prostatic regression *(45)*. In line with this, PRL has been shown to significantly inhibit the androgen withdrawal-induced apoptosis in DP and LP rat prostate cultures *(24)*. Furthermore, inhibition of activation of transcription factor Stat5, an important part of PRL-receptor mediated JAK/STAT signaling, induces cell death of human prostate cancer cells *(46)*. Interestingly, Li et al. recently demonstrated that activation of Stat5 in human prostate cancer is associated with high histological grade *(47)*. Furthermore, they showed that active Stat5 can distinguish prostate cancer patients whose disease is likely to progress earlier and, therefore, suggest the use of active Stat5 as a marker for selection of more individualized treatment (48).

Moreover, hyperprolactinemia has been shown to induce the synthesis of antiapoptotic Bcl-2 in rat prostate (49), which is in line with induced Bcl-2 expression in human BPH tissue as compared to control (44). Furthermore, PRL has been shown to inhibit TRAIL-induced apoptosis in the PC-3 prostatic cell line (50). TRAIL is a member of the TNF family that is known to induce apoptosis in prostate cells (51). This inhibition was suggested to be mediated by increased phosphorylation of Akt/PKB, a critical regulator of cell survival. The Akt pathway provides the survival signal that involves several proapoptotic proteins such as Bad (52, 53) and possibly other members of the Bcl-2 family. PRL has been shown to possess an antiapoptotic effect in the rat decidua, and this was shown to involve inhibition of caspase-3 activity mediated by the Akt-pathway (54). In that study, PRL was able to downregulate both caspase-3 mRNA levels as well as its activity. Taken together, our results suggest an importance of reduced apoptotic activity in the development of prostate hyperplasia in the PRL transgenic mice. The role of PRL in this regulation needs further investigation, as well as the involvement of diminished apoptotic activity for the development of prostate disease.

Indications of Tissue Remodeling and Activated Stroma

The broad molecular characterization made in the Pb-PRL transgenic mice gave us interesting clues of possible molecular mechanisms of importance for the prostate phenotype, including both the prostate hyperplasia and the dysplastic lesions resembling PIN I and PIN II displayed in this mouse model (32). Interestingly, numerous of the identified differentially expressed genes are directly associated with stromal cells and the extracellular matrix (ECM) proteins that are secreted from stromal cells, including fibronectin, vimentin, laminin, SPARC (osteonectin), calponin, and collagens. This may reflect altered activity of the prostatic stromal cells in the prostates of the transgenic mice. Interestingly, similar results have been reported in the DLP of the so-called ACI rat, a rat model that spontaneously develop prostate cancer (55). In addition to their structural role, ECM proteins have a pronounced influence on tissue remodeling regulating cell growth, differentiation, communication, and migration (56). Moreover, degradation of the ECM, mediated by a variety of proteolytic enzymes, such as matrix metalloproteinases (MMPs) and other proteases, have significant roles in normal and pathological tissue remodeling, including wound repair and tumorigenesis (57). Interestingly, a set of genes, including members of the families of cathepsins, MMPs, and TIMPs, was differentially regulated in the enlarged prostate of Pb-PRL transgenic mice as compared to controls. These genes may serve as potential actors that modify the tissue homeostasis, possibly by changing the reciprocal stromal–epithelial interactions, which eventually contributes to promote the pathological tissue growth observed in the model. The differential regulation of several of these candidate genes has previously been associated with prostate disorders in both humans and animal models indicating a relevance of these genes in the prostate pathogenesis of PRL transgenic mice. In addition to this gene expression study using cDNA microarray analysis, the gene expression analysis of the prostate of the Mt-PRL transgenic mice using the method of cDNA-RDA also indicated a potential importance and possible activation of the stromal compartment for the development and/or progression of the prostate phenotype *(36)*.

The histological resemblance of increased stroma/epithelial ratio, seen in both human BPH and the prostate hyperplasia of the PRL transgenic mice models, further supports the "embryonic reawakening theory" of BPH etiology *(10)*, where the prostate hyperplasia is proposed to be a disease of the prostatic stroma. This theory emphasizes that BPH represents a reawakening of the embryonic and inductive potential of prostatic stroma, which in turn induces hyperplastic changes in the epithelium through stromal–epithelial interactions. Several studies have proved the importance of the epithelial–stromal interactions both in normal prostate development as well as the influence of abnormal reciprocal interaction between epithelial cells and the embryonic mesenchyme or adult stroma in the progression of neoplastic growth in the human prostate gland *(58)*. Although the exact mechanisms of such tissue interactions are not fully understood, there is growing evidence that they may operate through cell–ECM interactions, remodeling of ECM and auto/paracrine growth factors *(56)*.

The phenomenon of tumorigenesis promotion by an activated stroma (generation of a so-called "reactive stroma") has previously been associated with prostate pathology and other human cancers *(59)*. The reactive stroma is characterized by ECM remodeling, elevated protease activity, increased angiogenesis, and an influx of inflammatory cells *(59)*. The list of differentially regulated genes in the Pb-PRL transgenic prostate *(32)* has much in common with the processes involved in what is in the literature described as characteristics of the reactive stroma [reviewed in *(59)*]. The upregulation of vimentin together with a downregulation of desmin suggests a myofibroblastic-like nature of the stroma cells, which is in line with the described phenotype of reactive stroma. In addition, reactive stroma cells typically express high levels of ECM components, such as collagen type I and III, fibronectin, and proteoglycans, as well as proteases that degrade the ECM observations that are in accordance with our present results. One possible hypothesis might be that PRL influences the initial induction of prostatic hyperplasia by modulating the stromal–epithelial interaction, resulting in an activation of the stroma with the phenotypical features of reactive stroma, which fits well with the stromal expansion as dominant feature of human BPH *(6, 8, 9)*.

CONCLUSION

The prostate phenotype of the two PRL transgenic mouse models (Mt-PRL and Pb-PRL transgenic mice) shares interesting histological characteristics with human BPH. The Pb-PRL transgenic model does resemble the situation in BPH better than the Mt-PRL transgenic model since the prostate hyperplasia develops in a mature gland under normophysiological androgen levels *(27, 28)*. This in combination with the molecular and histological similarities between the Pb-PRL transgenic model and human prostate pathology illustrates the potential further use of this model as a valuable tool in the study of BPH.

The use of differential gene expression technologies in our studies has enabled us to find molecular similarities between the prostate hyperplasia of the PRL transgenic mouse models and human prostate disorders. Of particular interest is the potential significance of reduced apoptosis for the development/progression of the prostate phenotype. Another interesting observation is the importance and possible activation of the stromal compartment for the development and/or progression of the prostate phenotype of the PRL transgenic mice. There is striking resemblance of the molecular pattern obtained in the PRL transgenic prostate to that previously described in the literature as the "embryonic reawakening theory" of BPH etiology and the theory of "reactive stroma" in prostate cancer etiology.

Overall, the differentially expressed genes identified in this study show many molecular similarities between the prostate hyperplasia of PRL-transgenic mice and human prostate pathology, including both BPH and prostate cancer.

REFERENCES

1. Boyle, P., Maisonneuve, P. & Steg, A. (1996) *J Urol* **155,** 176–80.
2. Oesterling, J. E. (1996) *Prostate Suppl* **6,** 67–73.
3. Kyprianou, N., Tu, H. & Jacobs, S. C. (1996) *Hum Pathol* **27,** 668–75.
4. Claus, S., Berges, R., Senge, T. & Schulze, H. (1997) *J Urol* **158,** 217–21.
5. Conner, P. & Fried, G. (1998) *Acta Obstet Gynecol Scand* **77,** 249–62.
6. Doehring, C. B., Sanda, M. G., Partin, A. W., Sauvageot, J., Juo, H., Beaty, T. H., Epstein, J. I., Hill, G. & Walsh, P. C. (1996) *Urology* **48,** 650–3.
7. Shapiro, E., Becich, M. J., Hartanto, V. & Lepor, H. (1992) *J Urol* **147,** 1293–7.
8. Robert, M., Costa, P., Bressolle, F., Mottet, N. & Navratil, H. (1995) *Br J Urol* **75,** 317–24.
9. Deering, R. E., Bigler, S. A., King, J., Choongkittaworn, M., Aramburu, E. & Brawer, M. K. (1994) *Urology* **44,** 64–70.
10. McNeal, J. E. (1978) *Invest Urol* **15,** 340–5.
11. Bosch, R. J. (1991) *Eur Urol* **20,** 27–30.
12. Reiter, E., Hennuy, B., Bruyninx, M., Cornet, A., Klug, M., McNamara, M., Closset, J. & Hennen, G. (1999) *Prostate* **38,** 159–65.
13. Costello, L. C. & Franklin, R. B. (1994) *Prostate* **24,** 162–6.
14. Nevalainen, M. T., Valve, E. M., Ingleton, P. M. & Harkonen, P. L. (1996) *Endocrinology* **137,** 3078–88.
15. Nevalainen, M. T., Valve, E. M., Ingleton, P. M., Nurmi, M., Martikainen, P. M. & Harkonen, P. L. (1997) *J Clin Invest* **99,** 618–27.
16. Nevalainen, M. T., Valve, E. M., Ahonen, T., Yagi, A., Paranko, J. & Harkonen, P. L. (1997) *FASEB J* **11,** 1297–307.
17. Ron, M., Fich, A., Shapiro, A., Caine, M. & Ben-David, M. (1981) *Urology* **17,** 235–7.
18. Negro-Vilar, A., Saad, W. A. & McCann, S. M. (1977) *Endocrinology* **100,** 729–37.
19. Schacht, M. J., Niederberger, C. S., Garnett, J. E., Sensibar, J. A., Lee, C. & Grayhack, J. T. (1992) *Prostate* **20,** 51–8.
20. Mori, T. & Nagasawa, H. (1984) *Acta Anat (Basel)* **120,** 180–4.
21. de Launoit, Y., Kiss, R., Jossa, V., Coibion, M., Paridaens, R. J., De Backer, E., Danguy, A. J. & Pasteels, J. L. (1988) *Prostate* **13,** 143–53.
22. Janssen, T., Kiss, R. & Schulman, C. (1995) *Acta Urol Belg* **63,** 7–14.
23. Janssen, T., Darro, F., Petein, M., Raviv, G., Pasteels, J. L., Kiss, R. & Schulman, C. C. (1996) *Cancer* **77,** 144–9.
24. Ahonen, T. J., Harkonen, P. L., Laine, J., Rui, H., Martikainen, P. M. & Nevalainen, M. T. (1999) *Endocrinology* **140,** 5412–21.
25. Assimos, D., Smith, C., Lee, C. & Grayhack, J. T. (1984) *Prostate* **5,** 589–95.
26. Wennbo, H., Kindblom, J., Isaksson, O. G. & Tornell, J. (1997) *Endocrinology* **138,** 4410–5.
27. Kindblom, J., Dillner, K., Ling, C., Tornell, J. & Wennbo, H. (2002) *Prostate* **53,** 24–33.
28. Kindblom, J., Dillner, K., Sahlin, L., Robertson, F., Ormandy, C., Tornell, J. & Wennbo, H. (2003) *Endocrinology* **144,** 2269–78.
29. Park, J. H., Walls, J. E., Galvez, J. J., Kim, M., Abate-Shen, C., Shen, M. M. & Cardiff, R. D. (2002) *Am J Pathol* **161,** 727–35.

30. Venter, J. C., Adams, M. D., Myers, E. W., Li, P. W., Mural, R. J., Sutton, G. G., Smith, H. O., Yandell, M., Evans, C. A., Holt, R. A., Gocayne, J. D., Amanatides, P., Ballew, R. M., Huson, D. H., Wortman, J. R., Zhang, Q., Kodira, C. D., Zheng, X. H., Chen, L., Skupski, M., Subramanian, G., Thomas, P. D., Zhang, J., Gabor Miklos, G. L., Nelson, C., Broder, S., Clark, A. G., Nadeau, J., McKusick, V. A., Zinder, N., Levine, A. J., Roberts, R. J., Simon, M., Slayman, C., Hunkapiller, M., Bolanos, R., Delcher, A., Dew, I., Fasulo, D., Flanigan, M., Florea, L., Halpern, A., Hannen-halli, S., Kravitz, S., Levy, S., Mobarry, C., Reinert, K., Remington, K., Abu-Threideh, J., Beasley, E., Biddick, K., Bonazzi, V., Brandon, R., Cargill, M., Chandramouliswaran, I., Charlab, R., Chaturvedi, K., Deng, Z., Di Francesco, V., Dunn, P., Eilbeck, K., Evangelista, C., Gabrielian, A. E., Gan, W., Ge, W., Gong, F., Gu, Z., Guan, P., Heiman, T. J., Higgins, M. E., Ji, R. R., Ke, Z., Ketchum, K. A., Lai, Z., Lei, Y., Li, Z., Li, J., Liang, Y., Lin, X., Lu, F., Merkulov, G. V., Milshina, N., Moore, H. M., Naik, A. K., Narayan, V. A., Neelam, B., Nusskern, D., Rusch, D. B., Salzberg, S., Shao, W., Shue, B., Sun, J., Wang, Z., Wang, A., Wang, X., Wang, J., Wei, M., Wides, R., Xiao, C., Yan, C., Yao A, Ye J, Zhan M, Zhang W, Zhang H, Zhao Q, Zheng L, Zhong F, Zhong W, Zhu S, Zhao S, Gilbert D, Baumhueter S, Spier G, Carter C, Cravchik A, Woodage T, Ali F, An H, Awe A, Baldwin D, Baden H, Barnstead M, Barrow I, Beeson K, Busam D, Carver A, Center A, Cheng ML, Curry L, Danaher S, Davenport L, Desilets R, Dietz S, Dodson K, Doup L, Ferriera S, Garg N, Gluecksmann A, Hart B, Haynes J, Haynes C, Heiner C, Hladun S, Hostin D, Houck J, Howland T, Ibegwam C, Johnson J, Kalush F, Kline L, Koduru S, Love A, Mann F, May D, McCawley S, McIntosh T, McMullen I, Moy M, Moy L, Murphy B, Nelson K, Pfannkoch C, Pratts E, Puri V, Qureshi H, Reardon M, Rodriguez R, Rogers YH, Romblad D, Ruhfel B, Scott R, Sitter C, Smallwood M, Stewart E, Strong R, Suh E, Thomas R, Tint NN, Tse S, Vech C, Wang G, Wetter J, Williams S, Williams M, Windsor S, Winn-Deen E, Wolfe K, Zaveri J, Zaveri K, Abril JF, Guigó R, Campbell MJ, Sjolander KV, Karlak B, Kejariwal A, Mi H, Lazareva B, Hatton T, Narechania A, Diemer K, Muruganujan A, Guo N, Sato S, Bafna V, Istrail S, Lippert R, Schwartz R, Walenz B, Yooseph S, Allen D, Basu A, Baxendale J, Blick L, Caminha M, Carnes-Stine J, Caulk P, Chiang YH, Coyne M, Dahlke C, Mays A, Dombroski M, Donnelly M, Ely D, Esparham S, Fosler C, Gire H, Glanowski S, Glasser K, Glodek A, Gorokhov M, Graham K, Gropman B, Harris M, Heil J, Henderson S, Hoover J, Jennings D, Jordan C, Jordan J, Kasha J, Kagan L, Kraft C, Levitsky A, Lewis M, Liu X, Lopez J, Ma D, Majoros W, McDaniel J, Murphy S, Newman M, Nguyen T, Nguyen N, Nodell M, Pan S, Peck J, Peterson M, Rowe W, Sanders R, Scott J, Simpson M, Smith T, Sprague A, Stockwell T, Turner R, Venter E, Wang M, Wen M, Wu D, Wu M, Xia A, Zandieh A, Zhu X. 1304–51. (2001) *Science* **291,**
31. Waterston, R. H., Lindblad-Toh, K., Birney, E., Rogers, J., Abril, J. F., Agarwal, P., Agarwala, R., Ainscough, R., Alexandersson, M., An, P., Antonarakis, S. E., Attwood, J., Baertsch, R., Bailey, J., Barlow, K., Beck, S., Berry, E., Birren, B., Bloom, T., Bork, P., Botcherby, M., Bray, N., Brent, M. R., Brown, D. G., Brown, S. D., Bult, C., Burton, J., Butler, J., Campbell, R. D., Carninci, P., Cawley, S., Chiaromonte, F., Chinwalla, A. T., Church, D. M., Clamp, M., Clee, C., Collins, F. S., Cook, L. L., Copley, R. R., Coulson, A., Couronne, O., Cuff, J., Curwen, V., Cutts, T., Daly, M., David, R., Davies, J., Delehaunty, K. D., Deri, J., Dermitzakis, E. T., Dewey, C., Dickens, N. J., Diekhans, M., Dodge, S., Dubchak, I., Dunn, D. M., Eddy, S. R., Elnitski, L., Emes, R. D., Eswara, P., Eyras, E., Felsenfeld, A., Fewell, G. A., Flicek, P., Foley, K., Frankel, W. N., Fulton, L. A., Fulton, R. S., Furey, T. S., Gage, D., Gibbs, R. A., Glusman, G., Gnerre, S., Goldman, N., Goodstadt, L., Grafham, D., Graves, T. A., Green, E. D., Gregory, S., Guigo, R., Guyer, M., Hardison, R. C., Haussler, D., Hayashizaki, Y., Hillier, L. W., Hinrichs, A., Hlavina, W., Holzer, T., Hsu, F., Hua, A., Hubbard, T., Hunt, A., Jackson, I., Jaffe, D. B., Johnson, L. S., Jones, M., Jones, T. A., Joy, A., Kamal, M., Karlsson, E. K., Karolchik D, Kasprzyk A, Kawai J, Keibler E, Kells C, Kent WJ, Kirby A, Kolbe DL, Korf I, Kucherlapati RS, Kulbokas EJ, Kulp D, Landers T, Leger JP, Leonard S, Letunic I, Levine R, Li J, Li M, Lloyd C, Lucas S, Ma B, Maglott DR, Mardis ER, Matthews L, Mauceli E, Mayer JH, McCarthy M, McCombie WR, McLaren S, McLay K, McPherson JD, Meldrim J, Meredith B, Mesirov JP, Miller W, Miner TL, Mongin E, Montgomery KT, Morgan M, Mott R, Mullikin JC, Muzny DM, Nash WE, Nelson JO, Nhan MN, Nicol R, Ning Z, Nusbaum C, O'Connor MJ, Okazaki Y, Oliver K, Overton-Larty E, Pachter L, Parra G, Pepin KH, Peterson J, Pevzner P, Plumb R, Pohl CS, Poliakov A, Ponce TC, Ponting CP, Potter S, Quail M, Reymond A,

Roe BA, Roskin KM, Rubin EM, Rust AG, Santos R, Sapojnikov V, Schultz B, Schultz J, Schwartz MS, Schwartz S, Scott C, Seaman S, Searle S, Sharpe T, Sheridan A, Shownkeen R, Sims S, Singer JB, Slater G, Smit A, Smith DR, Spencer B, Stabenau A, Stange-Thomann N, Sugnet C, Suyama M, Tesler G, Thompson J, Torrents D, Trevaskis E, Tromp J, Ucla C, Ureta-Vidal A, Vinson JP, Von Niederhausern AC, Wade CM, Wall M, Weber RJ, Weiss RB, Wendl MC, West AP, Wetterstrand K, Wheeler R, Whelan S, Wierzbowski J, Willey D, Williams S, Wilson RK, Winter E, Worley KC, Wyman D, Yang S, Yang SP, Zdobnov EM, Zody MC, Lander ES. (2002) *Nature* **420**, 520–62.

32. Dillner, K., Kindblom, J., Flores-Morales, A., Shao, R., Tornell, J., Norstedt, G. & Wennbo, H. (2003) *Endocrinology* **144**, 4955–66.

33. Tusher, V. G., Tibshirani, R. & Chu, G. (2001) *Proc Natl Acad Sci U S A* **98**, 5116–21.

34. Stoker, T. E., Robinette, C. L., Britt, B. H., Laws, S. C. & Cooper, R. L. (1999) *Biol Reprod* **61**, 1636–43.

35. Stoker, T. E., Robinette, C. L. & Cooper, R. L. (1999) *Toxicol Sci* **52**, 68–79.

36. Dillner, K., Kindblom, J., Flores-Morales, A., Pang, S. T., Tornell, J., Wennbo, H. & Norstedt, G. (2002) *Prostate* **52**, 139–49.

37. Hsu, S. Y., Kaipia, A., McGee, E., Lomeli, M. & Hsueh, A. J. (1997) *Proc Natl Acad Sci U S A* **94**, 12401–6.

38. Bruyninx, M., Hennuy, B., Cornet, A., Houssa, P., Daukandt, M., Reiter, E., Poncin, J., Closset, J. & Hennen, G. (1999) *Endocrinology* **140**, 4789–99.

39. Melkonyan, H. S., Chang, W. C., Shapiro, J. P., Mahadevappa, M., Fitzpatrick, P. A., Kiefer, M. C., Tomei, L. D. & Umansky, S. R. (1997) *Proc Natl Acad Sci U S A* **94**, 13636–41.

40. Ouyang, X. S., Wang, X., Lee, D. T., Tsao, S. W. & Wong, Y. C. (2001) *Carcinogenesis* **22**, 965–73.

41. Miyake, H., Nelson, C., Rennie, P. S. & Gleave, M. E. (2000) *Cancer Res* **60**, 170–6.

42. Thompson, T. C. & Yang, G. (2000) *Prostate Suppl* **9**, 25–8.

43. Xia, S. J., Xu, C. X., Tang, X. D., Wang, W. Z. & Du, D. L. (2001) *Asian J Androl* **3**, 131–4.

44. Cardillo, M., Berchem, G., Tarkington, M. A., Krajewski, S., Krajewski, M., Reed, J. C., Tehan, T., Ortega, L., Lage, J. & Gelmann, E. P. (1997) *J Urol* **158**, 212–6.

45. Kolbusz, W. E., Lee, C. & Grayhack, J. T. (1982) *J Urol* **127**, 581–4.

46. Ahonen, T. J., Xie, J., LeBaron, M. J., Zhu, J., Nurmi, M., Alanen, K., Rui, H. & Nevalainen, M. T. (2003) *J Biol Chem* **278**, 27287–92.

47. Li, H., Ahonen, T. J., Alanen, K., Xie, J., LeBaron, M. J., Pretlow, T. G., Ealley, E. L., Zhang, Y., Nurmi, M., Singh, B., Martikainen, P. M. & Nevalainen, M. T. (2004) *Cancer Res* **64**, 4774–82.

48. Li, H., Zhang, Y., Glass, A., Zellweger, T., Gehan, E., Bubendorf, L., Gelmann, E. P. & Nevalainen, M. T. (2005) *Clin Cancer Res* **11**, 5863–8.

49. Van Coppenolle, F., Slomianny, C., Carpentier, F., Le Bourhis, X., Ahidouch, A., Croix, D., Legrand, G., Dewailly, E., Fournier, S., Cousse, H., Authie, D., Raynaud, J. P., Beauvillain, J. C., Dupouy, J. P. & Prevarskaya, N. (2001) *Am J Physiol Endocrinol Metab* **280**, E120–9.

50. Ruffion, A., Al-Sakkaf, K. A., Brown, B. L., Eaton, C. L., Hamdy, F. C. & Dobson, P. R. (2003) *Eur Urol* **43**, 301–8.

51. Munshi, A., Pappas, G., Honda, T., McDonnell, T. J., Younes, A., Li, Y. & Meyn, R. E. (2001) *Oncogene* **20**, 3757–65.

52. Datta, S. R., Dudek, H., Tao, X., Masters, S., Fu, H., Gotoh, Y. & Greenberg, M. E. (1997) *Cell* **91**, 231–41.

53. del Peso, L., Gonzalez-Garcia, M., Page, C., Herrera, R. & Nunez, G. (1997) *Science* **278**, 687–9.

54. Tessier, C., Prigent-Tessier, A., Ferguson-Gottschall, S., Gu, Y. & Gibori, G. (2001) *Endocrinology* **142**, 4086–94.

55. Reyes, I., Reyes, N., Iatropoulos, M., Mittelman, A. & Geliebter, J. (2005) *Prostate* **63**, 169–86.

56. Liotta, L. A. & Kohn, E. C. (2001) *Nature* **411**, 375–9.

57. Mauch, C., Krieg, T. & Bauer, E. A. (1994) *Arch Dermatol Res* **287**, 107–14.

58. Hayward, S. W., Rosen, M. A. & Cunha, G. R. (1997) *Br J Urol* **79 Suppl 2**, 18–26.

59. Tuxhorn, J. A., Ayala, G. E. & Rowley, D. R. (2001) *J Urol* **166**, 2472–83.

INDEX

A

17-Aalpha- monooxygenase (hydroxylase)
(*CYP17A1*), 135
ACCβ downregulation, 55
Adipogenesis and microarray technology, 241–249
Adipose tissue, 240–241
Adipose tissue of type 2 diabetes gene transcript levels,
199–201
Adrenal adenomas and carcinomas genes expression,
137–140
Adrenal hyperplasia genes expression, 136–137
Adrenal specific transcripts, 135
Adrenocortical tumorigenesis, 131–132
Affymetrix Analysis Suit V 5.0, 72
Affymetrix GeneChip®, 192
Affymetrix gene expression analysis steps, 188
Affymetrix HG-U95Av2 arrays, 133, 266
Affymetrix human HG-U133+2 oligonucleotide
microarray, 134
AIS *See* Androgen insensitivity syndrome
Alendronate bisphosphonates drugs, 26
Alexa fluors 555 and 647, 189
Alpha-2-HS-glycoprotein, 56
Amphiregulin epidermal growth factor (EGF)-like
ligands, 29–30
Androgen-dependent signaling pathways, 92
Androgen insensitivity syndrome, 85, 87
Androgen-mediated protein metabolism, 100
Androgen receptor
domains and signaling, 84
and prostate cancer, 89
role in male sexual dimorphism and function, 85–87
stimulation in stromal compartment, 88
targets of, 98–101
Androgen response elements, 85
Androgen-responsive pathway in LNCaP cells, 99
Antiresorptives drugs, 26
Applied Biosystems Genome Survey arrays, 190
Applied Biosystems 5′ nuclease assay, 202
ARE *See* Androgen response elements
ArrayAssist payware versions, 193
ArrayExpress in Britain, 14
AR target genes from LNCaP model in prostate cancer,
102–104
ASCL3 genes, 99
Asoprisnil receptor modulators, 68
Axin 1 (*AXIN1*), 137
Axon GenePix commercial scanning software, 192

B

BD Biosciences Clontech Atlas™ Microarrays, 189
Benign prostatic hyperplasia, 85, 272
Benjamini-Hochberg false discovery rate testing, 7
3 Beta- and steroid delta-isomerase 2 (*HSD3B2*), 135
11-Beta-hydroxylase (*CYP11B1*), 135
BioConductor package, 12
Bioconductor tools for genomic data, 13
Bioinformatics and ChIP-chip experiments,
119–120
BioRad Icycler, 202
BMD. *See* Bone mineral density
Bone mineral density, 26
Bone remodeling, 25–26
Bone resorption rate, 26
BPH. *See* Benign prostatic hyperplasia
BRB-ArrayTools, 73
Breast cancers
c-MYC and ERα coregulation in, 123
and ERα signaling network in, 124–125
FoxA1 and ERα expression in, 123
glucocorticoid receptor (GR)activation of, 166
BTBR *ob/ob* diabetes susceptible mice, 253

C

CAIS patients with 46XY genotypes, 87
Calpain-10, 194
cAMP-GEF/Rap1/B-raf pathway, 27
cAMP-guanine nucleotide exchange factor, 27
cAMP-response element (CRE)-binding protein
(CREB), 31
Cancer subtype-specific signature profiling, 6
CAR. *See* Constitutive androstane receptor
Carney complex (CNC), 136
Casein kinase 1 (*CSNK1E*), 137
Casodex®, 89
Catenin-β1 (*CTNNB1*), 137
Catenin (cadherin-associated protein)- like 1
(*CTNNAL1*), 137
β-Catenin/T-cell factor complex, 140
CBR-1 agonist WIN inhibition by decidualization of
decidual fibroblasts, 158
CCAAT/enhancer binding protein (C/EBPβ), 52
CDB-2914 receptor modulators, 68
cDNA arrays, 5
cDNA microarray analysis steps, 187

cDNA representational difference analysis, 276
Chromatin immunoprecipitation analysis, 92
Chromatin immunoprecipitation analysis (ChIP-chip)
    applications of, 118–120
    diagram of, 117
Chromogranin B *(CHGB),* 135
cis-regulatory element CETS-1, 124
Class I cytokine receptor superfamily, 42
Cluster freeware applications, 193
CodeLink™ Bioarrays, 190
Commercial microarrays, 70
Comparative genomic hybridization (CGH)
    experiments, 131
Constitutive androstane receptor, 57
Cushing's syndrome, 136
Custom-spotted arrays, 187–188
Cyclin D1, 30
Cyclin-dependent kinases (CDKs), 30
CYP11B2 and *LHR* expression levels, 140
Cytochrome oxidase IV (COX IV) expression, 202
Cytotoxic cancer chemotherapeutic drugs, 30

D

Database for Annotation, Visualization, and Integrated
    Discovery (DAVID), 13
Ddehydroepiandrosterone, 220
Decidual cells decidualization, 147
    biology of, 148
    microarray studies of, 149, 152, 157
Decidual fibroblast differentiation, reprogramming
    scheme, 150–151
Decidualization-specific marker genes
    ETS1 oligo effect on, 158
    FOXO1A siRNA effects on, 159
Dehydroepiandrosterone (DHEA) sulfate, 135
Delta-like 1 homolog *(DLK1),* 135
Dex-regulated genes, 175
DHCR24 genes, 99
DHEA. *See* Ddehydroepiandrosterone
5-α-Dihydrotestosterone (DHT), 84
Dimensionality-reduction methods, 73
Distal Erα binding domains transcriptional
    enhancers, 122
DNA microarray analysis
    and global gene expression profiles study, 31–35
    and PTH in oteoblastic cells gene regulation, 27–28
    and PTH treatment of osteoporosis, 28–31
Dose-responsiveness demonstration, 7

E

ECM and auto/paracrine growth factors remodeling, 278
EGF receptor (EGFR/ErbB1), 29
Embryonic reawakening theory of BPH etiology, 278
Endocrine factor responsive gene, 6
Endocrine Pancreas Consortium, 186
Endocrinology and gene expression profiling studies, 5–7
Enlarged prostate, gene expression, 276–277

Enzo BioArray high-yield RNA transcript labeling
    kit, 71
Epac activation, 27
ERα binding and ERα-mediated transcription, 123
ERα-mediated transcription, 116
ERα signaling pathway, 123
ER α target gene identification and characterization,
    120–123
ERKs. *See* Extracellular signal-regulated kinases
Estrogen oestrogenic compounds drugs, 26
Estrogen receptor (ER) activation, 84
Estrogen-responsive genes in breast cancer, 123–124
European Bioinformatics Institute, 14–16
Expression signature, 6
Extracellular signal-regulated kinases, 27

F

False discovery rate (FDR), 193
FAS upregulation, 55
Fetuin. *See* Alpha-2-HS-glycoprotein
Fisher's Exact test, 13
Focal Adhesion Kinase (FAK), 45
Follicle stimulating hormone, 220
Forkhead protein FoxA1, 123
FSH. *See* Follicle stimulating hormone
Functional androgen response elements, 96
Functional Genomics of the Developing Endocrine
    Pancreas, 186

G

Gαq/phospholipase C (PLC)/protein kinase C (PKC)
    pathway, 27
Gαs/cAMP/protein kinase A (PKA) pathway, 27
Gastric inhibiting polypeptide, 136
GATA6 gene transcription, 229
Genbank Refseq database, 32
GeneChip® array from Affymetrix, 186
GeneChip® Operating System software, 192
Gene Expression Omnibus, 14–15
GeneFilters array, 189
Gene-gene interactions, 13
Gene ontology (GO), 28
Gene Ontology project (GO), 74
Genepix software, 72
Gene regulation by transcription factors combinatorial
    theory, 123
Genes co-functionality, 11
Gene Set Enrichment Analysis, 195
Gene Set Enrichment Assay, 194
GeneSifter commercial tool, 13
GeneSpring commercial tool, 13
GeneSpring GX 7.3 software package, 90
GeneSpring payware versions, 193
GeneSpring software package, 226
GenMAPP software, 74
GEO *See* Gene Expression Omnibus
GIP *See* Gastric inhibiting polypeptide

Global gene expression
  in leiomyoma, 69
  in myometrial smooth muscle cells, 69
Glucocorticoid receptor element (GRE), 166
Glucocorticoid receptor (GR), 52, 166
  induced cell survival, 174
  mediated breast cancer cell survival pathways,
    173–175
Glucocorticoids steroid hormones, 166
Glycogen synthase kinase-3β *(GSK3B)*, 137
Glycogen synthetase, 194
Gonadotropin releasing hormone agonists (GnRHa)
  therapy, 68
GOTree Machine, 194
G-protein-coupled receptor, 27
G-protein-coupled receptor's expression, 134
Growth hormone (GH)
  in liver and skeletal muscle, genes with DNA-binding
    activity, 46–47
  physiological effect of, 42
  regulated mechanism, 44
  secretion patterns, 44
  and STAT5 interaction, 52
  transcriptional actions, 44–45
  treatment effect on hypophysectomized male rats in
    hepatic expression of genes, 55
Growth hormone (GH) receptor (GHR), 42
  cytoplasmic domain, 42
  expression regulation, 43
  inactivating mutations, 42
GSEA *See* Gene Set Enrichment Analysis; Gene Set
  Enrichment Assay

H

HATs *See* Histone acetyltransferases
hGH replacement therapy on hypophysectomized (Hx)
  animals, 53
Histone acetyltransferases, 52
hMSCs *See* Human mesenchymal stem cells
Hormone activation and microarray data analysis,
  167–173
Hormone-induced expression alterations, 7–8
Hormone treatment, 7
Human Affymetrix GeneChip arrays, 71
Human decidualization *in vitro* models, 148
Human endometrial stromal cells, decidual cells, 147
Human mesenchymal stem cells, 240
  gene families of, 246
  models of, 245–247
Human obesity and insulin resistance studies,
  257–258
Human Survey Microarray version 2.0, 190
Human thyroid follicles suspension culture,
  208–210
Human uterine decidualization regulation, 159
3β-Hydroxysteroid dehydrogenase *(HSD3B)*, 135
Hypergeometric Distribution test, 13
Hypophysectomized male rats

GH regulation of hepatic gene expression in, 52–58
hepatic genes regulation by GH treatment in, 47–52

I

Ibandronate bisphosphonates drugs, 26
IGF-II and IGFBP-1 expression, 161
IGF-1 knockout mice, 42
IGF1 signaling, 6
Illumina platforms, 190
Incyte Genomics libraries, 186
Ingenuity Pathway Analysis, 194
Insulin-like growth factor binding protein-5
  (IGFBP-5), 34
Insulin-like growth factor 2 *(IGF2)*, 135
Insulin Receptor Substrate 1 (IRS-1), 45, 194
Insulin resistance syndrome, 11
Iodide effects on gene expression in thyroid follicles, 210
IP3K/Akt signaling, 232
IR-dependent signal transduction, 221

J

Jagged-1 expression, 30
Janus Kinase 2 (JAK2), 42
JNK-signaling pathway, 180
c-Jun *N*-terminal kinase (JNK), 178

K

KEGG *See* Kyoto Encyclopedia of Genes and Genomes
KEGG metabolic processes, 14, 75
Kennedy's disease, 87–88
KIAA0018 transcripts, 136
Kyoto Encyclopedia of Genes and Genomes, 74, 194

L

Laron syndrome, 42
Laser Capture Microdissection, 191
Lawrence Livermore National Lab cDNA clones, 189
LCM *See* Laser Capture Microdissection
Leiomyoma
  microarray platforms and analysis in research for, 78
  sampling from tumors for, 68–69
Leiomyoma isolation and culture
  data acquisition and analysis, 72–74
  gene expression profiling, 69
  ontology assessment, 74–75
  RNA preparation and hybridization, 70–72
  verification of, 75–77
Leiomyomas benign uterine tumors, 67–68
Leukemia inhibitory factor, 149
LIF *See* Leukemia inhibitory factor
LNCaP cells, hierarchical cluster analysis of genes, 91
LNCaP model for study of prostate cancer, 89
Luetinizing hormone (LH), 220

M

Mammalian GeneFilters ®, 189
Massive macronodular adrenocortical disease, 133
MCF10A-Myc cells, 180
MCF7 breast cancer cells, 120
McGill's Androgen Receptor Gene Mutations Database, 87
MCP-1 *See* Monocyte chemoattracant protein-1
Methimazole and amiodarone effects on thyroid follicles gene expression, 211–212
MIAMExpress, 15–16
Microarray assays, design of experiments, 7–10
Microarray-based gene expression studies, 6
Microarray data analysis
 complexicity and selection of stastical methods, 11–13
 data acquisition and quality assessment, 192–194
 downloadable datasets of, 15
 and extraction of biological information, 13–14
 limitations of, 16
 principles of, 10–11
 storage, standrads and exchange, 14–16
 valdation of, 16
 web sites of, 73
Microarray gene chip, 5
Microarray hybridization, 266
β-Microglobulin (*B2M*), 135
Mifepristone receptor modulators, 68
Mitogen-activated protein kinase (Mapk), 30, 45
MMAD *See* Massive macronodular adrenocortical disease
Molecular chaperone HSP22, 88
Monocyte chemoattracant protein-1, 161
21-Monooxygenase (hydroxylase) *(CYP21A2),* 135
Mouse PancChip, 186
mRNA expression and Leiomyoma, 75–76
Mrp2 gene, 55
Multiple endocrine neoplasia (MEN1)-associated neuroendocrine tumors, 11
Multiple neoplasia syndrome, 136
c-MYC transcription factor, 123
Myometrial smooth muscle cells
 data acquisition and analysis, 72–74
 gene expression profiling, 69
 ontology assessment, 74–75
 RNA preparation and hybridization, 70–72
 verification of, 75–77
Myosin light polypeptide 6 *(MYL6),* 135

N

NAD dehydrogenase IV expression, 202
National Institute of Diabetes and Digestive and Kidney Diseases, 186
National Institutes of Aging Array Analysis, 226
Nephroblastoma overexpressed gene, 135
NFκB-ligand (RANKL) receptor activator, 31

NF pituitary tumors and normal pituitary glands analysis study
 methods of, 265–268
 patients and tumor characterization, 264
 reagents in, 265
NGFIB orphan nuclear receptor, 136
n-Myc interactor (Nmi), 52
Nonfunctioning (NF) pituitary adenomas, 264
Normal human adrenals global expression profile, 134–136
Notch receptors, 30
NOV See Nephroblastoma overexpressed gene
Noxa and *DAB2* putative ELK-1 target genes, 180

O

Obese rat model and rodents, 257
Obesity, white adipose tissue and microarray technology, 239, 249–258
ob/ob and DIO mice models and rodents, 249–250, 253, 256
Off-the-shelf PCR-based gene expression assays, 190
Oligonucleotide arrays, 5
Oligonucleotide Human Genome U95Av2, 265
Open source R statistical programming language, 12–13
Osteoblasts bone-forming cells, 26
Osteoporosis metabolic bone disease, 25–26

P

PancChip cDNA microarray, 186
PAR-2 *See* Protease-activated receptor-2
Parathyroid hormone (PTH) anabolic agent, 26
Partek commercial tool, 13
Pb-PRL transgenic mice gene expression study, 273, 275
P21^cipl, 30
PCO, overian phenotype genes, 229
PCOS *See* Polycystic ovary syndrome
PCR-amplified cDNA clones, 186
Peptide hormone relaxin, 160
p85 expression, 56
PFAM conserved protein domains, 75
PFAM-conserved protein domains, 14
Phorbol 12-myristate-13-acetate (PMA), 29
Phosphatase-1 (Mkp-1), 30
PI3K/Akt activity and genes control, 231–232
PI-3-kinase-Akt pathways, 29
PIN *See* Prostatic intraepithelial neoplasia
Pituitary adenomas, 264
PLA2G2A genes, 99
Plasminogen-activator inhibitor 1 (PAI-1), 241
Plasminogen *(PLG),* 135
Polycomb Repressive Complex 2 (PRC2), 119
Polycystic ovary syndrome, 219
Polycystic ovary syndrome(PCOS) theca cells
 microarray assay of, 223–226
 steroidogenic profile of, 220
Power analysis problem, 9

PPNAD *See* Primary pigmented nodular adrenocortical disease

Pre-eclampsia, 161

Pregnancy, decidualization in pathological conditions, 160–161

Primary pigmented nodular adrenocortical disease, 133

PRL-receptor mediated JAK/STAT signaling, 276

Proapoptotic genes down-regulation, 276

Probabilistic graph models, 14

Proinflammatory cytokines effects on thyroid follicles gene expression, 211

Prolactin (PRL) action on prostate, 272–273

Prostate cancer, 88–89
   AR alterations and, 97–98

Prostate gland, functional genomics study, 273–274

Prostate-specific antigen (PSA) expression, 88

Prostate-specific rat probasin (Pb) promoter, 273

Prostatic hyperplasia mouse models and PRL genes, 273–274

Prostatic intraepithelial neoplasia, 85

Protease-activated receptor-2, 100

Proteasome activator PA28, 88

Protein expression and Leiomyoma, 76–77

Protein Kinase C (PKC), 45

Protein metabolism targets, 99–100

Protein-protein interactions, 13

PTH anabolic effect and Jagged-1, 30

PTH signaling and transcription factors, 33

PTH signaling in osteoblast, 31

PTH type I receptor (PTH1R), 26

Putative *cis*-regulatory modules in ERα-responsive promoters, 123

P450 xenobiotic-inducible superfamily, 56–57

Q

Q65 receptor group, 88

Q-RT-PCR *See* Quantitative real-time reverse transcription PCR

Quantitative real-time reverse transcription PCR, 134

QVALUE software packages, 73

R

Raloxifene oestrogenic compounds drugs, 26

Raloxifene receptor modulators, 68

RAP guaninenucleotide- exchange factor, 232

Ras-like GTPases, 45

Ras-Raf-MAP-kinase, 29

Rat and Mouse Survey microarrays, 190

RDA *See* cDNA representational difference analysis

Real-time reverse transcriptase PCR and diabetes research, 201–202

Relative gene expression level changes distributions, 9

Renal dysfunction and type 2 diabetes, 201

Retinoic acid genes metabolism, 229–230

Reverse transcriptase-real time quantitative polymerase chain reaction (RT-qPCR), 266–267

Risedronate bisphosphonates drugs, 26

S

SAGE *See* Serial analysis of gene expression

SAM *See* Significance Analysis of Microarrays

SBMA *See* Spinal and bulbar muscular atrophy

Scanalyze freeware applications, 192

SREBP cleavage-activating protein, 99

Selected microarray platforms merits and limitations, 187–191

Self-organizing maps, 193

Sentrix® Array Matrix, 190–191

Sentrix Beadchip arrays, 190–191

Serial analysis of gene expression, 16, 132

Signal Transducer and Activator of Transcription 5, 45

Significance Analysis of Microarrays, 193

Single-copy CGI promoters, 120

Single time point microarray data analysis, 167–168

Skeletal muscle of type 2 diabetes gene transcript levels, 195–199

Small heterodimer partner (SHP) expression, 54

SORD genes, 99

Specificity protein (Sp)1, 52

Spinal and bulbar muscular atrophy, 85, 88

Spotfire commercial tool, 13

SREBP-1/ADD1 expression in DIO mouse, 250

SREBP-1c *See* Sterol regulatory element binding protein-1c

SREBP1c gene expression, 55

SREBP cleavage-activating protein, 99

STAT5a/b *See* Signal Transducer and Activator of Transcription 5

STAT5b deficient mice analysis, 45

STAT5-mediated transactivation, 52

Steroidogenic acute regulator *(STAR)*, 135

Sterol regulatory element binding protein-1c, 202

Stromal-epithelial interactions, 88

SuperArray GEArray® bioscience, 189

SuperScript Choice system, 71

T

Tamoxifen oestrogenic compounds drugs, 26

TFBSs *See* Transcription factor binding sites

Thr-X-Tyr motif of MAP kinases, 30

Thyroid follicles gene expression
   iodide effects on, 210
      methimazole and amiodarone effects, 211–212
      proinflammatory cytokines effects on, 211

Thyroid follicles in suspension culture and monolayer culture, 208

Time course microarray data analysis, 168–173

Tissue biopsy, 10

Tissue transcriptional profiling, 131

3T3-L1 Cell line model, 241–245

3T3-L1 preadipocyte cell line, 240

TNF-related apoptosis inducing ligand, 157

Total RNA extraction and synthesis of biotin-labeled
   cRNA, 265–266
TRAIL *See* TNF-related apoptosis inducing ligand
TRAIL-induced apoptosis in PC-3 prostatic cell
   line, 277
Transcription and proliferation targets, 100–101
Transcription factor binding sites, 31
Transcription start site (TSS), 32
TRANSFAC database, 33
Transforming growth factor alpha (TGF-α), 29
Treatment-effect test, 7
TSH effect on thyroid function in human thyroid
   follicles, 209
Type 2 diabetes, gene transcript level alterations,
   194–201
Type V 17β-hydroxysteroid dehydrogenase, 220

U

Ubiquitin-mediated degradation, 88
UMR microarray data, 30
UMR 106-01, PTH-responsive rat osteosarcoma
   osteoblastic cell line, 27

V

Variance models analysis, 12
Vascular complications and type 2 diabetes, 201

W

White adipose tissue
   endocrine functions of, 241
   microarray technology and obesity, 249–258
Whole genome expression profile analysis, 5
Wilcoxon test, 193
WNT1-inducible signaling pathway protein 2 *(WISP2)*,
   137
Wnt signaling transduction pathway, 137
Wolff–Chaikoff effect in human thyrocytes, 210

X

xenograft model, 100

Y

Ying Yang-1 (YY1) transription factor, 52

Printed in the United States of America